New Frontiers in Production Engineering

New Frontiers in Production Engineering

Editors

José Carlos Sá
Francisco J. G. Silva
Gilberto Santos
Luís Pinto Ferreira

MDPI

Basel • Beijing • Wuhan • Barcelona • Belgrade • Novi Sad • Cluj • Manchester

Editors
José Carlos Sá
ISEP – School of Engineering,
Polytechnic of Porto
Porto, Portugal

Francisco J. G. Silva
ISEP – School of Engineering,
Polytechnic of Porto
Porto, Portugal

Gilberto Santos
Polytechnic Institute of
Cávado and Ave
Barcelos, Portugal

Luís Pinto Ferreira
ISEP – School of Engineering,
Polytechnic of Porto
Porto, Portugal

Editorial Office
MDPI
St. Alban-Anlage 66
4052 Basel, Switzerland

This is a reprint of articles from the Topical Collection published online in the open access journal *Sustainability* (ISSN 2071-1050) (available at: https://www.mdpi.com/journal/sustainability/topical_collections/New_Frontiers_in_Production_Engineering).

For citation purposes, cite each article independently as indicated on the article page online and as indicated below:

Lastname, A.A.; Lastname, B.B. Article Title. *Journal Name* **Year**, *Volume Number*, Page Range.

ISBN 978-3-0365-8216-0 (Hbk)
ISBN 978-3-0365-8217-7 (PDF)
doi.org/10.3390/books978-3-0365-8217-7

Cover image courtesy of Maria João Lopes Guerreiro Félix

Contents

About the Editors

José Carlos Sá

José Carlos Sá (Professor) has worked at ISEP—School of Engineering, Polytechnic of Porto, Porto, Portugal, since 2016 in the Mechanical Engineering Department. In 2015, he was awarded the title of Specialist in Quality Management by the Polytechnic Institute of Viana do Castelo. He has worked as a consultant in several companies (2004–2020) in the areas of occupational safety, quality management, lean and industrial engineering. He was the Director of Production and Quality in a multinational company. He has supervised more than 30 master's theses of ISEP-IPP students and has more than 60 indexed publications (WOS+SCOPUS) in the fields of quality management, occupational safety, Lean Six Sigma and continuous improvement. He has an h-index of 26 on Google Scholar (1934 citations) and 18 on SCOPUS (954 citations). He is a member of the Editorial Advisory Board of the journal *Technological Sustainability* (Emerald) and a member of the Editorial Board of the Journal of *Applied Research on Industrial Engineering*. He is also a member of the Review Board of the *International Journal of Environmental Research and Public Health* (MDPI). In 2015, he became a Senior Member of the Order of Engineers. He is one of the three founders of the International Conference on Quality Innovation and Sustainability (ICQIS), which is currently in its 5th edition.

Francisco J. G. Silva

Francisco J. G. Silva (Associate Professor with Habilitation) has worked at ISEP—School of Engineering, Polytechnic of Porto, Porto, Portugal, since 1995 in the Mechanical Engineering Department. He was the Director of the Bachelor's program in Mechanical Engineering and Master's program in Mechanical Engineering for 12 years. He has more than 280 papers published in SCOPUS- and CLARIVATE-indexed journals, 7 international written/edited books, has supervised more than 200 MSc students and 2 PhD students, and is currently supervising about 10 PhD students. He has an h-index of 51 on Google Scholar (7342 citations) and 37 on SCOPUS (4344 citations). He is a member of several Editorial Boards of indexed scientific journals, having reviewed more than 800 articles for more than 50 journals (Elsevier, Springer, Taylor Francis, MDPI, ASME, etc.). He has won several scientific awards, such as the Best Paper Award, Best Reviewers Award, IPP Scientific Merit Award, etc. He has also acted as an Evaluator of several scientific projects. Moreover, he has carried out some scientific projects related to manufacturing processes and their optimization.

Gilberto Santos

Gilberto Santos (Professor with Habilitation) has worked at Polytechnic Institute Cavado Ave, in Portugal, since 2006. He holds a Habilitation (DSc) from the New University of Lisbon (2019) in Industrial Engineering, and a PhD degree in Mechanical Engineering-Production Technologies from the University of Minho, achieved in 1999. He graduated in Mechanical Engineering from the Slovak University of Technology in Bratislava (STU).

He is a member of the ID+ Research Institute and has participated in about 20 R&D projects. He is one of the three founders of the International Conference on Quality Innovation and Sustainability (ICQIS), which is currently in its 5th edition. Gilberto participates as a speaker in several national and international conferences and he is the author of several publications in industrial engineering, namely, innovation, quality management, product development and integrated management systems (IMS)–QES (quality, environment and safety) (books, papers in international and national journals). He is a member of the research team of several IMS-QES research projects and has supervised about 70 Master's and 3 PhD theses. He has published about 250 papers (articles in journals, communications conferences, books and chapters of books). He is a reviewer for several international journals (Editorial Review Board—*TQM Journal*; *International Journal of Engineering and Industrial Management* (Portugal); *Revista Produção e Desenvolvimento/Journal Production and Development* (Brazil); *The International Journal of Engineering Technology and Scientific Innovation* (India); *Quality, Innovation, Prosperity* (Slovakia); *Sustainability* (MDPI); and *Standards* (MDPI).

Luís Pinto Ferreira

Luís Pinto Ferreira (Associate Professor) has worked at ISEP—School of Engineering, Polytechnic of Porto, Porto, Portugal, since 2016 in the Mechanical Engineering Department. He earned his PhD in Manufacturing Engineering from the University of Vigo (Spain), MSc in Industrial Engineering–Logistics and Distribution from the School of Engineering, University of Minho (Portugal), and his Bachelor's degree in Industrial Electronic Engineering from the School of Engineering, University of Minho (Portugal). He has published several scientifical papers, book chapters and communications in national and international events. He has been teaching in higher education institutions since 2000. He is currently the Deputy Director of the Master's program in Engineering and Industrial Management of the School of Engineering, Polytechnic of Porto (Portugal), and he has supervised more than 100 MSc students. He has more than 100 papers (ISI / SCOPUS) and 1 international book published. In 2004, he was awarded the APDIO/IO 2004 prize by the Operation Research Portuguese Association.

Preface

This Special Issue, entitled "New Frontiers in Production Engineering", focuses on the most recent developments in industrial engineering containing economic, environmental and social sustainability concerns. Its aim was to collect articles that promote the improvement of industrial processes and services, making them more sustainable in economic, environmental and social terms. These purposes are achieved through the dissemination of studies focused on human factors concerns, logistics, the digitalization of industrial processes and services, and the industrial management improvement of processes, targeting a more sustainable production. In fact, despite the countless developments that have been published in leading journals, these discoveries generate knowledge that is then translated into new developments, which constitutes an extra motivation for the collection, selection and publication of research that brings effective added value to the knowledge already disseminated in this area. This Special Issue is particularly aimed at professors, researchers and PhD students who seek to expand their knowledge in the field of industrial engineering in general, and in the field of sustainability in particular, as related to any of its three aspects. This Special Issue involves authors from different continents, showing that the concerns experienced by the scientific community are common regardless of where in the world the researchers may be based.

José Carlos Sá, Francisco J. G. Silva, Gilberto Santos, and Luís Pinto Ferreira
Editors

sustainability

Review

Digitalization as an Enabler to SMEs Implementing Lean-Green? A Systematic Review through the Topic Modelling Approach

Geandra Alves Queiroz [1,*], Paulo Nocera Alves Junior [2] and Isotilia Costa Melo [2]

1 Engineering Department, Production Engineering, Universidade do Estado de Minas Gerais, Passos 37900-106, Minas-Gerais, Brazil
2 Escuela de Ingeniería de Coquimbo (EIC), Universidad Católica del Norte (UCN), Coquimbo 1781421, Chile
* Correspondence: geandraqueiroz@gmail.com

Abstract: Small- and medium-sized enterprises (SMEs) play a fundamental role in the global economy. However, SMEs usually have different characteristics from larger enterprises, e.g., essential resource restrictions, lower performance, and higher environmental impacts. This requires them to search for strategies to be more competitive and sustainable. A possible solution relies on introducing Lean-Green practices. Previous research indicated that digitalization could be an enabler of Lean. Lean can also help to achieve increased environmental performance using the Lean-Green approach. In this study, this important yet under-studied area is investigated as we consider digitalization as an enabler for implementing lean in SMEs, with a focus on Lean-Green practices. A systematic literature review is executed, following a new framework based on topic modelling for extracting the papers. The topic modelling is executed through latent dirichlet allocation (LDA) which is a machine learning technique. In methodological means, this paper represents an example of the frontier of digitalization for research activities. Regarding the investigated focus, the main findings revealed that digitalization is an enabler to Lean and to Lean-Green. As digitalization supports information sharing, it consequently fosters performance measurement systems, improvements, and value chain integration.

Keywords: lean; lean-green; systematic literature review; topic modelling; industry 4.0; small and medium enterprise (SME); machine learning; artificial intelligence; digitization; digitalization; digital transformation; latent dirichlet allocation (LDA)

Citation: Queiroz, G.A.; Alves Junior, P.N.; Costa Melo, I. Digitalization as an Enabler to SMEs Implementing Lean-Green? A Systematic Review through the Topic Modelling Approach. *Sustainability* **2022**, *14*, 14089. https://doi.org/10.3390/su142114089

Academic Editors: José Carlos Sá, Francisco J. G. Silva, Gilberto Santos, Luís Pinto Ferreira and Manuel Pereira Lopes

Received: 21 September 2022
Accepted: 10 October 2022
Published: 28 October 2022

1. Introduction

Small- and medium-sized enterprises (SMEs) play a significant role in the global economy [1]. They face stiff competition in the globalized markets. Furthermore, these enterprises frequently present a lack of essential resources for organizational and technological changes as striking features [2]. They also tend to be less productive and pollute more than larger companies [3].

Remarkably, the Lean manufacturing (LM) paradigm is widely adopted in organizational strategy and is spread among several industries and service companies to improve productivity, quality, and delivery according to customer requirements by waste reduction and the minimum usage of resources [4]. In this context, to become more competitive, SMEs have tried to adopt the LM paradigm [5] and digital tools to improve the performance of their operations [6]. Digital tools are also fundamental pillars of the process of digital transformation (DT) which involves the combination of technology, organization models, and business processes aiming to create additional value for stakeholders within the changing digital environment [7].

Against this background, SMEs have been pushed to adapt their processes due to increased competitiveness in the manufacturing industry; in turn, they should embrace digital technologies and LM practices [8]. LM has become the essential approach for SME performance because of the competitiveness, high levels of quality, low cost, delivery time, and flexibility [9]. It is important that SMEs look for agile and lean methods to reinforce their processes [10] because they are in an increasingly competitive environment where large companies are undergoing the new industrial revolution [11]. This industrial revolution is sometimes also known as digitalization, digitization, digital transformation (DT), or industry 4.0 [12]. Information and communication technology (ICT) tools enable SMEs to integrate their business process, and from that, SMEs improve their operational efficiency and productivity [12]. Furthermore, ICT is considered a valuable tool that collaborates with SMEs to achieve a competitive advantage as it can facilitate management decisions, e.g., the use of social media for customer engagement improvement [13].

Implementing industry 4.0 principles as digital tools is critical for business performance [14]. However, it has been emphasized [14] that SMEs must face more barriers to using new technologies than larger companies [14]. Consequently, this requires some specific procedures for SMEs. In the same line of reasoning, the effectuation of LM in SMEs has been discussed as an imperative requirement and a challenge in the operations management literature [5,9]. Approximately 10% or less of SMEs can implement lean after one year [15]; however, SMEs cannot use many aspects of Lean due to their size restrictions, flat hierarchies, and weak strategy alignment [2]. SMEs also present limited financial resources, fear of adopting new technology, lack of top management commitment, and poor leadership. These are the most critical barriers to LM implementation in SMEs in India, and SME managers view adopting new technologies as an expense rather than a strategic investment [9,16].

Additionally, in the globalization era, SMEs must ensure at the same time sustainable profitability through cost savings while being environmentally conscious [16]. Despite these barriers, the possible solution for sustainable (operational and environmental) improvements in SMEs could be introducing new technologies and approaches such as Lean-Green [17]. At one time, not all Lean practices were in line with Green strategies; Lean-Green represents the evolution of Lean towards sustainability. Alongside being more aligned with the sustainable goals of the current world, it was demonstrated that the vital link between operational excellence and Lean-Green enables the construction of competitive advantage in a sustainability era [18].

In this sense, digitalization can be an enabler to LM and Lean-Green in SMEs. At the same time, LM can be a good foundation for DT in SMEs [19], as the proposal of LM using the Lean tool value stream map (VSM) to support industry 4.0 implementation by the understanding of changes in materials, equipment, processes, and information flows [20]. Nonetheless, SMEs lack resources, and finding the best solution to implement the LM is difficult.

However, these studies have not addressed the relationship between digitalization and Lean implementation or Lean-Green in SMEs. Thus, it is necessary to understand how DT can catalyze Lean and Lean-Green and SMEs' competitiveness. The literature on the combination of LM and DT is relatively novel. Considering this, this paper aims to answer the following questions:

RQ1. How can digitalization support Lean implementation in SMEs?

RQ2. How can digitalization be an enabler in implementing Lean-Green in SMEs?

The purpose of this paper is to identify by theoretical discussion how digitalization can support Lean implementation in SMEs and verify whether it can facilitate the integration of Lean and Green in SMEs. Alongside the thematic relevance, this paper also represents a methodological contribution. An explorative systematic literature review is executed through topic modelling based on latent dirichlet allocation (LDA). LDA is a machine learning technique. This approach can be a novelty for explorative literature reviews [21,22]. Among the advantages, it can be highlighted that the categories and mapping do not need to be known in advance, coding can be automated, and costs and time of pre-analysis, analysis, and post-analysis are low [22].

This article is organized as follows. In the next section, more details about the main concepts are presented. In Section 3, the research method is detailed. The Section 4 brings the results that are discussed in Section 5. Finally, Section 6 brings the conclusions and suggestions for further works.

2. Background

2.1. Digital Transformation (DT)

The terms "digital transformation," "digitalization," and "digitization" have no broadly accepted definition, but they may be synonymous in many cases [23–25]. According to Gong and Ribieri (2021) [26], "Digital transformation is a fundamental change process, enabled by the innovative use of digital technologies accompanied by the strategic leverage of key resources and capabilities, aiming to radically improve an entity and redefine its value proposition for its stakeholders." This definition emphasises the importance of the transformation reaching a strategic level based on the available resources and capabilities. Innovation is also directly associated with the idea of digital technologies. An SME is the entity of interest in the current context.

Another term usually associated with this context is "industry 4.0" [27]. Although sometimes mentioned as a synonym of industry 4.0, according to Culot et al. [27], the concept of DT differs from industry 4.0 because DT stresses the implications for strategy and business model innovation, underlines the emerging technologies on the business model, and in turn, highlights the rise of cross-industry ecosystems. On the contrary, the term "industry 4.0" is mainly restricted to the DT process in the manufacturing sector. Similarly, "digitization" is often limited to the service sector [27].

Given this reason, the terms "industry 4.0" and "digitization" were not used in this literature search because the focus was not restricted to the manufacturing and service sectors. For the literature search, it was assumed that DT and digitalization were synonyms, similar to what was assumed by a previous paper [28]. In this regard, although the four terms ("digital transformation," "digitalization," "digitization," or "industry 4.0") were not all used in the literature search, they are used in this paper as synonymous and their meaning is the process defined by Gong and Ribieri (2021) [26] as a DT [28].

2.2. Lean Manufacturing and Lean-Green

Lean Manufacturing (LM) has been considered a manufacturing management paradigm. It is also recognized as the philosophy to manage organizations, englobing manufacturing, and services [4]. Liker and Morgan (2006) [29] argue that LM goals are the pursuit of better quality, shorter lead times, lower cost, higher employee morale, and better security to apply to any activity and industry. Jasti and Kodali (2014) [30] show that results from previous empirical works highlight that LM enables inventory reduction, quality improvement, and value addition for the customers at the right time.

In summary, LM is taken as a set of principles and practices to eliminate all forms of waste within an organization [31]. Shah and Ward (2007, p. 791) [32] define LM as "an integrated socio-technical system whose main objective is to eliminate waste by concurrently reducing or minimizing supplier, customer, and internal variability." These wastes are categorized into defects, inventory, inappropriate processing, overproduction, waiting, handling, transportation, and talent [33].

A complete list of Lean tools does not exist. Still, the most used are 5S, kaizen, value stream map (VSM), just in time (JIT), single minute exchange of die (SMED), total productive maintenance (TPM), kanban, standardized work, visual management, 5 why's (root cause analysis), and A3 report [32]. This list includes the management routine, pull production, information boards, and the overall equipment effectiveness (OEE) indicator [34].

LM philosophy has evolved as a successful instrument because of its significant impact on waste reduction, efficient management of resources, and the firm's performance [35]. Nowadays, companies are integrating areas and tools in Lean. The recent trend includes the Green tools in Lean and digital tools or 4.0 in Lean. Worldwide, many organizations from the most diverse industries in the economy are embracing LM practices [36].

The Lean-Green has been considered an approach that supports the search for sustainable development of a production system's economic, environmental, and social pillars [37,38]. The Lean-Green focuses on waste reduction and the efficient use of resources [39,40]. Implementing Lean practices can offer significant advantages and synergies with a firm's environmental performance [37,41,42]. The empirical results from the study of Iranmanesh (2019) [43] show that Lean culture, involving process and equipment, product design, supplier relationships, and customer relationships positively and significantly affect sustainable performance. In the case of SMEs, Caldera, Desha, and Dawes (2019) [44] consider that the Lean-Green approach is an essential practice to support SMEs to achieve sustainability [44].

Additionally, digital tools have been discussed in the Lean literature as a very important dilemma in business [45]. The real-time information provided by digitalization is found to be very valuable in preparing an accurate VSM, which is the first stage in Lean implementation [46]. LM also offers great potential to implement innovative automation technologies in manufacturing [47], and digital tools can support manufacturing organizations to overcome the barriers of Lean implementation [48].

The literature shows that Lean and DT in SMEs can increase their competitiveness, exemplified by some Lean and digital tools. The study by Kolla, Minuferkr, and Plapper (2019) [49] shows that the combination of SMED "kaizen" and "heijunka" can help in SME product customization. As shown in a case study [50], in a food factory with the implementation of SMED, it was possible to achieve a 70% improvement in the overall OEE indicator. Digital tools, e.g., simulation software and 3D printing, can prototype and avoid the risk of product failure. This study also argues that ICT can support the integration of SMEs with their suppliers and customers. Furthermore, it can help to improve these relationships [49]. Another case resulted in improvements in the quality control process and Lean principles using the digitalization tools [51].

With regards to Lean-Green, the literature argues that it is essential to integrate the process information through the value chain [52,53] because the value chain integration is essential to the life cycle assessment (LCA) which contemplates a product life cycle "from the cradle to the grave." The information from the whole life cycle and value chain must be included in the environmental value stream map (E-VSM), a very important tool to integrate Lean and Green [53,54]. The proposal of E-VSM is that sustainability indicators, such as raw material consumption, water consumption, energy consumption, waste generation, and emissions, should be integrated into the traditional VSM [54,55]. The key issue for the implementation of Lean-Green is the inclusion of sustainability indicators in Lean systems [56,57]. Furthermore, this tool seeks to provide a wealth of information that can facilitate communication and management through Lean and Green indicators [56,57]. From this, the literature [3,18,54,56,57] argues about the importance of including Green performance indicators since this is what will allow integration with Lean systems and monitoring of environmental gains.

2.3. Small and Medium Enterprises (SMEs)

There is no globally standardized definition of small and medium enterprises (SMEs). The most common classifications are based on a financial measure and/or the number of employees, for example, the definitions from the European Union [58] and from South American countries such as Brazil [59] and Chile [60]. Some definitions, such as the Japanese, may specify whether the employees are temporary or permanent [61]. Even the same country may have different definitions of SME, depending on the sector, such as the USA [62]. In this study, it was assumed that all papers that used the acronym "SME" with the explanation "small and medium enterprises" and "small- and medium-sized enterprises" were referring to a comparable term. It is also considered that the term "SME" could encompass micro (usually defined as enterprises with less than 10 employees) and self-employed enterprises.

SMEs are often associated with lower productivity. This is attributed to their inability to achieve economies of scale, difficulties they face in accessing credit or investment, lack of appropriate skills, and their informality [63]. There are also some characteristics that may affect the performance of SMEs, such as the industrial sector, management, technology, technical competence in marketing, and innovation [64]. One study demonstrates that innovation driven by digitalization plays an important role in mediating the relationships between resources for business model experimentation as well as business model strategy implementation and the overall SME performance [28]. Another study [8] suggests that SME owners and managers should allocate more importance to innovation and DT for achieving higher levels of organizational performance. Lastly, another study demonstrated that the availability of digital technologies is not sufficient for improving SME performance. It is also necessary to develop digital technology and relational and innovative capabilities [65].

Regarding the Lean practices in SMEs passing through DT, few papers have been written and the body of knowledge is not yet established. The main pillars of Lean practices originate from total quality management, just-in-time management, and time-based competition, as well as supply chain and activity-based management [66]. In this regard, DT is expected to foster Lean and consequently support sustainability [67]. One paper proposed a Lean-thinking framework for supporting the open data-based DT of SMEs in the forest service in Finland [66]. The authors concluded that Lean thinking was enabled through a digital platform where the needs and wishes of different stakeholders of the ecosystem were communicated [66]. In this regard, the current paper expects to contribute by identifying how digitalization can support Lean implementation in SMEs and verifying whether it can help/support the integration of Lean and Green in SMEs.

2.4. Latent Dirichlet Allocation (LDA) Applied to Topic Modelling in SLRs

There are many methods for conducting a systematic literature review (SLR). A previous taxonomy [68] compared five methods, i.e., reading, human coding, dictionaries, supervised learning, and topic model. Regarding the assumptions, reading and topic modelling do not require the categories, category nesting (if any), and mapping to be known. Different from reading, topic modelling can automate coding, and the relevant text features are known, i.e., reducing subjectivity. Literature review costs can be divided into pre-analysis, analysis, and post-analysis. According to the mentioned taxonomy, all methods require high post-analysis costs [68]. Comparing human (reading and human coding) to more automated literature review methods (dictionaries, supervised learning, and topic model), the analysis costs of the automated methods are lower, specifically the costs of person-hours spent per text and the costs of the level of substantive knowledge. A more specific advantage of topic modelling through LDA is the lower pre-analysis cost, specifically the cost of person-hours spent conceptualizing and the cost of the level of substantive knowledge, even when compared to other automated methods (dictionaries and supervised learning).

However, topic modelling is rarely applied in an exploratory literature review [21,22]. Among topic modelling methods, the machine learning technique of latent dirichlet allocation (LDA) is the most used and is considered state-of-the-art [69]. Asmussen and Møller (2019) [22] proposed a framework for researchers from any field to apply LDA for topic modelling in an explorative literature review. This framework was followed and applied here. The LDA is a probabilistic method that extracts topics from a collection of papers. A topic is a distribution of words over a fixed vocabulary. The semantics and meaning of the sentences are not evaluated. However, LDA analyzes the words in each paper and calculates the joint probability distribution between the observed (words in the paper) and the unobserved (the hidden structure of topics). According to an example provided by the framework's creators [22], whether one of the topics in a paper is Lean, it can be assumed that the words Lean, just-in-time, and Kanban are more frequent compared to other papers that do not deal with Lean.

The LDA result is several topics, with the most prevalent topics grouped jointly. A probability for each paper is calculated for each topic, creating a matrix with the size of the number of topics multiplied by the number of papers [70]. While executing LDA, determining the number of topics (*k*) is a key parameter. As the framework proposed by Asmussen and Møller (2019) [22] is an unsupervised LDA, the relationship between the papers is not known before the model runs. Calculating the perplexity is normally used as cross-validation to estimate an adequate number of topics. Perplexity is a metric used to evaluate language models, where a low score indicates a better generalization. Lowering the perplexity is equivalent to maximizing the overall probability of papers being on a topic. Choosing the right number of topics is the art of balancing the right number of topics while keeping the perplexity at the lowest possible level.

3. Materials and Methods

This study adopted the combination of the SLR [71,72] and topic modelling through the LDA [22]. The primary phases of the SLR supported finding the studies that contained the keywords related to the generated topics. Alongside being beneficial to provide insights between the keywords and the topics, the topic modelling technique supported the paper's selection phase.

The SLR has been considered a replicable, scientific, and transparent literature review approach that minimizes bias. It is an iterative process used for identifying the extant literature about a research topic based on 5 steps: (i) RQs formulation; (ii) search strategy; (iii) selection and evaluation of relevant studies; (iv) analysis and synthesis of results; and (v) report the view [71]. The process framework used here is in Figure 1.

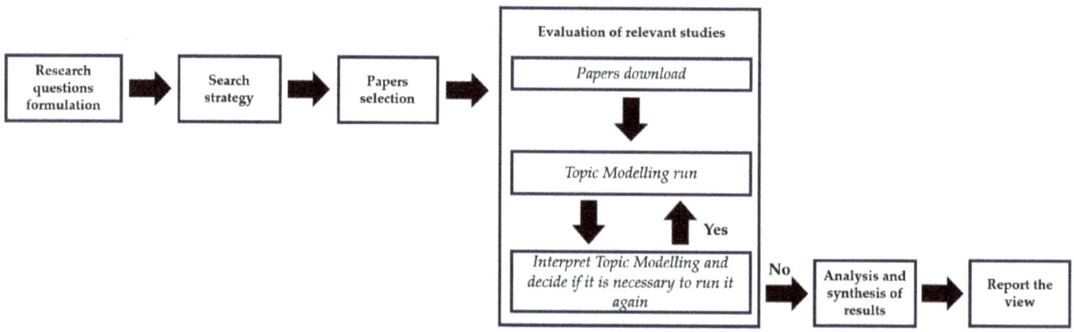

Figure 1. Schematic representation of the adopted framework, integrating topic modelling, based on latent dirichlet allocation (LDA), to the systematic literature review (SLR).

The first step defined the RQs, as presented in the introduction. Subsequently, the search strategy was designed and executed. Scopus and Web of Science were the databases chosen to find papers across reputable journals and conferences. These databases are regularly updated and have comprehensive international coverage [73]. It is important to emphasize that conference papers were selected because this research theme is new, so it is essential to capture the initial discussions held at scientific conferences [74]. The keywords were also identified among the 3 topics from a preliminary review on SMEs, Lean, and DT. The development of the search strings was done from the combination of the terms: small and medium enterprises (SMEs), digital transformation (DT), and Lean. The final search strings (available in Appendix A—the SLR Protocol) were defined after running tests to ensure reliable searches. After the search strings were established, the searches were undertaken in the databases. These searches resulted in 220 papers, but 47 of them were duplicated. They were excluded, resulting in 173 papers.

The step of paper evaluation (Figure 2) consisted of 3 phases for applying the topic modelling. The first phase consisted of the attempt to download the 173 resulting papers. Only 104 were available for download. The second phase consisted of the application of the topic modelling through LDA to perform the selection of the papers. One of the observed limitations of following the algorithm of Asmussen and Møller (2019) [22] is that the available algorithm is not able to read all papers. The reason why this happens is not clear. It is possibly due to some symbols and words non-identifiable by the algorithm. Topic modelling was applied to 92 papers, which the algorithm was able to read.

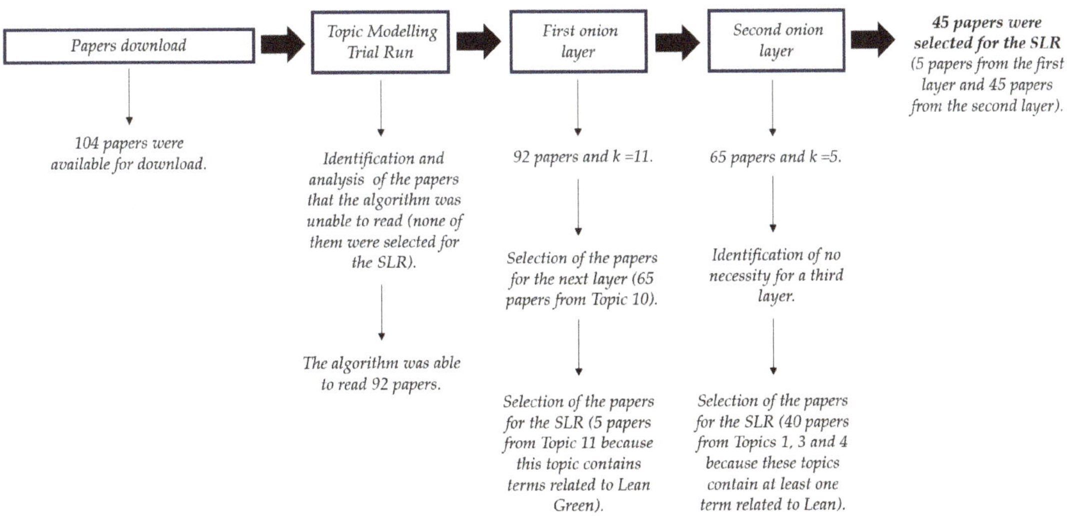

Figure 2. Flowchart of how topic modelling was applied for the selection of papers in the SLR and a detailed explanation of the number of selected papers per step.

The third phase of this step consisted of understanding the topic modelling results and judging whether it was necessary to run it again or not. It is expected that the larger the sample of paper, the more robust the results will be because of the lower perplexity levels. In general terms, 92 papers may be seen as a relatively small sample. Given this context, it was observed that above a certain threshold of topics (k), the LDA started to generate topics without any paper associated with it. Thus, it was decided to use the maximum number of topics (k) that still resulted in at least 1 paper associated with each topic. By observing the graphic of the number of topics (k) versus perplexity it was clear that, in the present case, the larger the k, the lower the perplexity. In other words, the model results had a better generalization.

As the objective is to use LDA for selecting papers, it is understood that the topics with fewer papers present a more discriminant list of words, which enables the analyst (a human being) to properly judge if these papers are pertinent to the RQ (or not). However, it was observed that in some cases, one topic has much more papers than the others. On the other hand, a topic with much more papers than the other will present a list of words that encompass many different themes, an obstacle to the adequate segregation of the papers. In such cases in this paper, the proposed solution is to run the LDA again, only with the papers on this topic, and to repeat this loop until the result does not present a concentration of papers on any topic. This is named the onion approach, and each loop is considered an onion layer. In the current case, it was necessary to perform 2 onion layers. Figure 2 details how this technique was used to filter out unwanted papers during the search for relevant literature. Further explanations are provided in the next section.

Finally, as is the case for any SLR, the final steps consist of analyzing and synthesizing the results (presented in Sections 4 and 5) and reporting the view on the theme (shown in the Discussion and summarised in the Conclusions).

4. Results

As presented in the previous section, the first onion layer had 92 papers. The maximum number of topics (k) that still resulted in at least one paper associated with each topic was k = 11. The topics are presented in Table 1. Topic 10 has 65 papers, and the other 27 papers were distributed among 10 topics. Hence, the papers from Topic 10 were used for the second onion layer (Table 2). In the first onion layer, it was decided to select the papers associated with Topic 11 (five papers) for careful reading, interpretation, and discussion in the SLR. Topic 11 was selected because it presents the words "lean", "sme", "green", and "environment". In this way, it is the only topic that has words directly related to the three keywords. It is also the only topic with two words related to "Green" (Lean-Green) and is one of the focuses of the RQs.

Table 1. Topics generated in the first onion layer (k = 11).

T1	T2	T3	T4	T5	T6	T7	T8	T9	T10	T11
lean	industri	servic	firm	data	var	vsm	suppli	learn	product	lean
implement	manufactur	construct	jit	erp	mainten	board	chain	construct	process	manufactur
startup	digit	agil	carbon	inform	cell	simul	manag	project	manag	smes
principl	factori	forest	adopt	bim	manufactur	director	strategi	train	system	manag
success	product	data	technolog	document	machin	audit	organis	plan	compani	implement
agre	procedia	softwar	malaysia	system	implement	govern	firm	cours	develop	perform
custom	autom	iot	cost	project	layout	twin	jit	technic	lean	factor
statement	smes	cloud	suppli	supplier	reduc	agent	technolog	elearn	research	green
disagre	logist	asset	relat	user	part	firm	supplier	educ	model	environment
compani	smart	transform	energi	order	engin	high	signific	engin	technolog	enabl

Table 2. Topics generated in the second onion layer (i.e., k = 5, considering the papers from the T10 from the previous layer).

T1	T2	T3	T4	T5
industri	design	lean	product	compani
product	system	manag	process	manag
manufactur	learn	implement	manufactur	technolog
digit	develop	smes	data	develop
smes	cost	product	time	inform
technolog	process	suppli	system	research
system	technic	manufactur	model	process
process	cell	chain	improv	model
lean	time	factor	vsm	project
custom	team	practic	sustain	andn

In the second onion layer, the maximum number of topics (k) that still resulted in at least one paper associated with each topic was k = 5. The topics are presented in Table 2. In this case, it was not observed the concentration of papers on one topic. It was therefore not necessary to perform a third onion layer. The papers from all topics that contained at least one word related to lean were selected for the SLR. The selected topics were 1, 3, and 4, and jointly they had 40 papers.

Given the discussed context, topic modelling through LDA supported the selection of 45 among 92 papers in an SLR (an extraction of around 49%), as shown in Figure 2. For achieving these results, there is no doubt that the time and cost invested were much smaller than those required when human beings execute the paper selection in an SLR. To guarantee that these faster results are reliable, measuring the perplexity is essential. However, it is also worth noting the empirical evidence of the current application. After reading the 45 papers, only five of them were discharged from the SLR (11%) and they were all from the second onion layer.

Among the discharged papers, one had the acronym "SME" for structural modelling equation, and another had the same acronym for the Society of Manufacturing Engineers. The third paper was an editorial commenting on another paper. The fourth one did not consider DT, and the fifth did not consider Lean. Given what was exposed above, in this paper it is argued the level of error for the application of this framework is tolerable for an SLR, given the practicality, time, and cost savings. However, as with any methodological approach, it can be continuously improved. For example, the current machine learning technique (LDA) is unsupervised. Further research is recommended to investigate the possibilities of a supervised LDA as well as other machine learning techniques.

Finally, Tables 3–6 summarise the 40 papers selected and analyzed to answer the research questions of this study. These papers are manually separated into four groups according to the focus of the current study (Figure 3). Further developments of the algorithm are recommended to consider grouping, ranking, and cycling selected papers automatically.

Table 3. Summary of the papers that discuss Lean implementation in SMEs.

Paper	Focus	Main Results
[15]	To provide and verify a step-by-step strategy from the low-to-high level of Lean application with appropriate tools and techniques to achieve the goals in each specific phase.	A roadmap for SMEs to successfully apply Lean in the context of limited resources.
[75]	This study identifies and prioritises critical success factors (CSF) based on implementing Lean Manufacturing for Thai SMEs.	"Technology resource" is the most important to enhance the implementation of Lean in SME organizations.
[76]	To identify the CSFs in managing maintenance (MM) activity in SMEs.	By identifying the CSF constructs, SMEs can utilize the information to improve their approach to a systematic MM program.
[77]	To provide an in-depth analysis of the implementation of Lean manufacturing in SMEs worldwide, to identify and present the critical difficulties that impact the undertaking, and to highlight the success factors of manufacturing firms.	The main barriers to Lean implementation are lack of leadership, commitment of top management, financial resources, resistance to change, training and knowledge about Lean tools, and know-how, skills, and expertise.
[1]	To provide a survey-based evaluation of applicability, benefits, and critical factors of Lean in SME product development and to identify how the Lean start-up approach operates in product development.	Lean methods can be introduced very quickly, promising high potential for improvements, and Lean methods combined with industry 4.0 technologies act as a booster for efficiency optimization in product development.
[78]	Based on the literature review, select the key success factors for Lean implementation in SMEs. These factors are then put in the context of industry 5.0 to explore the possibilities of Lean management as an enabler of industry 5.0.	Most of the success factors and tools are people-oriented, thus giving the human-centric approach to organizational and process improvement, as required by industry 5.0.
[79]	To identify key Lean practices for Indian automotive SMEs to reap the maximum benefits.	The findings show that customer involvement is the most important Lean practice, followed by problem identification and prevention, total productive maintenance, and others.
[80]	To identify and suggest improvements in two areas crucial for implementing Lean principles in an aluminium semi-permanent casting small to medium enterprise (SME).	This paper builds an analytical model to measure production performance and presents a case study of an aluminium foundry. The case study reveals that using two Lean principles of bottleneck identification and Lean buffering could allow for a timely response to customer requirements.
[81]	To explore how Lean principles can improve the entrepreneurial ecosystem in Malaysian SMEs.	Using Lean principles could be more cost-effective because they avoid lengthy implementations of unproven strategies and investments in the entrepreneurial ecosystem.
[82]	To identify the steps SMEs can follow in searching for a plant distribution model and applying layout improvement to increase productivity.	Implementing the tools for plant distribution and 5S was proposed to reduce waste in the production process since this generated a high total cycle time.

Table 3. *Cont.*

Paper	Focus	Main Results
[83]	To analyse which aspects of the JIT philosophy also apply to small- and medium-sized enterprises (SMEs).	Barriers include a lack of supplier cooperation and partnerships, an inability to develop the necessary technologies and methodologies to reduce or eliminate waste, difficulties in managing demand fluctuations, a lack of capital to acquire advanced technologies, quality control problems, and inadequate employee training and development. Enablers include the ability to empower employees, reduce JIT implementation time, overcome employee resistance to change, and receive various forms of governmental support.
[84]	To investigate the opposing factors in adopting Lean manufacturing in Indian SMEs and systematically evaluate causal/effect barriers by the Grey-DEMATEL technique.	Nine out of fourteen barriers belong to the casual group, and five barriers belong to the effect group. This study revealed that fear of adopting new technology is a high influencing barrier among all barriers.
[9]	To investigate the interrelationships among LM adoption barriers in Indian SMEs.	The findings show that limited financial resources, fear of adopting new technology, lack of top management commitment, and poor leadership quality are the most critical barriers to LM diffusion in Indian SMEs.
[85]	To analyse the contextual relationship and dependency amongst enablers for LM implementation in Bulgarian SMEs.	The findings demonstrated that "leadership and commitment by management," "human resource management," "customer relation management," "supplier relation management," and "information technology system" are the most significant enablers for Lean implementation in Bulgarian SMEs.

Table 4. Summary of the papers that discuss the integration of DT and Lean.

Paper	Focus	Main Results
[20]	To explore the integration of the VSM with simulation to help industry 4.0 initiatives.	VSM combined with simulation can help enterprises understand changes in materials, equipment, processes, and information flows associated with industry 4.0 application scenarios.
[86]	To understand how effective Turkish SMEs use enterprise resource planning (ERP) systems to assess their adherence to Lean.	The effective usage of specific ERP modules can contribute toward applying Lean principles and vice versa.
[11]	To propose a digital twin-enabled VSM approach for SMEs.	A theoretical framework which combines IoT data-driven production process planning and simulation methods.
[87]	To propose that entrepreneurial orientation (EO), participative management style, supplier relations, resource management, JIT, and technology utilization are several drivers of an effective management decision-making approach.	EO, supplier relations, resource management, JIT methodology, and TU positively impact agile supply chain management.
[45]	To understand how Lean and industry 4.0 integration can contribute to increased flexibility and productivity.	Implementation of LP and industry 4.0 concepts in Angola are not being uniformed in terms of company size and geography. Five S is the leanest tool implemented.
[88]	To investigate business intelligence (BI) tools that help an SME to improve its supplier order fulfilment management.	SMEs can benefit from BI tools at an affordable cost, joining the emergent trend of deploying business analytics to make systematic innovations and collaborating with supplier relationships to implement a just-in-time strategy.
[7]	To discuss the problem of digital transformation of supply chain management (SCM) due to the changing business environment and the desire to meet customer demand in Thailand.	SCM processes are altered, such as reduction in product design and a manufacturing period, faster delivery of products to customers, easily meeting of the preferences and demands of customers, and faster and effective decision-making supported by big data and analytical decision techniques.
[49]	To provide state-of-the-art literature on existing assessment models and consequently map Lean and industry 4.0 components to the specific characteristics of manufacturing SMEs.	Development of a hybrid model including Lean and industry 4.0 features suitable for manufacturing SMEs.

Table 4. *Cont.*

Paper	Focus	Main Results
[89]	To document the positive experience of one enterprise in jointly auditing and improving data quality and IT infrastructure, which better aligned its XPS with sustainability objectives.	The paper provides a management supporting tool to promote change in the natural evolution of businesses, for example, the BI panels. Regarding the link between sustainability and business objectives via big data and BI tools, the company detected one weakness concerning data analysis.
[19]	To demonstrate the state-of-art regarding Lean production and digitization and to present an approach based on the consistent opinion of the reviewed literature, which formulates digitization as the next step of Lean management in production systems.	This study reveals the demand for a methodological approach in an SME environment that quantifies the profitability of implementing digital technologies in Lean. It formulates digitization as the next step of Lean in production systems.
[90]	This paper aims to understand the impact of digitalization on international, Lean, and global start-up speeds (LGS).	Digitalization allows LGS to increase decision-making efficiency and optimize international market evaluation strategies and processes.
[14]	This paper presents the value stream mapping analysis results supported by the overall equipment effectiveness (OEE) coefficient analysis.	It was found that the partial digitization of the one production operations in SMEs has a positive effect on the course of the process. Creating a running chart for the variability of Total Productive Maintenance (TPM) indicators helps to improve the organization and efficiency of work.
[91]	To present the outcomes of expert workshops to identify requirements of SMEs in the field of smart logistics management.	SMEs will only benefit from industry 4.0 by following customized implementation strategies, approaches, concepts, and technological solutions.
[92]	To evaluate how Lean can support industry 4.0 in pursuit of greater customer value and manufacturing excellence.	This paper finds that the pursuit of SMEs toward process efficiency and waste reduction can be best achieved through a focus on foundational digitalization and data management, then taking a stepwise approach towards the cyber-physical systems of industry 4.0.
[93]	To investigate Lean manufacturing and healthcare logistics 4.0 concepts, methods, and tools.	The framework elaborated for implementing industry 4.0 and logistics 4.0 concepts in SMEs, in opposition to many architectures of the literature, considers sustainability as the kernel of industry 4.0 concept implementation in SMEs.
[94]	To examine driving factors of Lean supply chain management and major supply chain and information technology solutions applied in SMEs.	There is a positive relationship between the successfulness of Lean supply chain practices and IT solutions adopted, and the Lean supply chain would increase. It is also shown that these supply chain and information technologies play crucial roles in helping small companies transform into leaner organizations.
[95]	To present a case study about a Chinese company that constructed an intelligent Lean system.	China's small- and medium-sized manufacturing enterprises can promote intelligent Lean through the path of "standardization, lean, digital, intelligent" concepts.
[96]	To investigate the relationship between Logistics 4.0 concepts and technologies and logistics performance indicators in manufacturing enterprises.	The implementation of smart and Lean concepts has a major impact on logistics performance, whereas information and communication technologies, as well as autonomous logistics systems and vehicles, are not completely implemented and exploited yet.
[12]	To analyze the relationship between information and communication technologies (ICT), industry 4.0, and agile manufacturing.	ICT plays a key role, but it is not a goal itself. They are a prerequisite for the implementation of industry 4.0. Still, they serve to achieve agility in the manufacturing system and, as a result, achieve a competitive advantage for enterprises operating in turbulent and unpredictable environments.

Table 5. Critical success factors (CSFs) to DT adoption in Lean SMEs.

Paper	Focus	Main Results
[97]	A theoretical association of advances in manufacturing technology and tools.	Framework with the technologies and process improvement methodologies to point out the difficulties of SMEs to obtain technological advancements because of limited resources and lack of finances available to them.
[66]	To study the functional areas which can potentially leverage industry 4.0 technologies and help India's SMEs to adopt digital technologies for the identified functional areas.	Manufacturers want to change their design and manufacturing strategies based on performance metrics. Therefore, they need first to capture real-time machine data, analyze it, and then incorporate the resulting improvements in manufacturing and design decisions in that order.
[98]	In this paper, the analysis of the SMEs, and automation integrators of the project, identify a correlation between the challenges, age, and size of the SMEs.	It is presented that a strategic focus on production with "simple" smart technology concepts can enable SMEs to become more adaptable to the changing and dynamic environment. For example, collaborative robots and AGVs (automatic guided vehicles) in a reconfigurable environment can adapt to changing environments.
[99]	To show global challenges, such as digital transformation, are urging companies to become more dynamic and flexible.	The Lean start-up approach in large manufacturing companies examined the feasibility of accelerating product development while it makes innovation processes cheaper, more flexible, and more reliable.

Table 6. DT and Lean-Green.

Paper	Focus	Main Results
[100]	This paper proposes a method to drive process innovation toward the increase in efficiency of a production plant.	Correct data management permits to plan the best practices to improve processes and systems involved, in terms of environmental and economic impacts, meaning a process of sustainable innovation.
[17]	To analyse the Lean-Green performance of Indian manufacturing SMEs by investigating the influential relationships of various factors and the firm's set of Lean and Green practices.	Enterprises need to decrease their operational sizes to improve operational and environmental performance. The possible alternative and more practical strategy could be introducing new technology innovation and holistic adoption of manufacturing excellence initiatives, such as Lean-Green.
[16]	To identify the drivers for integrated Lean and Green manufacturing from the combined support of existing literature and expert opinions.	The results reveal that top management commitment, technology up-gradation, current legislation, green brand image, and future legislation are the five most important drivers for implementing integrated Lean-Green in Indian manufacturing SMEs.

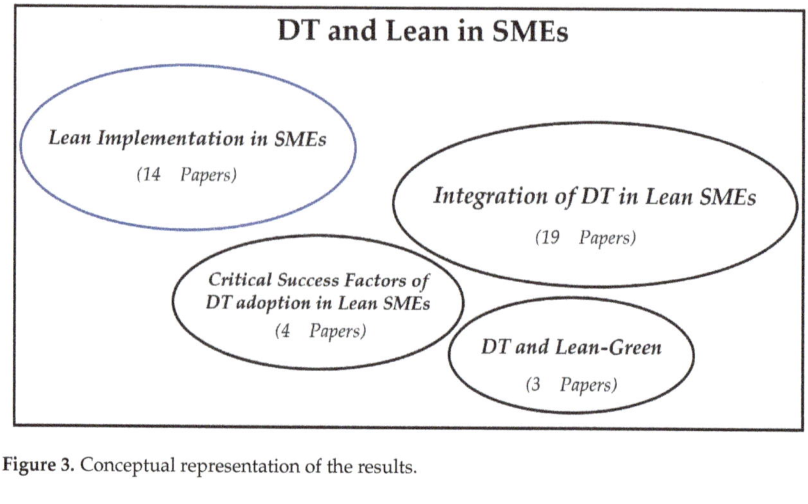

Figure 3. Conceptual representation of the results.

Tables 3–6 present the objective (focus) and the main results of each group of papers shown in Figure 3. Table 3 is the group of 14 papers that discuss Lean implementation in SMEs. Different from the papers in other groups, these papers bring DT as a remarkable secondary point but not as the focus of the investigation. Table 4 brings the 19 papers that focus on DT in Lean SMEs. Table 5 brings four papers that discuss the critical success factors (CSFs) of DT adoption in Lean SMEs. Finally, Table 6 brings three papers that discuss DT and Lean-Green in SMEs, which is also one of the RQs of the current SLR.

5. Discussion

5.1. How Can Digitalization Support Lean Implementation in SMEs?

From these 40 studies, it was possible to answer the RQs and find some propositions about Lean, SMEs, and Digitalization. The first point observed from the studies is that DT can collaborate on information sharing, which is essential for a successful Lean implementation. The lack of resources for advanced technologies can hinder Lean in SMEs. For example, the information technology system was highlighted as one of the most important facilitators for implementing Lean in Bulgarian SMEs [85]. The findings indicated that information sharing from digitalization could be the enabler, mainly for VSM, JIT, Lean Supply Chain, performance indicators (e.g., OEE), TPM program, and ERP systems.

The VSM tool needs information from the process, such as lead-time, efficiency cycle time, etc., to analyse and improve the process [101]. This is exemplified by a case [14] in which the partial digitalization of the operations of an SME helped to visualize the efficiency data measured by the OEE. These data helped in the construction and visualization of the current situation in the VSM. Thus, it was possible to trace a VSM of the future situation based on concrete data, which led to satisfactory results of efficiency increase in availability, performance, and quality. At the same time, the information flow achieved through VSM can assist the implementation of technologies from industry 4.0, entering a virtuous cycle where digitalization fosters Lean practices and vice versa [20].

Lu, Liu, and Min (2021) [11] detail the integration of DT in the VSM in production line optimization. This study shows that from digital information through a radio frequency identification (RFID) chip, a new VSM solves the problems where the standard VSM method is powerless. The RFID reader enabled managers to give directly quantized answers to the problems and to be notified about other possible circumstance changes and problems after each implemented optimization plan, e.g., the daily output capacity, average takt and reasonable work in progress (WIP) inventory, etc. Therefore, these data are a timestamp table which allows the managers to see every change in the process in a faster way and make more accurate calculations for the implementation of improvements, such as a Takt matrix, which can be used for modelling and making an overall process simulation of mixed series production.

The case of JIT paradigm requires suitable technologies that allow information sharing to adequately plan the flow of materials between the production chain. As exemplified in [87], the process information in SMEs informing JIT supplier partnerships aim towards order fulfilment cycle time reduction. Another example is shown through empirical results from the implementation of Lean integrated with concepts of industry 4.0 or digitalization. The integration results in a higher goods security level and a real-time data sharing policy [102]. Therefore, there is a positive correlation between the effectiveness of Lean supply chain practices, such as JIT, and the level of information technology solutions implemented. This is also essential to help SMEs implement Lean [93]. Information and communication technologies and Lean implementation have a great impact on logistics performance [95]. In agreement with these studies, in the study of Dowlatshahi and Taham (2009) [83], the hypothesis arises that advanced and relevant manufacturing technologies and methodologies are necessary for enhancing JIT implementation in SMEs.

The research of Gandhi, Thanki, and Thakkar (2019) [16] argues that SMEs present an environment with high uncertainty and low flexibility, so, from this aspect, the adoption of internet and communication technology can provide a competitive advantage. This is because, as pointed out by Alshamaila et al. (2013) [103], if SMEs adopt cloud computing technology they can receive benefits in terms of cost-saving, access to a large pool of hardware resources, enhanced information sharing, improved supply chain collaboration, and a faster time to market. Another finding relates to the use of ERP. The research [74] pointed out the effective use of some ERP modules collaborated to implement Lean.

The use of a great amount of data in ERP systems can provide support to build collaborative supplier relationships, which is the foundation of JIT in SMEs [87]. However, SMEs are often unable to acquire the advanced technologies needed for the effective implementation of JIT due to limited financial resources [83], but it was found that BI tools can be used at affordable costs [87].

Alongside Lean supply chain practices, other Lean practices can be supported by DT tools in SMEs. Some studies show that DT tools in SMEs can have a positive impact on TPM, visual management, and performance indicators [14,76]. The data for the TPM indicators collaborate to visualize the variability and motivate the workers to improve the process efficiency. Therefore, they help to visualize the process information and are a good way to implement new management concepts in SMEs [14,76]. These cases [76] show that visualization in maintenance management collaborates to improve productivity, obtaining superior product quality, optimizing operation costs, improving the delivery of products to customers, and empowering employees. A survey of Indian SMEs indicated that manufacturers that aim to make some improvements in their operations must capture real-time machine data, analyze them, and incorporate the resulting improvements in manufacturing and design decisions in this order [66].

Another critical issue is that the DT allows Lean to optimize the decision-making process contributing to internationalization as the international managers work to integrate forces and mediate between local market realities and corporate goals [89]. Another consequence of increasing information sharing in SMEs through DT tool implementation is supply chain collaboration, because it is vital to place SMEs in the current context of market trends and the needs of their stakeholders [7]. In short, based on the results of this literature review, it is clear that information is the main way in which digital tools collaborate for Lean in SMEs.

Furthermore, it was found that the use of simulation tools supports the viability of Lean improvements in SMEs. It was demonstrated [82] that layout changing and 5S reduce waste in the production process. Simulation software to test and validate the improvements before the implementation is most effective in reducing the total cycle time. It can also reduce the risk and help to find better solutions. Related to the first RQ, it was also found that the Lean start-up approach can help SMEs to accelerate the innovation process and makes it cheaper, more flexible, and more reliable despite the smaller resources than for large companies [10,99].

In short, based on the results of this literature review, it is possible to understand that digital technologies can be an enabler of Lean implementation. Firstly, it is clear that information sharing from digitalization is the main way in which digital tools collaborate for Lean in SMEs. It was noted that for making decisions or implementing improvements, it is necessary to visualize the data. In this case, to Lean work properly, collaborative supplier relationships are essential. Technologies such as simulation software also allow the process to better understand and reduce project risks.

5.2. How Can Digitalization Support Lean-Green Implementation in SMEs?

This RQ is quite unexplored in the literature, but it was possible to obtain some findings/deductions. Similar to the first RQ, information sharing from digital tools is the central method in enabling the Lean-Green in SMEs. The study of Papetti et al. (2016) [99] argues that correct data management allows the best practice implementations to improve sustainable innovation of the operations, that is, in terms of environmental and economic impacts.

It is essential to highlight that process and supply chain information is more critical when environmental impacts must be considered [52,53,55]. This information allows the data to properly allocate consumption and, consequently, identifies the causes of a specific consumption trend (e.g., under/overproduction, increase in energy consumption, etc.) [99]. The study on Indian manufacturing SMEs found that technology up-gradation (utilization of energy resource efficiency advanced technology) is one of the five most important drivers behind Lean and Green manufacturing integration [16].

A case study driven by a building material supplier located in Hong Kong showed that data provided by BI tools helped the SME to detect weaknesses in data analysis and allocated sustainability-specific data to improve the efficiency of the entire production unit. The method used is the basis for a web application tool [87]. Another example of process data was demonstrated in a study that proposed a method to increase the efficiency of a production plant in terms of energy consumption [99]. These cases are in line with the Lean-Green integration proposals that highlight the importance of including sustainability indicators. In this instance, it is necessary to have access to process information. Another issue related to Lean-Green is the increase in their sustainable performance. It has been demonstrated that collaborations with big companies and investing in technology improved communication between employees and suppliers. It can also reduce process lead times and achieve better operational performance [17].

On these grounds, it is possible to infer that DT can collaborate with Lean-Green in a similar but more relevant way than Lean. This proposition is supported by the life cycle approach of environmental aspects. As understood by the Lean-Green literature which shows the direct effect that information sharing has on Lean decisions, Lean-Green needs even more integration of the data from the whole value chain. In this way, the environmental impact data from all value chains and during the product life cycle are the foundation for sustainable improvements. Thus, when digitalization enables the Lean-Green implementation in SMEs, these firms can obtain more advantages over their competitors and collaborate for sustainable global targets since they are more unproductive and more pollutant than larger companies.

6. Conclusions

This paper aimed to investigate the relationship between the DT in the Lean and Lean-Green in SMEs using a topic modelling systematic review approach. With this approach, it was possible to select 40 papers and find theoretical evidence fast, effectively, and at a low cost, proving that this approach is a viable option for the realization of SLRs. Thus, the method helped to find the state-of-art effectively and from a vast number of papers and to understand relationships among the studied constructs. Nowadays, it is possible to find many papers in databases, and the time for paper selection is usually the limiting factor for completing SLRs.

Based on the findings, it was possible to understand that the DT can enable Lean and Lean-Green, and that both can improve SME competitiveness. The main point highlighted in the studies was the information sharing obtained from the DT. The results showed that the integration and visualization of the process data collaborate with the performance measurement system, helping to make decisions. It was also demonstrated that the implementation of the tools JIT, ERP, VSM, TPM, and E-VSM could be better supported by the DT, which provides the process data. In the case of JIT and E-VSM, the DT provides the information from the value chain. Therefore, regardless of Lean-Green, obtaining more data from the value chain is emphasized as an essential requirement for sustainable improvement. Despite some relationships found in the literature, this area still requires more research, mainly regarding the DT and Lean-Green in SMEs. Thus, future studies are recommended to quantitatively investigate how much Lean-Green and DT implementation improve SME sustainability performance.

As with all research, the current study is not free from limitations. It is necessary to consider that some papers were unavailable for download. Thus, some contributions from these unavailable works may have been missed. The adopted algorithm could not read all papers and it should be improved to enable the reading of all. Another limitation refers to the fact that only one filter could be replaced by the automatic solution. Further development of this framework could directly search the papers, check for duplicated ones, exclude them, and download the remaining papers. Alongside solving these issues, the algorithm should be able to read some two- and three-word expressions as one word for generating a topic. For example, the expression "supply chain" would be considered as two words "suppl" and "chain" in a topic by the current algorithm. It should be considered "supply chain" as one word in a topic. Further developments of the algorithm should also consider integrating visualization tools, such as a cloud of words, ranking, or cycling, enabling a faster interpretation of the results.

Finally, it is important to mention that the selection was based on an unsupervised LDA, which implies that the selected papers might not always be the same for all times the model is running. Therefore, the findings here represent a theoretical approach, and any generalization must be cautious. For this reason, other automatic techniques should be tested, and the differences between the results should be investigated.

Author Contributions: Conceptualization, G.A.Q., P.N.A.J. and I.C.M.; methodology, P.N.A.J. and I.C.M.; validation, G.A.Q. and I.C.M.; formal analysis, G.A.Q. and I.C.M.; investigation, G.A.Q. and I.C.M.; resources, G.A.Q. and I.C.M.; writing—original draft preparation, G.A.Q. and I.C.M.; writing—review and editing, G.A.Q., P.N.A.J. and I.C.M.; funding acquisition, P.N.A.J. All authors have read and agreed to the published version of the manuscript.

Funding: The Universidad Católica del Norte (UCN) partially supported the payment of the Article Processing Charge.

Institutional Review Board Statement: Not applicable.

Informed Consent Statement: Not applicable.

Data Availability Statement: Not applicable.

Conflicts of Interest: The authors declare no conflict of interest.

Appendix A. SLR Protocol

Literature Review Protocol	
Objective	To understand how digitalization enables Lean in SMEs.
Research Question	RQ1. How can digitalization support Lean implementation in SMEs? RQ2. How can digitalization be an enabler in implementing Lean-Green in SMEs?
Keywords and Synonyms	digital transformation; digital transition; digital innovation; digitalization; industry 4.0; SME; small- and medium-sized enterprises; digital media; social media; social network; internet; technology; ICT; small and middle compan*; small and middle firm; lean; Toyota production system; just in time.
Source Selection Criteria Definition	Criteria: The sources should be available and globally recognized as high-quality sources. Studies Language: English. Source Search Methods: The sources should be available and globally recognized as high-quality sources. Source List: Web of Sciences and Scopus.
Studies Type Definition	Research published in journals, books, and conferences.
Studies Initial Selection:	3 August 2022.
Studies Quality Evaluation:	The quality is defined by the databases selected.

Literature Review Protocol		
Search string	Web of Science (38 papers returned): TS = ((((((“digital transformation” or “digital transition” or “digitalization” or “industry 4.0” or “digital media” or “social media” or “social network” or “internet” or “digital” or “technology” or “ICT”) and (“lean” OR “toyota production system” OR “Just in time ”) and (“sme” or “small and medium firm” or “small and medium compan*” or “small and medium-sized enterprise”)))))) Scoupus (182 papers returned): TITLE-ABS-KEY (((“digital transformation” OR “digital transition” OR “digitalization” OR “industry 4.0” OR “digital media” OR “social media” OR “social network” OR “internet” OR “digital” OR “technology” OR “ICT”) AND (“lean” OR “toyota production system” OR “Just in time”) AND (“sme” OR “small and medium firm” OR “small and medium compan*” OR “small and medium-sized enterprise”)))	
Data Extraction Form Fields	Filter 1—Exclusion of duplicate studies. Filter 2—Download of papers. Filter 3—Topic modelling application for papers extraction.	
	Inclusion Criteria	**Exclusion Criteria**
Study Selection Criteria (Filter 4), executed through human reading	• The paper addresses some aspects of digital transformation and Lean in SMEs. • The paper addresses some aspects of digital transformation and Lean-Green in SMEs.	• The paper does not consider digital transformation. • The paper does not consider SMEs. • The paper does not consider any Lean practice. • The paper is not in English (only the abstract is in English).
Studies analysis	Forty studies were analyzed, focusing on answering the RQs.	

References

1. Rauch, E.; Dallasega, P.; Matt, D.T. Critical Factors for Introducing Lean Product Development to Small and Medium Sized Enterprises in Italy. *Procedia CIRP* **2017**, *60*, 362–367. [CrossRef]
2. Dombrowski, U.; Crespo, I.; Zahn, T. Adaptive Configuration of a Lean Production System in Small and Medium-Sized Enterprises. *Prod. Eng.* **2010**, *4*, 341–348. [CrossRef]
3. Siegel, R.; Antony, J.; Garza-Reyes, J.A.; Cherrafi, A.; Lameijer, B. Integrated Green Lean Approach and Sustainability for SMEs: From Literature Review to a Conceptual Framework. *J. Clean. Prod.* **2019**, *240*, 118205. [CrossRef]
4. Womack, J.P.; Jones, D.T. *Lean Thinking: Banish Waste and Create Wealth in Your Corporation*; Free Press: New York, NY, USA, 2003.
5. Da Silva, F.W.S.; Alves, A.C.; Figueiredo, M.C.B. Lean Production in Small and Medium Sized Companies from the Free Economic Zone of Manaus: A Reality or Just Fiction? *Gestao e Producao* **2019**, *26*, 1–14. [CrossRef]
6. AlMulhim, A.F. Smart Supply Chain and Firm Performance: The Role of Digital Technologies. *Bus. Process Manag. J.* **2021**, *27*, 1353–1372. [CrossRef]
7. Khalid, B.; Naumova, E. Digital transformation SCM in view of COVID-19 from Thailand SMEs perspective. *Glob. Chall. Digit. Transform. Mark.* **2021**, 49–66.
8. AlMujaini, H.; Hilmi, M.F.; Abudaqa, A.; Alzahmi, R. Corporate Foresight Organizational Learning and Performance: The Moderating Role of Digital Transformation and Mediating Role of Innovativeness in SMEs. *Int. J. Data Netw. Sci.* **2021**, *5*, 703–712. [CrossRef]
9. Jaiswal, P.; Singh, A.; Misra, S.C.; Kumar, A. Barriers in Implementing Lean Manufacturing in Indian SMEs: A Multi-Criteria Decision-Making Approach. *J. Model. Manag.* **2021**, *16*, 339–356. [CrossRef]
10. Marinho, B.F.D.; Costa Melo, I. Fostering Innovative SMEs in a Developing Country: The ALI Program Experience. *Sustainability* **2022**, *14*, 13344. [CrossRef]
11. Lu, Y.; Liu, Z.; Min, Q. A Digital Twin-Enabled Value Stream Mapping Approach for Production Process Reengineering in SMEs. *Int. J. Comput. Integr. Manuf.* **2021**, *34*, 764–782. [CrossRef]
12. Mohammed, I.K.; Trzcielinski, S. The Interconnections Between ICT, Industry 4.0 and Agile Manufacturing. *Manag. Prod. Eng. Rev.* **2021**, *12*, 99–110. [CrossRef]
13. Rozak, H.; Adhiatma, A.; Fachrunnisa, O.; Rahayu, T. Social Media Engagement, Organizational Agility and Digitalization Strategic Plan to Improve SMEs' Performance. *IEEE Trans. Eng. Manag.* **2021**, 1–10. [CrossRef]
14. Klimecka-Tatar, D.; Ingaldi, M. Digitization of Processes in Manufacturing SMEs-Value Stream Mapping and OEE Analysis. *Procedia Comput. Sci.* **2022**, *200*, 660–668. [CrossRef]
15. Minh, N.D.; Kien, D.T. A Four-Phase Framework for Lean Implementation in Small and Medium Enterprises. *Management* **2021**, *25*, 259–277. [CrossRef]

16. Gandhi, N.S.; Thanki, S.J.; Thakkar, J.J. Ranking of Drivers for Integrated Lean-Green Manufacturing for Indian Manufacturing SMEs. *J. Clean. Prod.* **2018**, *171*, 675–689. [CrossRef]
17. Thanki, S.; Thakkar, J.J. An Investigation on Lean–Green Performance of Indian Manufacturing SMEs. *Int. J. Product. Perform. Manag.* **2020**, *69*, 489–517. [CrossRef]
18. Sá, J.C.; Reis, M.; Dinis-Carvalho, J.; Silva, F.J.G.; Santos, G.; Ferreira, L.P.; Lima, V. The Development of an Excellence Model Integrating the Shingo Model and Sustainability. *Sustainability* **2022**, *14*, 9472. [CrossRef]
19. Hoellthaler, G.; Braunreuther, S.; Reinhart, G. Digital Lean Production-An Approach to Identify Potentials for the Migration to a Digitalized Production System in SMEs from a Lean Perspective. *Procedia CIRP* **2018**, *67*, 522–527. [CrossRef]
20. Ferreira, W.D.P.; Armellini, F.; de Santa-Eulalia, L.A.; Thomasset-Laperrière, V. Extending the lean value stream mapping to the context of Industry 4.0: An agent-based technology approach. *J. Manuf. Syst.* **2022**, *63*, 1–14. [CrossRef]
21. Saha, B. Application of Topic Modelling for Literature Review in Management Research. In *Interdisciplinary Research in Technology and Management*; CRC Press: Boca Raton, FL, USA, 2021; pp. 249–256. ISBN 9781003202240.
22. Asmussen, C.B.; Møller, C. Smart Literature Review: A Practical Topic Modelling Approach to Exploratory Literature Review. *J. Big Data* **2019**, *6*, 93. [CrossRef]
23. Riedl, R.; Benlian, A.; Hess, T.; Stelzer, D.; Sikora, H. On the Relationship between Information Management and Digitalization. *Bus. Inf. Syst. Eng.* **2017**, *59*, 475–482. [CrossRef]
24. Mertens, P.; Wiener, M. Riding the Digitalization Wave: Toward a Sustainable Nomenclature in Wirtschaftsinformatik. *Bus. Inf. Syst. Eng.* **2018**, *60*, 367–372. [CrossRef]
25. Legner, C.; Eymann, T.; Hess, T.; Matt, C.; Böhmann, T.; Drews, P.; Mädche, A.; Urbach, N.; Ahlemann, F. Digitalization: Opportunity and Challenge for the Business and Information Systems Engineering Community. *Bus. Inf. Syst. Eng.* **2017**, *59*, 301–308. [CrossRef]
26. Gong, C.; Ribiere, V. Developing a Unified Definition of Digital Transformation. *Technovation* **2021**, *102*, 102217. [CrossRef]
27. Culot, G.; Nassimbeni, G.; Orzes, G.; Sartor, M. Behind the Definition of Industry 4.0: Analysis and Open Questions. *Int. J. Prod. Econ.* **2020**, *226*, 107617. [CrossRef]
28. Bouwman, H.; Nikou, S.; de Reuver, M. Digitalization, Business Models and SMEs: How Do Business Model Innovation Practices Improve Performance of Digitalizing SMEs? *Telecommun. Policy* **2019**, *43*, 101828. [CrossRef]
29. Liker, J.K.; Morgan, J.M. The Toyota Way in Services: The Case of Lean Product Development. *Acad. Manag. Perspect.* **2006**, *20*, 5–20. [CrossRef]
30. Jastia, N.V.K.; Kodali, R. Lean Production: Literature Review and Trends. *Int. J. Prod. Res.* **2015**, *53*, 867–885. [CrossRef]
31. Mostafa, S.; Dumrak, J.; Soltan, H. A Framework for Lean Manufacturing Implementation. *Prod. Manuf. Res.* **2013**, *1*, 44–64. [CrossRef]
32. Shah, R.; Ward, P.T. Defining and Developing Measures of Lean Production. *J. Oper. Manag.* **2007**, *25*, 785–805. [CrossRef]
33. Liker, J.K. *Toyota Way: 14 Management Principles from the World's Greatest Manufacturer*; McGraw-Hill Education: New York, NY, USA, 2004.
34. Arlbjørn, J.S.; Freytag, P.V.; de Haas, H. Service Supply Chain Management: A Survey of Lean Application in the Municipal Sector. *Int. J. Phys. Distrib. Logist. Manag.* **2011**, *41*, 277–295. [CrossRef]
35. Jaiswal, P.; Kumar, A.; Prasad, K. Assessment of Drivers to Implement Lean Manufacturing in Indian SMEs Using Intuitionistic Fuzzy Based TOPSIS Approach. *Int. J. Mod. Manuf. Technol.* **2017**, *9*, 30–38.
36. Losonci, D.; Demeter, K. Lean Production and Business Performance: International Empirical Results. *Compet. Rev.* **2013**, *23*, 218–233. [CrossRef]
37. Teixeira, P.; Sá, J.C.; Silva, F.J.G.; Ferreira, L.P.; Santos, G.; Fontoura, P. Connecting Lean and Green with Sustainability towards a Conceptual Model. *J. Clean. Prod.* **2021**, *322*, 129047. [CrossRef]
38. Bhattacharya, A.; Nand, A.; Castka, P. Lean-Green Integration and Its Impact on Sustainability Performance: A Critical Review. *J. Clean. Prod.* **2019**, *236*, 117697. [CrossRef]
39. Garza-Reyes, J.A. Lean and Green-a Systematic Review of the State of the Art Literature. *J. Clean. Prod.* **2015**, *102*, 18–29. [CrossRef]
40. Cobra, R.L.R.B.; Guardia, M.; Queiroz, G.A.; Oliveira, J.A.; Ometto, A.R.; Esposto, K.F. "Waste" as the Common "Gene" Connecting Cleaner Production and Lean Manufacturing: A Proposition of a Hybrid Definition. *Environ. Qual. Manag.* **2015**, *25*, 25–40. [CrossRef]
41. Cherrafi, A.; Elfezazi, S.; Chiarini, A.; Mokhlis, A.; Benhida, K. The Integration of Lean Manufacturing, Six Sigma and Sustainability: A Literature Review and Future Research Directions for Developing a Specific Model. *J. Clean. Prod.* **2016**, *139*, 828–846. [CrossRef]
42. Dieste, M.; Panizzolo, R.; Garza-Reyes, J.A. Evaluating the Impact of Lean Practices on Environmental Performance: Evidences from Five Manufacturing Companies. *Prod. Plan. Control* **2020**, *31*, 739–756. [CrossRef]
43. Iranmanesh, M.; Zailani, S.; Hyun, S.; Ali, M.; Kim, K. Impact of Lean Manufacturing Practices on Firms' Sustainable Performance: Lean Culture as a Moderator. *Sustainability* **2019**, *11*, 1112. [CrossRef]
44. Caldera, H.T.S.; Desha, C.; Dawes, L. Evaluating the Enablers and Barriers for Successful Implementation of Sustainable Business Practice in 'Lean' SMEs. *J. Clean. Prod.* **2019**, *218*, 575–590. [CrossRef]
45. Charrua-Santos, F.; Santos, B.P.; Enrique, D.V.; Alberto, A.; Bibete, H.; Osório, G.J.; Lima, T.M. An Overview of Lean Production and Industry 4.0 in Different Context. In Proceedings of the ICITM 2020—2020 9th International Conference on Industrial Technology and Management, Oxford, UK, 11–13 February 2020; pp. 69–72. [CrossRef]

46. Chen, J.C.; Chen, K.M. Application of ORFPM System for Lean Implementation: An Industrial Case Study. *Int. J. Adv. Manuf. Technol.* **2014**, *72*, 839–852. [CrossRef]
47. Kolberg, D.; Zühlke, D. Lean Automation Enabled by Industry 4.0 Technologies. *IFAC-PapersOnLine* **2015**, *48*, 1870–1875. [CrossRef]
48. Sanders, A.; Elangeswaran, C.; Wulfsberg, J. Industry 4.0 Implies Lean Manufacturing: Research Activities in Industry 4.0 Function as Enablers for Lean Manufacturing. *J. Ind. Eng. Manag.* **2016**, *9*, 811. [CrossRef]
49. Kolla, S.; Minufekr, M.; Plapper, P. Deriving Essential Components of Lean and Industry 4.0 Assessment Model for Manufacturing SMEs. *Procedia CIRP* **2019**, *81*, 753–758. [CrossRef]
50. Garcia-Garcia, G.; Singh, Y.; Jagtap, S. Optimising Changeover through Lean-Manufacturing Principles: A Case Study in a Food Factory. *Sustainability* **2022**, *14*, 8279. [CrossRef]
51. Garcia-Garcia, G.; Coulthard, G.; Jagtap, S.; Afy-Shararah, M.; Patsavellas, J.; Salonitis, K. Business Process Re-Engineering to Digitalise Quality Control Checks for Reducing Physical Waste and Resource Use in a Food Company. *Sustainability* **2021**, *13*, 12341. [CrossRef]
52. Souza, J.P.E.; Alves, J.M. Lean-Integrated Management System: A Model for Sustainability Improvement. *J. Clean. Prod.* **2018**, *172*, 2667–2682. [CrossRef]
53. Vinodh, S.; Ben Ruben, R.; Asokan, P. Life Cycle Assessment Integrated Value Stream Mapping Framework to Ensure Sustainable Manufacturing: A Case Study. *Clean Technol. Environ. Policy* **2016**, *18*, 279–295. [CrossRef]
54. Faulkner, W.; Badurdeen, F. Sustainable Value Stream Mapping (Sus-VSM): Methodology to Visualize and Assess Manufacturing Sustainability Performance. *J. Clean. Prod.* **2014**, *85*, 8–18. [CrossRef]
55. Kaswan, M.S.; Rathi, R. Green Lean Six Sigma for Sustainable Development: Integration and Framework. *Env. Impact Assess. Rev.* **2020**, *83*, 106396. [CrossRef]
56. Lucato, W.C.; Vieira, M.; Da Silva Santos, J.C. Eco-Six Sigma: Integration of Environmental Variables into the Six Sigma Technique. *Prod. Plan. Control.* **2015**, *26*, 605–616. [CrossRef]
57. Helleno, A.L.; de Moraes, A.J.I.; Simon, A.T.; Helleno, A.L. Integrating Sustainability Indicators and Lean Manufacturing to Assess Manufacturing Processes: Application Case Studies in Brazilian Industry. *J. Clean. Prod.* **2017**, *153*, 405–416. [CrossRef]
58. European Commission SME Definition. Available online: https://ec.europa.eu/eurostat/web/structural-business-statistics/small-and-medium-sized-enterprises (accessed on 11 October 2022).
59. OECD Brazil | Financing SMEs and Entrepreneurs 2020: An OECD Scoreboard | OECD ILibrary. Available online: https://www.oecd-ilibrary.org/sites/8153da8d-en/index.html?itemId=/content/component/8153da8d-en (accessed on 5 April 2022).
60. Library of Chilean Congress [Biblioteca del Congreso Chileno] Estatuto de Las PYMES—Ley Fácil—Biblioteca Del Congreso Nacional de Chile. Available online: https://www.bcn.cl/leyfacil/recurso/estatuto-de-las-pymes (accessed on 7 January 2022).
61. Government of Japan Small and Medium Enterprise Agency. Available online: https://www.chusho.meti.go.jp/sme_english/outline/07/01.html (accessed on 5 April 2022).
62. U.S. Small Business Administration (SBA). *Table of Size Standards*; U.S. Small Business Administration (SBA): Washington, DC, USA, 2022.
63. World Trade Organization. *Levelling the Trading Field for SMEs*; World Trade Organization: Geneva, Switzerland, 2017.
64. Kim, S.-S. Sustainable Growth Variables by Industry Sectors and Their Influence on Changes in Business Models of SMEs in the Era of Digital Transformation. *Sustainability* **2021**, *13*, 7114. [CrossRef]
65. Troise, C.; Corvello, V.; Ghobadian, A.; O'Regan, N. How Can SMEs Successfully Navigate VUCA Environment: The Role of Agility in the Digital Transformation Era. *Technol. Forecast. Soc. Chang.* **2022**, *174*, 121227. [CrossRef]
66. Kankaanhuhta, V.; Packalen, T.; Väätäinen, K. Digital Transformation of Forest Services in Finland—A Case Study for Improving Business Processes. *Forests* **2021**, *12*, 781. [CrossRef]
67. Dutta, G.; Kumar, R.; Sindhwani, R.; Singh, R.K. Digital Transformation Priorities of India's Discrete Manufacturing SMEs—A Conceptual Study in Perspective of Industry 4.0. *Compet. Rev.* **2020**, *30*, 289–314. [CrossRef]
68. Quinn, K.M.; Monroe, B.L.; Colaresi, M.; Crespin, M.H.; Radev, D.R. How to Analyze Political Attention with Minimal Assumptions and Costs. *Am. J. Pol. Sci.* **2010**, *54*, 209–228. [CrossRef]
69. Lancichinetti, A.; Sirer, M.I.; Wang, J.X.; Acuna, D.; Körding, K.; Amaral, L.A.N. High-Reproducibility and High-Accuracy Method for Automated Topic Classification. *Phys. Rev. X* **2015**, *5*, 011007. [CrossRef]
70. Blei, D.M. Probabilistic Topic Models. *Commun. ACM* **2012**, *55*, 77–84. [CrossRef]
71. Denyer, D.; Tranfield, D. Producing a Systematic Review. In *the Sage Handbook of Organizational Research Methods*; Buchanan, D.A., Bryman, A., Eds.; American Psychological Association: Washington, DC, USA, 2009; pp. 671–689.
72. Moher, D.; Liberati, A.; Tetzlaff, J.; Altman, D.G. Preferred Reporting Items for Systematic Reviews and Meta-Analyses: The PRISMA Statement. *PLoS Med.* **2009**, *6*, e1000097. [CrossRef] [PubMed]
73. Aghaei Chadegani, A.; Salehi, H.; Md Yunus, M.M.; Farhadi, H.; Fooladi, M.; Farhadi, M.; Ale Ebrahim, N. A Comparison between Two Main Academic Literature Collections: Web of Science and Scopus Databases. *Asian Soc. Sci.* **2013**, *9*, 18–26. [CrossRef]
74. Levy, Y.; Ellis, T.J. A Systems Approach to Conduct an Effective Literature Review in Support of Information Systems Research. *Inf. Sci. Int. J. Emerg. Transdiscipl.* **2006**, *9*, 181–212. [CrossRef]
75. Kiatcharoenpol, T.; Laosirihongthong, T.; Chaiyawong, P.; Hierarchy, A.A.; Ahp, P. A Study of Critical Success Factors and Prioritization by Using Analysis Hierarchy Process in Lean Manufacturing Implementation for Thai SMEs. In Proceedings of the 5th

International Asia Conference on Industrial Engineering and Management Innovation; Atlantis Press: Paris, France, 2015; pp. 10–13. [CrossRef]
76. Bakri, A.; Alkbir, M.F.M.; Awang, N.; Januddi, F.; Ismail, M.A.; Aizat Ahmad, A.N.; Zakaria, I.H. Addressing the Issues of Maintenance Management in Smes: Towards Sustainable and Lean Maintenance Approach. *Emerg. Sci. J.* **2021**, *5*, 367–379. [CrossRef]
77. Salma, A.; Anas, C.; Mohammed, E.H. Bibliographic Study on the Difficulties Encountered by SMEs during the Implementation of Lean Manufacturing. *J. Adv. Manuf. Syst.* **2021**, *20*, 163–190. [CrossRef]
78. MLADINEO, M.; Cubic, M.; Nikola, G.; Marina, C.Z. Human-Centric Approach of the Lean Management as an Enabler of Industry 5.0 in SMEs. In Proceedings of the International Conference "Mechanical Technologies and Sctrucutural Materials", Split, Croatia, 23–24 September 2021.
79. Tiwari, R.K.; Tiwari, J.K. Identification of Key Lean Practices within Indian Automotive SMEs Environment. *Int. J. Ind. Syst. Eng.* **2019**, *33*, 17–37. [CrossRef]
80. Wadhwa, R.S. Implementing Continuous Improvement in Metal Casting: Case Study of an Aluminum Foundry. *Adv. Mat. Res.* **2013**, *622*, 433–436. [CrossRef]
81. Anuar, A.; Sadek, D.M.; Abdull Rahman, N.L.; Othman, N. Lean Principles Improving Entrepreneurial Ecosystem in Malaysian SMEs. *Int. J. Supply Chain. Manag.* **2020**, *9*, 73–77.
82. Ruiz, S.; Simón, A.; Sotelo, F.; Raymundo, C. Optimized Plant Distribution and 5S Model That Allows SMEs to Increase Productivity in Textiles. In Proceedings of the 17th LACCEI international Multi-Conference for Engineering, Education and Technology: "Industry, Innovation, And Infrastructure for Sustainable Cities and Communities", Montego Bay, Jamaica, 24–26 July 2019. [CrossRef]
83. Dowlatshahi, S.; Taham, F. The Development of a Conceptual Framework for Just-In-Time Implementation in SMEs. *Prod. Plan. Control.* **2009**, *20*, 611–621. [CrossRef]
84. Jaiswal, P.; Kumar, A. Analyzing Barriers of Lean Manufacturing Adoption in Indian SMEs Using an Integrated Approach of Grey Decision Making Trial and Evaluation Laboratory (DEMATEL). *Lect. Notes Eng. Comput. Sci.* **2016**, *2224*, 688–691.
85. Prasad, S.; Baltov, M.; Neelakanteswara Rao, A.; Lanka, K. Interdependency Analysis of Lean Manufacturing Practices in Case of Bulgarian SMEs: Interpretive Structural Modelling and Interpretive Ranking Modelling Approach. *Int. J. Lean Six Sigma* **2020**, *12*, 503–535. [CrossRef]
86. Iris, C.; Cebeci, U. Analyzing Relationship between ERP Utilization and Lean Manufacturing Maturity of Turkish SMEs. *J. Enterp. Inf. Manag.* **2014**, *27*, 261–277. [CrossRef]
87. Maryam, M.; Rezaei, S.; Shahijan, M.K. Agile Supply Chain Management (ASCM): A Management Decision Making Approach. *Asia Pac. J. Mark. Logist.* **2017**, *29*, 1–17. [CrossRef]
88. Ho, D.C.K.; Mo, D.Y.W.; Wong, E.Y.C.; Leung, S.M.K. Business Intelligence for Order Fulfilment Management in Small and Medium Enterprises. *Int. J. Internet Manuf. Serv.* **2019**, *6*, 169–184. [CrossRef]
89. Goti, A.; de la Calle, A.; Gil, M.J.; Errasti, A.; Bom, P.R.D.; García-Bringas, P. Development and Application of an Assessment Complement for Production System Audits Based on Data Quality, IT Infrastructure and Sustainability. *Sustainability* **2018**, *10*, 4679. [CrossRef]
90. Neubert, M. Digitalization's impact on lean global, start-up firms' internationalization speed. *Innov. Manag. Entrep. Sustain.* **2018**, *volume*, 765–775.
91. Dallasega, P.; Woschank, M.; Ramingwong, S.; Tippayawong, K.Y.; Chonsawat, N. Field Study to Identify Requirements for Smart Logistics of European, US and Asian SMEs. In Proceedings of the International Conference on Industrial Engineering and Operations Management, Bangkok, Thailand, 5–7 March 2019; Volume 2019, pp. 844–855.
92. Mofolasayo, A.; Young, S.; Martinez, P.; Ahmad, R. How to Adapt Lean Practices in SMEs to Support Industry 4.0 in Manufacturing. *Procedia Comput. Sci.* **2022**, *200*, 934–943. [CrossRef]
93. Dossou, P.E.; Rafael, P.; Cristiane, S.; Joao, C.J. How to Use Lean Manufacturing for Improving a Healthcare Logistics Performance. *Procedia Manuf.* **2020**, *51*, 1657–1664. [CrossRef]
94. Bin Zhou, S.; Ji, F.X. Impact of Lean Supply Chain Management on Operational Performance. *Int. J. Bus. Anal.* **2015**, *2*, 1–19. [CrossRef]
95. Peng, C.; Zhong, L. Intelligent Transformation of Small and Medium-Sized Manufacturing Enterprise in China-Case Study of Dongguan Taiwei Electronics. In Proceedings of the 2021 International Conference on Computer Engineering and Artificial Intelligence, Tianjin, China, 23–26 April 2021; pp. 511–518. [CrossRef]
96. Woschank, M.; Dallasega, P. The Impact of Logistics 4.0 on Performance in Manufacturing Companies: A Pilot Study. *Procedia Manuf.* **2021**, *55*, 487–491. [CrossRef]
97. Adeyeri, M.K.; Ayodeji, S.P.; Akinnuli, B.O.; Farayibi, P.K.; Ojo, O.O.; Adeleke, K. *Development of SMEs Coping Model for Operations Advancement in Manufacturing Technology*; Elsevier Inc.: Amsterdam, The Netherlands, 2018; ISBN 9780081024140.
98. Grube, D.; Malik, A.A.; Bilberg, A. Generic Challenges and Automation Solutions in Manufacturing SMEs. In *Proceedings of the Annals of DAAAM and Proceedings of the International DAAAM Symposium, Zadar, Croatia, 8–11 November 2017*; Volume 28, pp. 1161–1169. [CrossRef]
99. Jesemann, I.; Beichter, T.; Constantinescu, C.; Herburger, K.; Rüger, M. Investigation of the "Lean Startup" Approach in Large Manufacturing Companies towards Customer Driven Product Innovation in SMEs. *Procedia CIRP* **2021**, *99*, 711–716. [CrossRef]
100. Papetti, A.; Marilungo, E.; Gregori, F.; Germani, M. Driving Process Innovation: A Structured Method for Improving Efficiency in SMEs. *Procedia CIRP* **2016**, *50*, 448–453. [CrossRef]

101. Rother, M.; Shook, J. *Learning to See: Value Stream Mapping to Add Value and Eliminate MUDA.*; Lean Enterprise Institute, Ed.; Lean Enterprise Institute: Boston, MA, USA, 1999; Volume 1, ISBN 9780966784305.
102. Frontoni, E.; Rosetti, R.; Paolanti, M.; Alves, A.C. HATS Project for Lean and Smart Global Logistic: A Shipping Company Case Study. *Manuf. Lett.* **2020**, *23*, 71–74. [CrossRef]
103. Alshamaila, Y.; Papagiannidis, S.; Li, F. Cloud computing adoption by SMEs in the north east of England. *J. Enterp. Inf. Manag.* **2013**, *26*, 250–275. [CrossRef]

sustainability

Article

Fostering Innovative SMEs in a Developing Country: The ALI Program Experience

Bruno Francisco Diniz Marinho [1,*] and Isotilia Costa Melo [2]

1 Luiz de Queiroz College of Agriculture, University of São Paulo (USP), Piracicaba 13418-445, SP, Brazil
2 Escuela de Ingeniería de Coquimbo (EIC), Universidad Católica del Norte (UCN), Coquimbo 1781421, Chile
* Correspondence: bruno.marinho@mjv.com.br

Abstract: Small and medium enterprises (SMEs) play an essential economic role through income and employment generation as well as reducing inequalities. In this regard, continuous innovation is a pillar for guaranteeing SMEs' survival worldwide. In Brazil, the ALI Program (Portuguese acronym for Local Innovation Agent) trains groups of SMEs to implement a continuous innovation process based on agile methodologies. This paper applied focus group methodology to investigate whether, after participating in the program, SMEs practice some innovation processes in their business and the difficulties perceived by entrepreneurs in incorporating innovation processes into their ventures. Based on the extant literature and on the focus group, it was observed that SMEs see innovation as a risk, fear innovating because of tradition, family, and/or generational context, and do not see innovation as a process suitable to be systematized with agile tools. In addition, factors such as the age of the SMEs' leaders, the age of the SMEs, the gender of the SMEs' leaders, and the nature of the SMEs (family business or not) may affect their openness to innovation. Recommendations are stated for practitioners (such as ALIs) to improve their training quality, policymakers to improve and create similar programs, and researchers interested in future research directions.

Keywords: small and medium-sized enterprises (SMEs); micro and small businesses (MSBs); Brazilian support service for micro and small enterprises (SEBRAE); local innovation agent ALI program; organizational agility; entrepreneurship; lean startup; agile methods; agile tools

Citation: Marinho, B.F.D.; Costa Melo, I. Fostering Innovative SMEs in a Developing Country: The ALI Program Experience. *Sustainability* **2022**, *14*, 13344. https://doi.org/10.3390/su142013344

Academic Editors: José Carlos Sá, Francisco J. G. Silva, Gilberto Santos, Luís Pinto Ferreira and Manuel Pereira Lopes

Received: 15 July 2022
Accepted: 15 September 2022
Published: 17 October 2022

1. Introduction

Small and medium-sized enterprises (SMEs) deserve specific attention as they represent a significant share of global business. SMEs account for 90% of all industries and 50% of employment globally [1]. Their relevance to Latin American countries is evidenced by the fact that they represent 99% of companies in the region and 70% of the job positions [2]. Given this context, innovation is considered the main factor of social and economic development and the source of competitiveness of regions and countries [3].

Olavarrieta and Diaz [4] pointed out that high-quality replication studies in Latin American and emerging markets are frequently desk rejected because they are seen as not helping the knowledge-building process. However, the authors listed five problems that may arrive from this usual editorial practice. Among them, this process creates the wrong incentive for the top-trained researchers in emerging nations, turning their focus on novel and very specific problems, but ones that are not relevant and necessary for their business communities (such as innovation in SMEs).

Consequently, this also raises concerns about the applicability of business theories for SMEs in Latin American and emerging markets, such as organizational agility [3,5]. Moreover, the extant literature demonstrates how SMEs in Brazil are not mature enough regarding innovation. One of the studies indicates that only 3.65% of SMEs are occasional innovators [6]. Another study demonstrates that SMEs have no systematic innovation processes [7]. On the other hand, a study case demonstrated preliminary positive results of the implementation of innovation management in a technology-based SME [8]. In a more recent paper carried out after the COVID-19 pandemic, it was indicated that SMEs need to innovate in their business models, even partially [9].

Some studies work on using agile methodologies in emerging enterprises [10], and others bring the possibility of innovation in microenterprises [11]. However, few studies investigate the cases of applying agile methodologies in SMEs in Brazil, the largest Latin American country. Since 1972, SEBRAE (Portuguese acronym of the Brazilian Support Service for Micro and Small Enterprises), a non-profit organization, has worked to stimulate entrepreneurship and enable the competitiveness and sustainability of micro and small businesses (MSBs) in Brazil [12]. SEBRAE operates in all Brazilian states offering courses, consultancy, and projects to support entrepreneurship and innovation [12]. As well as previous authors investigating SEBRAE, here, SME is assumed as a synonym of MSB [13,14].

Among the SEBRAE programs, the Local Innovation Agent Program (known by the Portuguese acronym ALI Program) was founded in 2008. It is a government-sponsored program and is also the largest innovation fostering program focused on SMEs in Brazil [14]. The ALI Program's objective "is to revolutionize the small business through innovation" [15]. To achieve this purpose, the ALI Program hires local innovation agents (ALIs) from the business sector. Each ALI becomes responsible for performing eight-month training with a group of 20 SMEs on innovation and skills identified as necessary for SMEs' survival, based on organizational agility. An eligible ALI candidate must have graduated from a university in the last ten years, prove experience in the business sector, and be approved. After being approved in this initial phase and before training SMEs in the ALI Program, the ALIs receive training on innovation and agility that is also eliminatory [14]. Hence, the ALI Program training for SMEs is based on and inspired by agile and innovation methodologies.

Like Olavarrieta and Diaz [4] observed for replication research in Latin America, most publications about the ALI Program are limited to journals with a regional scope. Carvalho et al. (2020) [14] executed a literature review about the ALI Program considering Web of Sciences, Scopus, and *Revista de Administração e Inovação* (RAI), the most important Brazilian innovation journal. The authors found 34 papers and identified that most of them only focused on SMEs from a unique Brazilian state. The program was investigated in only 10 of the 27 Brazilian states. The most investigated state is Paraná (seven papers). Of the papers, 13 (38%) were descriptive, 4 (12%) were qualitative, and 17 (50%) were quantitative. Regardless of the research limitations, the authors concluded that SMEs in the ALI Program were able to improve innovation over the program independent of previous management and innovation levels, which is evidence of the relevance of conducting research about the ALI Program and other, similar programs in different Latin American countries and emerging markets.

Along these lines, the current paper aims to register and discuss the understanding of five SMEs in different industries after receiving the eight-month innovation training in the ALI Program in São Paulo between 2019 and 2020. From this point of view, the current paper contributes to bringing a more recent period of investigation to the literature in a broader spread journal. Remarkably, the ALI Program had two versions: one from 2008 to 2018 and another between 2019 and 2020 [14]. Hence, it is important to note whether, during this period, the results differ from the previous versions. The current registration and discussion may be a useful tool for improving and building new versions of the program in the future.

São Paulo is the Brazilian state with the largest population, four times larger than the population of the state of Paraná [16], the most-investigated state. Assuming the number of SMEs is proportional to the number of people, there is a gap in the investigation of the context of the ALI Program application in São Paulo. In this regard, this current paper is mainly directed to practitioners, such as ALI agents, ALI Program leaders, and policymakers, not only from Brazil, but also those interested in applying for similar programs in similar contexts, e.g., other developing countries. The recording and discussion of the

ALI Program experience are important for the improvement of the program as well as for developing similar innovation-fostering programs for SMEs in other countries, especially in Latin American and emerging markets.

Based on the perspective of organizational agility [3,5], the current research also aims to investigate (1) whether participating SMEs practice some innovation processes in their business; and (2) the difficulties perceived by entrepreneurs in incorporating innovation processes into their ventures. Comparing the observations with the extant literature (especially about similar programs in Latin America), the evidence is discussed, considering the particularity of the studied cases and what can be considered for future research directions. In this regard, this paper is also directed to researchers to whom the recording and interpretation of an experience of fostering innovation in SMEs in a less-investigated context (Brazil) may be relevant for building scientific advances. Future research directions are proposed along the text and summarized in the conclusion.

2. Materials and Methods

2.1. Theoretical Background

Like many management practices, adopting agility through the development and promotion of agile methods has been almost entirely driven by consultants and practitioners, with little participation from the academic community in the first stages of evolution [17]. Later, in the context of information system development, a definition of agility was proposed as follows: "continual readiness of a method to rapidly or inherently create change, proactively or reactively embrace change, and learn from change while contributing to perceived customer value (economy, quality, and simplicity), through its collective components and relationships with its environment" [17]. In this regard, agility is different from flexibility and leanness. Flexibility also corresponds to the ability to create change and proactively or reactively embrace change. However, unlike agility, flexibility does it in a timely manner and through its internal components (and relationships with its environment). Moreover, leanness is defined as "the contribution to perceived customer value (economy, quality, and simplicity)" [17]. Thus, flexibility and leanness can be seen as the basis for achieving agility, which, differently from both, also encompasses the creation through its collective components.

Organizational agility refers to the ability of organizations (here, SMEs) to create a new value by adapting their resources and strategies [5]. Organizational agility is also seen as the ability of an SME to cope with the volatile, uncertain, complex, and ambiguous (VUCA) changes and thrive in a competitive environment of continually and unpredictably changing opportunities. Organizational agility is divided into reactive, proactive, and innovative. Reactive agility responds to a change in the market after identifying it. Proactive agility identifies a new customer's desire in the market and changes the organizational strategy to attend to that desire and achieve the maximum return. Innovative agility is centered on developing new products and markets that customers did not know they desired [18]. In this regard, agility is seen as a key ability for SMEs to compete and survive [3,19]. Specifically, for SMEs, digital technologies capability, relational capability, and innovation capability contribute to building organizational agility. In turn, agility has a positive impact on performance [3].

Consequently, agility is also now seen as a fundamental pillar for research in the innovation context. Agility improves the innovation performance of innovative firms by facilitating ambidexterity [20]. Ambidexterity refers to the organizational exploration and exploitation orientation. On the one hand, exploitation is focused on current internal capabilities, knowledge, and well-established processes. Exploitation is usually associated with short-term success and profitability. On the other hand, exploration is focused on discovering new capabilities, learning new knowledge, and creating new ways of doing business. Exploration is associated with uncertain outcomes, high autonomy, and long-term success [21]. Although there is evidence that agility and innovation are connected, i.e., agility affects innovation performance in innovative firms [20], and innovation capability affects agility in SMEs [3], it also raises other questions for further investigations, such as which kind of agility impacts most the innovation performance. SMEs also have inherent characteristics that differentiate them from larger enterprises, such as limited resources and specialization capabilities, as well as more flexibility for adapting to new circumstances [3,22,23]. These characteristics reflect in the innovation [22,23]. For example, it was demonstrated that decision-makers in more innovative SMEs are more likely to adopt social media and public cloud computing [24]. SMEs may also be recommended to take different innovative routes to improve their performance when digital transformation is changing their business model [25].

Given this context, questions such as "does agility affect innovation performance differently in SMEs than in other enterprises?" or "which kind of agility impacts most the innovation performance in SMEs?" may arise as recommended for further investigation. This answer is useful for designing more effective innovation-fostering programs. Innovation has been pointed out as a driver for improving SMEs' performance [25–27]. Additionally, innovative SMEs contribute most significantly to a country's economy (particularly regarding job creation) [3]. Consequently, they have been the center of attention in many studies [3,25,27–30].

Parallel to the concept of agility, there are different intensities of innovation, e.g., incremental, frugal, radical, or disruptive. Incremental innovation often refers to a series of small improvements made to an existing product or service. Frugal innovation is an emerging concept defined as the process of designing a novel product architecture at reasonable prices that provide consumers with the latest applications, as opposed to the existing solution [31]. In this regard, the agile tool Minimum Viable Product (MVP) taught by Ries in his book [32] represents an essential approach for SMEs to achieve frugal innovation. A study stated that frugal innovation achievement is a very tricky and complex task for SMEs [33].

Radical innovation requires creating new internal knowledge within an organization that generates the successful commercialization of a novel product or service. The concept of disruptive innovation is defined by Bower and Christensen [34] as a discovery that enables a significant breakthrough in capabilities. In this regard, disruptive innovation causes changes to the extent that traditional competencies are rendered irrelevant or trivial [35], and existing rules of the competition are disturbed [36]. Searching to answer how SMEs achieve the agility to respond to disruptive digital innovations in their ecosystem, a study concluded that mitigating organizational rigidity is enabled by the mechanism of achieving boundary openness, while developing innovative capability is enabled by the mechanism of achieving organizational adaptability [19].

Interpreting innovation from the perspective of organizational agility suggests that frugal and incremental innovations are more related to reactive agility, while radical innovation is more related to proactive agility. Innovative agility seems to be related to disruptive agility. However, deeper investigations are recommended.

Besides proactive, reactive, and innovative agility, intellectual agility is a common term in the literature. Intellectual agility is often considered a synonym for the wider concept of organizational agility [5]. However, intellectual agility is rooted in intellectual capital. Hence, intellectual agility is about creating an environment within the organizations in which employees can invest their efforts in the formulation of responses to organizational challenges through modifications in the existing product, process, and management, and creating innovative strategies [5].

Knowledge understood as intellectual capital has been pointed out as essential for explaining the cumulative growth of innovative SMEs. Intellectual capital has three components, i.e., human capital, structural capital, and relational capital. Among them, human capital seems to be unable to directly influence growth if not through social capital and, to a very low degree, through relational capital. Thus, the key to growth seems to be fostering SMEs' capability for transforming knowledge from human capital into social capital (i.e., organizational value) [29].

Finally, based on 110 SMEs in Serbia, a paper [5] investigated the nexus between entrepreneurial leadership, human capital, and innovation. The authors concluded that the intellectual agility of employees positively influences the innovativeness of SMEs, but this effect is strongly mediated through entrepreneurial leadership. In this regard, the relationship between intellectual capital and intellectual agility is an open field for further investigations in SMEs.

2.2. *Innovation Fostering Programs Focused on SMEs in Latin America*

Multiple searches for similar programs in Latin America (i.e., focused on fostering innovation in SMEs) were executed in Scopus in August 2022. Eleven papers were identified: four in Spanish and seven in English. This may be evidence that the Latin American context is investigated less frequently. The papers studied programs in Mexico, Argentina, Chile, Colombia, Costa Rica, and Venezuela. The most investigated country was Mexico [37–41]. Four papers did not mention any program, but rather recommendations for program creation [42–45]. It is worth noting that none of the mentioned programs were based on agility, such as the ALI Program.

The extant literature about the programs can be divided between those programs that directly finance innovation [37–39,46] and those that finance indirectly [40,41,47], e.g., through training and consulting, such as the ALI Program. Similarly, there were multi-sectorial programs [37–39,46] and specific programs, e.g., focused on SMEs in agriculture [40,48]. The ALI Program has no sectorial focus. Some programs also promote innovation to improve SMEs' exporting [43,45,46] or create jobs [47]. The ALI Program promotes innovation to increase SMEs' competitiveness, guaranteeing survival.

Among the analyzed papers, one [38] investigated the collaboration between SMEs and academia regarding open innovation. A government program in Mexico sponsored this collaboration. This program relates to what is proposed by the ALI Program regarding innovation learning. In the Mexican case, the SMEs participated in the funding program with projects to develop innovation and collaborated with a research center or university. The authors applied an absorptive capacity approach to understanding the process of developing new knowledge for achieving innovation.

Drawing on the ambidexterity literature, the authors [38] interpret the organizational exploration and exploitation orientation results. Although the sample of the Mexican program was heterogeneous and limited (16 SMEs), it was possible to conclude that the exploitation of new knowledge is a complex dimension for creating value from collaboration, which makes the outcome difficult to measure using traditional means. In this case, the authors argued that ambidexterity for developing new knowledge for innovation is based on exploiting characteristics as an iterative learning process between SMEs and universities, while exploring characteristics is integrally delegated to the university [38].

On this subject, the orientation of the ALI Program is not clear. Before becoming an ALI, the selected candidates receive training. After this training, they conduct an eight-month training with groups of 20 SMEs (a cycle). After each cycle, the ALI is instructed to write a scientific paper under the supervision of an experienced academic. However, as a search in Scopus can reveal, absolutely no paper from an ALI has been published in English yet.

It is possible to assume that the ALI Program attempts to connect SMEs to universities and research centers. However, this attempt is weak and underdeveloped. Other strategies should be investigated and benchmarked beyond the Latin American context to strengthen this connection between SMEs and academia. Moreover, researchers could investigate more about the ambidexterity development of the SMEs in the ALI or similar programs.

Being a university graduate is a requirement for any candidate to become an ALI [13]. It should be ideal that during the training to become an ALI, candidates receive a scientific education equivalent to a master's degree so that ALIs could deliver more in terms of developing new knowledge for innovation based on exploring characteristics and strengthening ambidexterity.

2.3. Brazilian Context

The ALI Program is an important program that promotes innovation among SMEs in Brazil. It started in 2008 with pilot versions in the Federal District and the state of Paraná. Between 2008 and 2010, the program was conducted by SEBRAE and adopted by some Brazilian states. From December 2010, the program began to be conducted at the national level and promoted in partnership with SEBRAE and the National Council for Scientific and Technological Development (CNPQ). Remarkably, the ALI Program had two versions: one from 2008 to 2018, and another between 2019 and 2020 [14]. In 2019 and 2020, the program served 5840 SMEs in the state of São Paulo (the largest Brazilian state in terms of population) from different industries.

In Brazil, the classification of SMEs follows the parameters of the National Statute of Microenterprises and Small Businesses, created by Complementary Law nº. 123/2006. The purpose of this statute is to simplify and differentiate this sector, stimulating the generation of employment and reducing informality in the economy [49].

According to the statute, SMEs are divided into:

- Individual micro entrepreneur (MEI): gross revenue of up to BRL 81,000.00 per year.
- Micro entrepreneur (ME): gross revenue of up to BRL 360,000.00 per year.
- Small business (EPP): gross revenue greater than BRL 360,000.00 and equal to or less than BRL 4,800,000.00 per year.

The ALI Program attends the three categories, and they are considered SMEs in the current paper. In 2019 and 2020, 146 local innovation agents (ALIs) were selected for the program. Within the program, ALIs become responsible for applying a pre-established methodology that accompanies entrepreneurs (SMEs' representatives) in identifying their customers' problems, creating a value proposition, developing new solutions, and planning their implementation. The objective of the ALI Program is to transform SMEs through innovation processes, and for this, it works with accredited agents (ALIs) who visit SMEs, applying its innovative methodology designed by the CERTI Foundation (CERTI is a Portuguese acronym Reference Centers for Innovative Technologies) [15] and based on the organizational agility.

Among other books, the ALI Program's recommended bibliography presents the books *Scaling Lean: Mastering the Key Metrics for Startup Growth* [50] by Maurya and *Competing Against Luck: A Story of Innovation and Customer Choice* [51] by Christensen. So, the choice for textbooks of the ALI Program represents the state-of-art of scientific discussion on the impacts of organizational agility on innovation for SMEs, translated and adapted to a comprehensive language for the SMEs' practitioners.

At the end of the program, the ALIs have to write a scientific paper reflecting on their experience, but it is not necessary to publish it. This paper is born within the scope of the paper from an ALI in the city of São José dos Campos, São Paulo.

2.4. ALI Program Methodology

The ALI selection process consists of three phases and a training session. In the first stage, a curricular and documentary evaluation of the candidate is carried out. The ALI candidate must have higher education. In the second stage, a knowledge test is carried out involving, among other topics, innovation, business management, entrepreneurship, and knowledge about innovation ecosystems. In the third stage, an individual interview is carried out with the local program managers. After this process, the candidates participate in a qualifying face-to-face training, where the methodology of the program is applied in groups and ends with an application of part of the methodology of each candidate to a real SME. The training has an eliminatory character and is conducted by SEBRAE's consultants. It serves as an evaluation of the behavioral characteristics of the candidates.

When starting their work, the ALIs had the function of applying the methodology to 40 SMEs of different sizes and industries over a period of one and a half years. This service was divided into two eight-month cycles, where each agent was responsible for 20 SMEs per cycle.

The ALI Program has no cost to the SME, and the SME is free to decide whether they will implement the proposed tools or not. The methodology consists of eleven face-to-face meetings with the responsible ALI and a representative of the SME in a sequence of presentations and application of tools designed for the program.

The first meeting is the only group meeting with the ALI and the representatives of all SMEs. At this meeting, the ALI presents the program and all its stages and the main concept of innovation. After this meeting, representatives are invited to fill in a tool called "Innovation Radar" [52]. This is a self-assessment where each representative chooses a score from 1 to 5, measuring SMEs' points defined as follows: Productivity and Cost Reduction, Culture of Innovation, Capital, Network, Processes, Technology, Customer Experience, Opportunities, Presence, Revenue with Innovation, and Market and New Markets. After this assessment, each representative receives an individualized diagnosis of their maturity to innovate and the answer as to whether they are approved to participate in the ALI Program. The individualized diagnosis categorizes the SMEs according to their score as beginners, aspirants, eventual innovators, and impact innovators.

Although this is not a definitive criterion, in general, the SMEs categorized as beginners are indicated for more basic programs offered by SEBRAE, the aspiring and eventual innovators are invited to participate in the ALI Program, and the impact innovators are indicated for a more advanced program.

From the second meeting, the meetings are individual, with each representative of the SMEs. In this last meeting, the ALI leads the representative to define a business model implementation plan, defining cost estimates, objectives, metrics, actions, and deadlines for the work carried out in the program to be implemented and its results to be measured. The representative receives a certificate of completion of the follow-up.

Although the SME is free to decide whether they will implement the proposed tools or not, after 6–12 months of participation in the program, the ALI returns to accompany and reassess the SME. Reassessing the innovation radar is mandatory and reassessing the management is optional. This is the way the success of the program may be evaluated [53].

2.5. Methodological Procedures

A focus group is established in qualitative research. Although there are many possible variations of the method, a focus group basically consists of an informal discussion among selected individuals about specific topics. Thus, a focus group enables access to participants' own language, concepts, and concerns. It fosters the production of more fully articulated conceptualizations and enables the observation of the collective sense-making process [54].

A focus group is usually one or more discussion groups. Each group has 6–8 (and rarely more than 12 participants) from a pre-existing group (here, representatives/leaders of the SMEs that participated in the ALI Program). Participants focus collectively upon a topic selected by the researcher (ALI) and presented to them. The most common way of presenting a topic is through a set of questions [54].

In the current application, the questions were freely elaborated by the ALI based on the experience in the ALI Program. They were:

- What is innovation to you?
- How do you practice innovation in your enterprise?
- What are the obstacles for your enterprise to be more innovative?
- What opportunities do you see that innovation can bring?

It is recommended that focus group participants be homogeneous, particularly concerning "status" factors such as occupation, social class, or age. The focus group study was carried out after the end of the first cycle of the ALI Program [54]. This cycle had the registration of 20 SMEs, with 19 enterprises participating until the end of the cycle. The event was optional for participants, and all SMEs were invited. Five representatives of the SMEs showed up, and they were investigated as one focus group. They were considered to be homogeneous once they all had leadership positions at their SMEs. The focus group was recorded, a recommended practice for this research method [54].

Three of the representatives were women and two men. The names of the representatives and SMEs have been changed to preserve the identity of the participants:

- Alexander, manager of Confection Prime, a textile SME.
- Arielle from Accounting Speedy, an accounting SME.
- Cindy from Coffee Daisy, an SME in the commercial industry that is a coffee shop and flower shop.
- Denzel from Emporium Fresh, a natural products emporium.
- Nikki, manager of Business Intellect, a consultant network.

Focus groups are an adequate methodological choice when the purpose is to elicit people's understandings, opinions, and views. One of the advantages of the focus group method is its flexibility and consequent potential breadth of use. However, the disadvantages lie in its limited reliability and validity and various forms of moderator and respondent bias [54].

Deciding on using a focus group method can be based on three considerations: the purpose of the research, the kind of output desired, and the practical aspects of conducting focus groups [54]. When carrying out work that investigates changes in mentality, the confrontation with thoughts, emotions, dilemmas, and other subjective aspects of the respondents is assumed; Rey [55] states that qualitative research is the best approach to bring a more authentic expression of the subjects involved. The emergence of the subject is legitimated, considering research as a communication process that intends to facilitate the authentic expression of the studied subjects [55].

The purpose of this research was wholly aligned with the focus group method to better understand how people feel or think about a problem, idea, product, or service [56]. Given the restrictions of being one ALI in one city studying a unique cycle and the infrastructure availability (a meeting room with recording equipment), the research should have a qualitative, descriptive, and exploratory nature. Thus, the focus group was judged an adequate method.

The event was held on 11 November 2020, in the service space of the SME called Coffee Daisy. At first, the ALI presented some concepts about innovation to the participants and then started a discussion about the application of innovation in the participating SMEs. After this presentation, the agent encouraged the participants to talk about how innovation can be applied in their SMEs, asking them the directive questions. Based on the question, the entrepreneurs raised their points of view and exchanged information with the other members of the group. The agent also recorded the findings of the event in the format of a paper, which was corrected and supervised by an experienced academic.

3. Results

The ALI systematically compiled the information at the event in Table 1. The names of the SMEs and their representative participants have been changed for privacy reasons. This table presents statements made by entrepreneurs regarding their understanding of innovation and management practices, identifying opportunities for how innovation can improve their enterprises, and the risks associated with implementing innovative business models.

The information collected in the focus group shows that SMEs are motivated and recognize the importance of making innovations in their businesses. However, the following considerations can be made about the practice of innovation processes in their business and the perceived difficulties in incorporating innovation processes into their ventures.

Table 1. Summary of results of the focus group with entrepreneurs.

Representative/SME	SME Classification	Gender	Statements	Opportunities	Risks
Alexander/Confection Prime	*EPP*	Man	Innovation is related to the insights that the entrepreneur has daily.	Innovation must always be exercised.	One difficulty in innovating is overcoming the fear of the risks involved.
				The entrepreneur can overcome the fears involved in the "act" of innovating.	In family businesses, implementing innovations can be more difficult compared to the way of managing previous generations.
Arielle/Accounting Speedy	*EPP*	Woman		Recently changed working tools (software).	
				Accounting is a hypercompetitive market, so it is important to do it differently.	
				Changes in the physical space proved to be efficient in terms of the relationship of employees and in the way of working.	
				The enterprise changed its approach to customer relations, carrying out closer and more individualized communication.	
				Changes in laws make accounting enterprises need to adapt to new trends constantly.	
Cindy/Coffee Daisy	*ME*	Woman	The enterprise focuses more on the purpose and desire of businesswomen than on customer needs.	She talks a lot with her customers informally.	Afraid of "doing different".
Denzel/Emporium Fresh	*MEI*	Man			Due to their more traditional worldview, it is difficult for entrepreneurs from another generation to innovate.
					It is difficult to undertake due to financial insecurity in a business.
Nikki/Business Intellect	*MEI*	Woman		Launching new products and services can be a good solution to financial stagnation.	
				Observing substitute competitors can help to understand unmet customer needs and improve your offering.	

The observation "The enterprise focuses more on the purpose and desire of businesswomen than on customer needs" indicates that the Coffee Daisy is very far from the customer-centric approach. Customer centricity is a pillar of the agile philosophy. During the ALI Program, the SMEs have repeatedly come in contact with the concept of "Jobs to be done". This concept emphasizes the importance of listening to the customer to understand what results they are looking for. In addition, the SMEs conducted several interviews with customers during the program. However, this SME failed to see value in listening to its customers and creating a systematic opportunity discovery through systematically summarizing informal conversations. Although encouraged to build minimum viable products (MVP) (i.e., incremental and frugal innovation) in a validated learning process with constant iterations (organizational proactive and reactive agility), the SME was unable to overcome the fear of "doing it differently", as mentioned by the businesswoman.

This same fear appears in Denzel's statement, who fears innovation due to financial risks, even having used the construction of MVP, which is a tool precisely to mitigate risks and validate the product on a small scale before investing a lot. The generational statement where the fear of innovating due to a more traditional worldview reveals an aspect that Dabic and colleagues [5] already stated. While investigating intellectual agility, these authors observed that younger employees could bring innovation to SMEs and concluded that hiring younger employees into the team is a good recruitment practice [5]. In this regard, executing future focus groups with participants from the same generation may be interesting.

Businesswoman Arielle demonstrates a greater sensitivity to the results expected by her customers, claiming to have made changes in the service of her clients, making her interactions more individualized. The businesswoman did not state whether this improvement arose from an opportunity arising from the observations of her customers' needs. The absence of this statement may demonstrate a devaluation of the importance of listening to the customer's voice. In addition, the businesswoman points to specific changes such as work software and room layout, which can improve deliveries. Still, as they do not involve continuous processes, they do not guarantee a substantial change around a more innovative performance in the market. Changes in legislation are mentioned as influencing the company's performance. Although they bring value to the customer, they are conditional changes for the continuity of an accounting business. The enterprise's tradition is mentioned as an impediment to growth.

Businesswoman Nikki sees the launch of new products and services as a solution to periods of financial stagnation. Again here, innovation is not seen from a systematic bias, but as an individualized action for a specific problem. The representative also points out the importance of observing indirect competitors who serve their customers better. However, it does not mention a defined and continuous process in its enterprise that allows the understanding that it is a method incorporated into the business practice.

In his speech, Alexander showed that he sees innovation as spontaneous insights from everyday life, which may indicate a lack of understanding of the processes and tools that encourage entrepreneurs and employees to develop new ideas (i.e., systematic innovation based on agile tools). Some statements reinforce the view of innovation as spontaneous ideas rather than systems and methodologies applied in enterprise management when referring to innovating as the "act of innovating". The entrepreneur reveals fears of innovating and the need to overcome them, and the generational obstacle appears again when the difficulties of innovating in a family business are mentioned.

From this discussion stimulated by the focus group, it is possible to see some persistent points in the statements of SMEs:

1. The view of innovation as a risk.
2. The fear of innovating because of tradition, family, and/or generational contexts.
3. Innovation is not seen as a process/system.

Although the ALI Program brings these SMEs a vision that is more aligned with the best practices and consolidated concepts in the area of innovation, this is not enough to internalize in its participants the need to change processes and work systematically in the search for opportunities with customers, in the collaboration between the team and an anticipatory vision of the future, and exploring new opportunities, as mentioned by Dabic and colleagues [5] regarding intellectual agility as an essential role of leaders in SME innovation.

4. Discussion

Once most of the representatives (19 out of 20) attended the program until the end of the training (eight months), they believed in the importance of developing human capital through continuous education. Human capital has been pointed out as the most important pillar of intellectual capital for promoting the cumulative growth of innovative SMEs [29]. Thus, this also indicates that SMEs' representatives may have a practical perception of what the drivers of growth actually are. In the observed cases, this perception agrees with what was observed in the previous literature [29].

The literature also indicated that government programs focused on developing marketing and innovation skills are more effective in promoting SMEs' survival [30]. In this regard, the SME representatives mentioned these skills as being the most important of the training period. However, it is worth noting that the observed program was executed before the pandemic and did not focus on digital transformation. This should have become a critical skill for survival during the pandemic. Hence, future investigations could focus on understanding the impacts of government programs on the survival of SMEs through improving digital transformation and innovation.

It is worth noting that although innovative SMEs are frequently associated with an innovative geographical hub [28], no representative of SMEs mentioned that aspects pertinent to their geographical location were affecting their business and possibilities for innovation. However, Arielle from Accounting Speedy mentioned the impact of national laws on her business, and Denzel from Emporium Fresh mentioned financial insecurity that may be related to the financial conditions of the country (besides the SME financial performance).

Among the analyzed SMEs, two are MEIs (the smallest possible category in terms of revenue, also classified as self-employment), one is ME (intermediary category), and two are EPPs (the largest category). There was no observed tendency to understand innovation in certain ways depending on the SME's size.

On the other hand, three of the five representatives of the SMEs were women (one in each size category of SME). This is a very small sample for making non-gender-biased assumptions. However, none of the women described understanding innovation as a process, while the men showed they understood innovation as a process, but at the same time considered innovation as a risk. In this regard, the association of innovation as a risk may be an obstacle to women in adopting innovation as a process in their SMEs because the female social role construction relates women with non-risky behaviors in many cultures [57,58].

Previous findings pointing out that family SMEs led by women tend to be less engaged in innovation through digital transformation is intriguing [59]. It is also known that gender-based roles are influenced by the local culture [57,58]. In this regard, future investigations should focus on understanding the specific particularities and necessities for equally promoting innovation in SMEs independently of the leaders' and representatives' gender. The gender-focused perspective on future investigations is relevant globally once SMEs are responsible for more than 50% worldwide [1] and most female jobs [60].

On this theme, another correlated topic is the family SME. Two representatives (Alexander from Confection Prime and Denzel from Emporium Fresh, both men) mentioned that their SMEs are family businesses, which is an obstacle to implementing innovation as a process. Hence, further investigation should also replicate the study [59] to confirm the results (i.e., family SMEs led by women tend to be less engaged in innovation) in different cultural contexts. Apparently, in the currently observed context, the expected results may be confirmed.

The age of the leadership and the age of the SME also seem to affect the acceptance of agile and innovation methodologies. Denzel and Alexander mention generational conflict as an obstacle to innovation. Arielle mentioned the "tradition" of the SME as an obstacle, which may be related to the age of the enterprise or to a generational conflict. The literature expects those younger employees to be more open to intellectual agility [5]. Moreover, based on the technological acceptance model (TAM) and social cognitive theory (SCT), a previous paper [61] concluded that, in a family business, younger and older generations have different preferences when adopting smartphones and tablets at work. Generational considerations should be taken into account while training SMEs.

In summary, six factors were discussed as possibly affecting the adoption of innovation and organizational agility. They were the age of the SMEs' leaders, the age of the SMEs, the gender of the SMEs' leaders, the nature of the SMEs (family business or not), geographical position, and the size of the SMEs. Except for the last two, the other four seemed to affect in the following way: older leaders, older SMEs, SMEs led by women, and family SMEs tend to be more resistant to adopting organizational agility and systematization of innovation.

5. Conclusions

The ALI Program is a government program to promote innovation in SMEs. The ALI Program offers eight-month innovation training based on agile methodologies to achieve the innovation objective. One local innovation agent (ALI) is responsible for training 20 SMEs in each eight-month training cycle. After each cycle, the ALI is instructed to write a scientific paper under the supervision of an academic. Here, the experience of an ALI in São Paulo between 2019 and 2020 is presented and discussed based on the perspective of organizational agility.

Regardless of the identified obstacles, the actions of the ALI Program can be understood as a driver for enterprises to understand the need to innovate in a context of uncertainty (VUCA). It is observed that, even for well-established enterprises, the need to innovate and the understanding of what innovation really is raised a debate that had not been carried out by the ALI Program team before. In this sense, the actions of the ALI Program are effective in reinforcing the entrepreneurs' awareness of the importance of innovating and motivating them to do things differently. On the other hand, the SMEs studied here still understand innovation as one-off actions and have not incorporated innovation management into their businesses.

The findings showed that even after finishing the ALI Program, SMEs see innovation as a risk, fear innovating because of tradition, family, and/or generational contexts, and do not see innovation as a process suitable to be systematized with agile tools. Many reasons can be investigated regarding the reasons for the program's unsuccess, and they should be the focus of further research. It is suggested here that the age of SMEs' leaders, the age of the SMEs, the gender of the SMEs' leaders, and the nature of the SMEs (family business or not) could be explicative factors for the observed innovation resistance. Finally, based on what was observed in the literature from similar programs in Latin America, it is recommended that policymakers associate the ALI Program with a financing program so that the SMEs participating in the program have more resources for investing in innovation.

Like all research, this study is not without limitations, and the limitations represent opportunities for further research. First, the sample size (five SMEs) is minimal and does not enable general and robust conclusions. This happens because the participation in the research was completely free and limited to the experience of one ALI during one cycle in one specific region (São José dos Campos, São Paulo state).

To overcome this limitation, it is recommended to the ALI agents and ALI Program leaders to promote coordination and collaboration for data collection among different ALIs and cycles so that the results can be analyzed considering a broader view. This also enables more research with other methods, such as quantitative and mixed approaches, thus improving the strength and the implications of the findings.

It is also recommended that policymakers create strategies for promoting the collaboration between SMEs, and academia for innovation should be built and tested. Specifically, it is recommended that the ALIs receive scientific training, equivalent to a master's degree, before training SMEs. It is assumed that once ALIs have more scientific capacitation, they will be able to show the relevance of scientific knowledge and research to SMEs' businesses. Consequently, more SME representatives will be interested in participating in research.

As pointed out in the introduction, there is a lack of replication studies for Latin American and emerging markets. Once more data are feasible, it is recommended to replicate in Brazil the paper that investigated the nexus between entrepreneurial leadership, human capital, and innovation in Serbia [5]. It is relevant to confirm whether the relationship between the intellectual agility of employees and the innovativeness of SMEs, strongly mediated through entrepreneurial leadership, is confirmed in other contexts for building and improving the theoretical knowledge on the theme.

Second, although the best practices for focus groups were followed [54], the guiding questions were created based on the ALI's experience. Because no guiding questions on the theme were found, it is recommended that researchers and/or ALI agents systematically develop guiding questions based on a literature review and expert validation.

Third, although the focus group was assumed to be homogeneous once all members had occupied similar leadership positions, the findings and previous literature indicate that generation and gender may affect the results, probably due to cultural reasons. So, it is recommended that researchers and/or ALI agents execute focus groups with all members of the same gender. Then, the results from focus groups with different genders could be compared and discussed. The same procedure is recommended for investigating generational differences.

Regarding necessary replication studies, a study about French family SMEs pointed out that gender differences affect the role of IT in SME performance and the importance of family harmony [59]. It is recommended that researchers replicate this study in Latin American and emerging markets once the cultural view of gender roles may affect the results. In addition, it is recommended to policymakers and/or ALI Program leaders to consider specific particularities and necessities for equally promoting innovation in SMEs. These particularities may be based on gender, generation, regional culture, or the nature of the SME (family or non-family business) and may require different educational approaches to reach the same innovative performance. Given this context, promoting research on the theme and generating knowledge may be seen as the first step toward creating adequate policies and educational approaches.

This research work is just the beginning of a better understanding of how the concept of organizational agility can contribute to the innovation of SMEs in practice. It is hoped that this work could contribute a piece to the puzzle.

Author Contributions: Conceptualization, B.F.D.M.; methodology, B.F.D.M.; validation, I.C.M.; formal analysis, B.F.D.M. and I.C.M.; investigation, B.F.D.M. and I.C.M.; resources, B.F.D.M.; writing—original draft preparation, B.F.D.M. and I.C.M.; writing—review and editing, B.F.D.M. and I.C.M.; funding acquisition, I.C.M. All authors have read and agreed to the published version of the manuscript.

Funding: The Universidad Católica del Norte (UCN) partially supported the payment of the Article Processing Charge (APC) with the regular grant of the Escuela de Ingeniería de Coquimbo (EIC). The authors partially supported the APC with their own resources.

Institutional Review Board Statement: Ethical review and approval were waived for this study because all participants were adults, capable, and completely able to exercise their freedom. The questions were asked publicly in an open place. No question or questions were judged to be sensitive. The full answers will not be publicly divulged.

Informed Consent Statement: Informed consent was obtained from all subjects involved in the study.

Data Availability Statement: Not applicable.

Conflicts of Interest: The full answers given by the SME representatives will not be provided because this information is strategical for their business.

References

1. World Bank. World Bank SME Finance: Development News, Research, Data. Available online: https://www.worldbank.org/en/topic/smefinance (accessed on 8 January 2022).
2. Fernández, I.A.; Rosas, M.d.L.A.C. El Desarrollo Sostenible Como Imperativo Estratégico: El Contexto de La Pequeña y Mediana Empresa Latinoamericana. *Rev. Lasallista Investig* **2020**, *16*, 28–43. [CrossRef]
3. Troise, C.; Corvello, V.; Ghobadian, A.; O'Regan, N. How Can SMEs Successfully Navigate VUCA Environment: The Role of Agility in the Digital Transformation Era. *Technol. Forecast. Soc. Change* **2022**, *174*, 121227. [CrossRef]
4. Olavarrieta, S.; Diaz, D. The Strong Need for Extended Research and Replications in Latin American and Emerging Markets. *J. Bus. Res.* **2021**, *127*, 384–388. [CrossRef]
5. Dabić, M.; Stojčić, N.; Simić, M.; Potocan, V.; Slavković, M.; Nedelko, Z. Intellectual Agility and Innovation in Micro and Small Businesses: The Mediating Role of Entrepreneurial Leadership. *J. Bus. Res.* **2021**, *123*, 683–695. [CrossRef]
6. Berne, D.F.; Coda, R.; Krakauer, P.; Donaire, D. The Innovation Challenge in Micro and Small Enterprises (MSE). *Innov. Manag. Rev.* **2019**, *16*, 235–252. [CrossRef]
7. Vasconcelos, R.; Oliveria, M. Does Innovation Make a Difference? *Innov. Manag. Rev.* **2018**, *15*, 137–154. [CrossRef]
8. De Souza, G.; Melo, I.C.; Amaral, D.C. Evaluating the Transition for Self-Managed Teams through Analysis of Roles in Agile Product Development Process in a Technology-Based Startup. *J. Technol. Manag. Appl. Eng.* **2020**, *36*, 2–14.
9. Rodrigues, C.D.; de Noronha, M.E.S. What Companies Can Learn from Unicorn Startups to Overcome the COVID-19 Crisis. *Innov. Manag. Rev.* **2021**. [CrossRef]
10. Nardes, F.B.S.; Miranda, R.C.D.R. Lean Startup e Canvas: Uma Proposta de Metodologia Para Startups. *Rev. Bras. Adm. Cient.* **2014**, *5*, 252–272. [CrossRef]
11. Silva, G.; Dacorso, A.L.R. Inovação Aberta Como Uma Vantagem Competitiva Para a Micro e Pequena Empresa. *Rev. Adm. Innov. RAI* **2013**, *10*, 251–269. [CrossRef]
12. SEBRAE. O Que é SEBRAE? Available online: https://www.sebrae.com.br/sites/PortalSebrae/canais_adicionais/o_que_fazemos#:~{}:text=O%20Servi%C3%A7o%20Brasileiro%20de%20Apoio%20%C3%A0s%20Micro%20e,apoio%20aos%20pequenos%20neg%C3%B3cios%20de%20todo%20o%20pa%C3%ADs (accessed on 11 July 2022).
13. De Carvalho, G.D.G.; Vosgerau, D.S.R.; de Carvalho, H.G.; Candido, R.; Cavalcante, M.B. Competências Dos Agentes Locais de Inovação: Análise de Conteúdo Utilizando Atlas TI. *Rev. Espac.* **2016**, *37*, 1–20.
14. De Carvalho, G.D.G.; de Resende, L.M.M.; de Carvalho, H.G.; Pontes, J.; Correa, R.O. The Local Innovation Agents Program: A Literature Review on the Largest Brazilian Small Business Innovation Support Program. *Int. J. Innov. Sci.* **2020**, *12*, 565–588. [CrossRef]
15. SEBRAE. Programa ALI. Available online: https://www.sebrae.com.br/sites/PortalSebrae/ufs/pe/sebraeaz/ali-agente-local-de-inovacao,f137767836fc7710VgnVCM100000d701210aRCRD (accessed on 11 July 2022).
16. IBGE—Brazilian Institute of Geography and Statistics Cities and States [Cidades e Estados]. Available online: https://ibge.gov.br/cidades-e-estados.html (accessed on 11 July 2022).
17. Conboy, K. Agility from First Principles: Reconstructing the Concept of Agility in Information Systems Development. *Inf. Syst. Res.* **2009**, *20*, 329–354. [CrossRef]
18. Najrani, M. The Endless Opportunity of Organizational Agility. *Strateg. Dir.* **2016**, *32*, 37–38. [CrossRef]
19. Chan, C.M.L.; Teoh, S.Y.; Yeow, A.; Pan, G. Agility in Responding to Disruptive Digital Innovation: Case Study of an SME. *Inf. Syst. J.* **2019**, *29*, 436–455. [CrossRef]
20. Martinez-Sanchez, A.; Perez-Perez, M.; Vicente-Oliva, S. Absorptive Capacity and Technology: Influences on Innovative Firms. *Manag. Res. J. Iberoam. Acad. Manag.* **2019**, *17*, 250–265. [CrossRef]
21. Cenamor, J.; Parida, V.; Wincent, J. How Entrepreneurial SMEs Compete through Digital Platforms: The Roles of Digital Platform Capability, Network Capability and Ambidexterity. *J. Bus. Res.* **2019**, *100*, 196–206. [CrossRef]
22. Chen, Y.-Y.K.; Jaw, Y.-L.; Wu, B.-L. Effect of Digital Transformation on Organisational Performance of SMEs. *Internet Res.* **2016**, *26*, 186–212. [CrossRef]
23. González-Varona, J.; López-Paredes, A.; Poza, D.; Acebes, F. Building and Development of an Organizational Competence for Digital Transformation in SMEs. *J. Ind. Eng. Manag.* **2021**, *14*, 15. [CrossRef]
24. Hassan, S.S.; Reuter, C.; Bzhalava, L. Perception or Capabilities? An Empirical Investigation of the Factors Influencing the Adoption of Social Media and Public Cloud in German SMEs. *Int. J. Innov. Manag.* **2021**, *25*, 2150002. [CrossRef]
25. Bouwman, H.; Nikou, S.; de Reuver, M. Digitalization, Business Models, and SMEs: How Do Business Model Innovation Practices Improve Performance of Digitalizing SMEs? *Telecomm. Policy* **2019**, *43*, 101828. [CrossRef]
26. AlMujaini, H.; Hilmi, M.F.; Abudaqa, A.; Alzahmi, R. Corporate Foresight Organizational Learning and Performance: The Moderating Role of Digital Transformation and Mediating Role of Innovativeness in SMEs. *Int. J. Data Netw. Sci.* **2021**, *5*, 703–712. [CrossRef]

27. Ardito, L.; Raby, S.; Albino, V.; Bertoldi, B. The Duality of Digital and Environmental Orientations in the Context of SMEs: Implications for Innovation Performance. *J. Bus. Res.* **2021**, *123*, 44–56. [CrossRef]
28. Simmie, J. Knowledge Spillovers and Reasons for the Concentration of Innovative SMEs. *Urban Stud.* **2002**, *39*, 885–902. [CrossRef]
29. González-Loureiro, M.; Dorrego, P.F. Intellectual Capital and System of Innovation: What Really Matters at Innovative SMEs. *Intang. Cap.* **2012**, *8*, 239–274. [CrossRef]
30. Najib, M.; Rahman, A.A.A.; Fahma, F. Business Survival of small and medium-sized restaurants through a Crisis: The Role of Government Support and Innovation. *Sustainability* **2021**, *13*, 10535. [CrossRef]
31. Yousaf, Z.; Radulescu, M.; Sinisi, C.I.; Serbanescu, L.; Păunescu, L.M. Towards Sustainable Digital Innovation of SMEs from the Developing Countries in the Context of the Digital Economy and Frugal Environment. *Sustainability* **2021**, *13*, 5715. [CrossRef]
32. Ries, E. *The Lean Startup: How Today's Entrepreneurs Use Continuous Innovation to Create Radically Successful Businesses*, 1st ed.; Crown Publishing: New York, NY, USA, 2011.
33. Nassani, A.A.; Sinisi, C.; Mihai, D.; Paunescu, L.; Yousaf, Z.; Haffar, M. Towards the Achievement of Frugal Innovation: Exploring Major Antecedents among SMEs. *Sustainability* **2022**, *14*, 4120. [CrossRef]
34. Bower, J.L.; Christensen, C.M. Disruptive Technologies: Catching the Wave. *Harv Bus. Rev.* **1995**, *73*, 1–43.
35. Schumpeter, J. *The Theory of Economic Development*; Harvard University Press: Boston, MA, USA, 1934.
36. Lyytinen, K.; Rose, G.M. The Disruptive Nature of Information Technology Innovations: The Case of Internet Computing in Systems Development Organizations. *MIS Q.* **2003**, *27*, 557. [CrossRef]
37. Góngora, G.; García, D.; Madrid, A. Efecto Del Apoyo Público Sobre El Comportamiento Innovador y El Rendimiento En PYMES. *Rev. Cienc. Soc. RCS* **2010**, *16*, 400–417. [CrossRef]
38. Avalos-Quispe, G.; Hernández-Simón, L. Open Innovation in SMEs: Potential and Realized Absorptive Capacity for Interorganizational Learning in Dyad Collaborations with Academia. *J. Open Innov. Technol. Mark. Complex.* **2019**, *5*, 72. [CrossRef]
39. Van Hoof, B.; Lyon, T.P. Cleaner Production in Small Firms Taking Part in Mexico's Sustainable Supplier Program. *J. Clean. Prod.* **2013**, *41*, 270–282. [CrossRef]
40. Donovan, J.; Rutsaert, P.; Domínguez, C.; Peña, M. Capacities of Local Maize Seed Enterprises in Mexico: Implications for Seed Systems Development. *Food Secur.* **2022**, *14*, 509–529. [CrossRef]
41. Solleiro, J.L.; Gaona, C.; Castañón, R. Políticas Para El Desarrollo de Sistemas de Innovación En México. *J. Technol. Manag. Innov.* **2014**, *9*, 98–109. [CrossRef]
42. Morantes, M.E.L.; Ferrer, N.J.L.; Parra, O.D.S.B.; de Hernández, V.E.V. Gestión Tecnológica En Pymes Del Sector Textil Del Municipio Maracaibo-Estado Zulia—Venezuela. *Rev. Venez. Gerenc.* **2018**, *23*, 314–335.
43. Escandón-Barbosa, D.; Ochoa, A.E. Success Factors in the Internalisation of Export Companies in Colombia. *Econ. Reg.* **2021**, *17*, 570–581. [CrossRef]
44. Albarracín, E.J.G.; Lema, D.G.P. De Impacto de La Innovación Sobre El Rendimiento de La Mipyme: Un Estudio Empírico En Colombia. *Estud. Gerenc.* **2012**, *28*, 11–27. [CrossRef]
45. Alvarez, R. Sources of Export Success in Small- and Medium-Sized Enterprises: The Impact of Public Programs. *Int. Bus. Rev.* **2004**, *13*, 383–400. [CrossRef]
46. Monge-González, R.; Rodríguez-Alvarez, J.A.; Leiva, J.C. An Impact Evaluation of a Fund to Finance Innovation in SMEs. *Acad. Rev. Latinoam. Adm.* **2016**, *29*, 20–43. [CrossRef]
47. Castillo, V.; Maffioli, A.; Rojo, S.; Stucchi, R. The Effect of Innovation Policy on SMEs' Employment and Wages in Argentina. *Small Bus. Econ.* **2014**, *42*, 387–406. [CrossRef]
48. Junior, P.N.A.; Melo, I.C.; Yamanaka, L.; Severino, M.R.; Rentizelas, A. Supporting the Bidding Decisions of Smallholder Farmers in Public Calls in Brazil. *Agriculture* **2021**, *12*, 48. [CrossRef]
49. SEBRAE. Lei Geral Da Micro e Pequena Empresa. Available online: https://www.sebrae.com.br/sites/PortalSebrae/artigos/lei-geral-da-micro-e-pequena-empresa,46b1494aed4bd710VgnVCM100000d701210aRCRD (accessed on 11 July 2022).
50. Maurya, A. *Scaling Lean: Mastering the Key Metrics for Startup Growth*; Penguin Random House: New York, NY, USA, 2016.
51. Christensen, C.M.; Dillon, K.; Hall, T.; Duncan, D.S. *Competing Against Luck: The Story of Innovation and Customer Choice*; Harper Business: New York, NY, USA, 2016.
52. Bachmann, D. *Perfil Do Grau de Inovação Das MPEs Do Paraná*; SEBRAE: Curitiba, Brazil, 2009.
53. De Carvalho, G.D.G.; de Carvalho, H.G.; Cardoso, H.H.R.; Gonçalves, A.D. Assessing a Micro and Small Businesses Innovation Support Programme in Brazil: The Local Innovation Agents Programme. *J. Int. Dev.* **2018**, *30*, 1064–1068. [CrossRef]
54. Wilkinson, S. Focus Group Methodology: A Review. *Int. J. Soc. Res. Methodol.* **1998**, *1*, 181–203. [CrossRef]
55. Gonzalez Rey, F.L. *Pesquisa Qualitativa e Subjetividade: Os Processos de Construção Da Informação*, 1st ed.; Cengage Learning: Boston, MA, USA, 2001.
56. Krueger, R.A. *Casey Focus Groups: A Practical Guide for Applied Research*, 4th ed.; SAGE Publications: Thousand Oaks, CA, USA, 2008.
57. Bussey, K.; Bandura, A. Social Cognitive Theory of Gender Development and Differentiation. *Psychol Rev.* **1999**, *106*, 676–713. [CrossRef]
58. Eagly, A.H.; Karau, S.J. Role Congruity Theory of Prejudice toward Female Leaders. *Psychol Rev.* **2002**, *109*, 573–598. [CrossRef]
59. Dutot, V.; Bergeron, F.; Calabrò, A. The Impact of Family Harmony on Family SMEs' Performance: The Mediating Role of Information Technologies. *J. Fam. Bus. Manag.* **2021**. [CrossRef]

60. Lopez-Nicolas, C.; Nikou, S.; Molina-Castillo, F.-J.; Bouwman, H. Gender Differences and Business Model Experimentation in European SMEs. *J. Bus. Ind. Mark.* **2020**, *35*, 1205–1219. [CrossRef]
61. Apergis, E. Who Is Tech Savvy? Exploring the Adoption of Smartphones and Tablets: An Empirical Investigation. *J. High Technol. Manag. Res.* **2019**, *30*, 100351. [CrossRef]

sustainability

MDPI

Article

Increasing the Sustainability of Manufacturing Processes in Plastic Injection: Recovering Out-Of-Service Robots to Eliminate Manual Assembly Operations

Rúben Costa [1], Vitor Sousa [1], Francisco J. G. Silva [1,2,*], Raul Campilho [1,2], José C. Sá [1,2], Arnaldo Pinto [1] and João Pereira [1]

1 ISEP–School of Engineering, Polytechnic of Porto, 4200-072 Porto, Portugal
2 INEGI–Institute of Science and Innovation in Mechanical and Industrial Engineering, 4200-465 Porto, Portugal
* Correspondence: fgs@isep.ipp.pt

Abstract: In the 20th century, there was a burst concerning the development of the automobile industry, which has become an essential asset for society. With its evolution, this industry created a foundation that based its competitiveness on satisfying people's needs with the highest possible quality and always respecting the delivery deadlines. With the growth in demand, the improvement of certain processes was needed to achieve the desired production goals, accomplished through automation and robotics, as production and assembly lines increasingly used fully automated processes. In plastic injection lines, production is constant and carried out quickly, so it is desirable to perform component assembly steps that immediately support the output of the injection mould parts. This work consists of adapting an obsolete robotic cell to be implemented in one of the production lines to insert components into the injected parts, replacing labour work. Through a mechanical project and an automation design, the equipment was concluded and is currently in production, fulfilling the necessary requirements and improving the process' cycle time. This proves that it is possible to recover old equipment, which is able to improve current tasks and common needs in modern industry, increasing the economic sustainability of the processes and saving resources.

Keywords: plastic injection; sustainability; robotics; automation; manufacturing cell

Citation: Costa, R.; Sousa, V.; Silva, F.J.G.; Campilho, R.; Sá, J.C.; Pinto, A.; Pereira, J. Increasing the Sustainability of Manufacturing Processes in Plastic Injection: Recovering Out-Of-Service Robots to Eliminate Manual Assembly Operations. *Sustainability* **2022**, *14*, 12300. https://doi.org/10.3390/su141912300

Academic Editor: Antonis A. Zorpas

Received: 1 August 2022
Accepted: 25 September 2022
Published: 27 September 2022

1. Introduction

The plastics injection industry is an extremely challenging sector as it dedicates its production to other types of extremely competitive industries, such as the automotive segment and the sector of household items and packaging, among many others. Therefore, the plastics industry constantly seeks to increase its productivity, as innovation is a determining factor in the survival of this sector [1,2], which can be viewed from different angles. The sector is constantly looking to broaden its influence through the launch of new products where the weight factor is extremely important, and plastics can be a solution due to their characteristically low density. This factor assumes an even greater magnitude when the manufactured parts take on even greater proportions, where the replacement of materials and the development of new technologies can allow the industry to reach market segments that are still little-explored [3]. Given the cost structures normally linked to this industry, it is easy to see that there are two fundamental factors which can affect the competitiveness of the process: the moulds' durability and cycle time management, with particular emphasis on the cooling/solidification time of the injected components. The durability of moulds has been a permanent target for many studies, especially when abrasive reinforcements are injected together with plastics [4,5]. On the other hand, cycle time has also been studied in terms of technology [6,7] and management [1]. However, the plastic injection process is not confined to the injection operation itself, especially in more complex parts commonly supplied to the automotive industry, requiring its extraction from the mould and interconnected operations, such as the inspection and assembly of other components. In this perspective, the help of automation and robotics has been proven to be fundamental [8].

Depending on the operations' complexity and the desired flexibility, there are cases where automation is considered the best solution, especially for large series of the same type of product, whereas robotics is employed when greater flexibility is required. The literature is vast in examples in which either one solution or the other, or a combination of both, have led to excellent results in terms of increased productivity, improved levels of repeatability and quality, and even a drastic reduction of repetitive tasks that usually cause health problems for operators [9,10]. Seeber et al. [11] studied the physical interactions between the different resources existing in a workstation, trying to establish methodologies that help Process Engineers in the design of these workstations. Silva et al. [8] developed an automatic solution for inserting previously shaped wires into a mould, where plastic over-injection was performed on the ends of these wires. In this case, the authors opted for automation without robotics, as the setup time needed to change production from one reference to another was small, and changes in the product to be manufactured were infrequent. Considering these factors, the authors argued that a solution based on automation was significantly more economical and perfectly sufficient to make the process much more productive and competitive. Taking into account the previous process of this same product, i.e., the bent wire, Magalhães et al. [12] developed an innovative system based solely on automation in order to guide, inspect, and prepare the product for the next phase. In this case, the developed system could be manufactured at a significantly lower cost than an industrial robot. Furthermore, the system occupied less space, allowed for a faster cycle time than a robot, could be perfectly adjustable to the cycle time of the equipment responsible for forming the wire, and sorted the bent wires, placing them in the supply position for the next process, developed in [8]. Regarding the savings achieved by this solution, the return-on-investment calculated for the new equipment was less than 9 months, which clearly shows the advantages provided by the carried-out study. Araújo et al. [13] developed a new subproduct-transport concept in the manufacturing process of the cushions and suspension mats used in car seat structures. The developed system, based solely on traditional automation, allowed for the reduction of two previously necessary jobs, guaranteeing higher quality assurance and greater productivity, by achieving an efficiency gain of about 40%. Quality assurance was also the main motivation for the work carried out by Costa et al. [14] based on a transmission set usually integrated in the windshield wiper system of automobiles. The initial system allowed for the possibility of a wrong system assembly, which was only semi-automatic, due to the incorporation of human labour. For this reason, the level of defectively assemblies was high enough to cause concern about the process' sustainability. Thus, a new fully automatic equipment was designed from scratch, based only on conventional automation, which allowed for the solving of quality problems in a totally effective way. The return-on-investment calculated was around 2 years, which is perfectly compatible with the investments normally made in the automotive components industry. The Bowden cables manufacturing process has also been the subject of several productivity improvements based on the use of conventional automation. Moreira et al. [15] and Martins et al. [16] developed systems that allowed for the aggregation of different processes in a single piece of equipment, thus reducing the management of semi-products between different stages of the Bowden cables' manufacturing process, with perfectly quantified benefits. Several other studies [17–19] have been developed defending that, in certain situations, the concept to solve certain problems can only go through solutions based on conventional automation to the detriment of robotics, making the investment more easily amortized and allowing for some flexibility [20,21].

Contrary to the previously reported situations, many other investigations have been carried out in the field of robotics, often in close-collaboration with conventional automation. The use of robotics in manufacturing cells must be scrupulously studied, with a view on reducing the displacement time of robots and the energy consumed as much as possible, as described by Gadaleta et al. [22] and Gultekin et al. [23]. Robotics can be affected by inaccuracies of the robotic arm in relation to the jig where the different parts to be joined or assembled are placed. Aware of this possibility, Millington et al. [24] developed a system for real-time positional correction, which, despite its simplicity, proved to meet the requirements initially established in terms of the accuracy of the robotic arm related to the jig, being supported by complex real-time mathematical calculations. Beyond the positioning of the parts by the robotic arm into the jig, robotics presents other challenges, namely possible collisions of the robot with jigs or even with the parts to be worked on; as such Mutti et al. [25] developed their own algorithms to avoid this type of collision. Another challenge normally related to the application of robotic systems was also explored by Honarpardaz et al. [26], namely the design of grippers necessary for handling parts or tools. This study aimed to simplify the design approach of these grippers, taking into account whether they perform only one operation, or if they need to perform multiple ones. As in most solutions based only on conventional automation, solutions based on robotics allow for the real-time control of industrial operations, which leads to a greater interaction between the different production cells, optimizing logistical operations and providing information on the real-time workflow [27]. Séguin et al [28] carried out a study to minimize the number of robots added to a given complex process, minimizing investment and avoiding collisions, without the cycle time being severely impaired. Chernweno and Torn [29] developed a framework capable of decomposing the different tasks performed for a given manual process, with the objective on the study and successful implementation of a robotic system capable of replacing all operations performed manually. In order to validate the framework, a case study was developed that worked as a 'proof-of-concept'. In addition to all the frameworks and models developed around robotics and automation, some case studies have also been presented referring to specific applications, demonstrating with real cases the advantages offered by the implementation of these systems. Matenga et al. [30] successfully used robotics for the rewinding of the stators of electric motors, a task that is traditionally performed manually, given the complexity and diversity of the dimensions of these stators. This solution allowed for the rewinding task to be performed six times faster, with the advantage of allowing the collection of data that can be made accessible by the manufacturer and the customer. Silva et al. [31] developed a robotic system for the inspection and packaging of suspension mats and cushions used to support foam blocks in car seat structures. Through artificial and robotic vision, each manufactured product is handled by a gripper specially designed for this purpose, which is located at the end of the line, promoting the phased inspection of each area most prone to manufacturing defects and then sorting these products inside a large box, where six product stacks are produced. The solution essentially aimed to increase the reliability in the analysis of the manufactured products and to reduce the manual labour associated with the process, with a view on avoiding repetitive work and generating occupational diseases.

Nevertheless, robotization is also no longer seen as an automation free from human work around it, instead starting to consider the integration of manual labour with the robotization of some tasks. Ruiz et al. [32] called this trend hybridization; however, it is necessary to ensure the safety of employees who need to work within the volumetry defined for each robot, or for a set of robots, when that is the case. Johansen et al. [33] defended the use of collaborative robots when production is based on a high diversity of products produced in relatively low quantities, due to a high degree of customization. A similar situation has been presented by Andronas et al. [34]. Notwithstanding, there has still been a relatively low rate of application of this type of robots, with safety reasons being those that are most often cited as the cause for this reduced use [35]. This problem is currently the subject of several studies and publications, thus revealing its importance [36–39]. In spite of this, there are already studies on the implementation of collaborative robotic systems in SMEs (small and medium enterprises), hence demonstrating that their adoption may be slow, but will certainly reach higher proportions in the relatively near future [40]. However, the successive evolution of robots becomes worrying in industrial terms, as companies tend to discard first-generation robots in favour of more modern and agile ones with greater programming capacity and connection to peripheral systems. This constitutes an environmental problem that urgently needs to be resolved [41]. Fortunately, although still in a tenuous way, studies are beginning to arise which aim to adapt first-generation robots to tasks previously performed only through conventional automation or through manual labour, and for which these robots have sufficient requirements, thus allowing their reuse, as an alternative to its complete disposal [42].

This work aims to describe the framework developed in order to implement robotic solutions in the assembly of components. To make a 'proof-of-concept', the work also describes a case study in which the concept is successfully applied. This concept has been developed around a case study based on an injected plastic component used in automobiles, which in some components need to be assembled later, an operation previously performed manually. With the point to increase the precision and repeatability of the process, as well as improving the competitiveness of the product, a robot from one of the first generations was recovered, which was reprogrammed and adapted to the process. Furthermore, conditions were created in terms of peripheral components so that it could carry out the desired operations with the intended success. Thus, this work presents as its main novelty the development a solution able to improve the productivity by reusing previous devices already out-of-service, but good enough to help in improving the performance of manufacturing processes. The developed concept introduces some novel interesting mechanical solutions, which can be taken as examples regarding other needs in the same activity sector, or even in other kind of industries, helping by this way in overcoming difficult problems with simple and cheap solutions.

2. Materials and Methods

The main target of this work was the manufacturing and assembly process of components in plastics parts for the automotive sector obtained by injection moulding, namely the insertion of clips on finished injection moulded parts. The specific target to develop the concept was the production line of the mould MO.9163, which is responsible for producing the parts of the upper B-pillar of a well-known automotive vehicle, as shown in Figure 1.

Figure 1. Upper B-pillar of the well-known automotive vehicle, highlighted in yellow.

The targeted part is manufactured in Polypropylene at high pressure in the mould and removed by mechanical extraction. The part has two versions that differ only in the indication or not of the air bag. The mould used to produce these two versions is the same; however, when there is a change of version, it is necessary for the SMED's (single-minute exchange of die) team to change the mould insert, i.e., the place that engraves 'air bag' on the part. These parts fit into the car body through two clips that are inserted by operators during the production process. The area where the clips are housed is called the clip holder, and can be seen in Figure 2 (left). Since the component under study exists on both sides of the car (right B-pillar and left B-pillar), two parts have to be produced in a single injection, one for each side, as shown in Figure 2 (right), with the clips being at the bottoms of the parts.

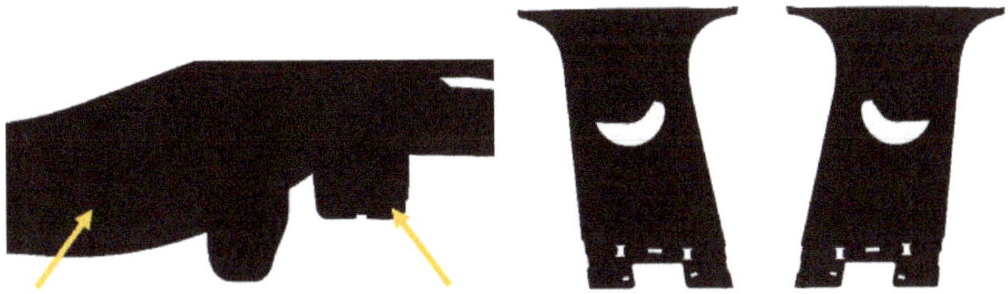

Figure 2. Clip holders for parts (**left**) and MO.9163 parts (**right**).

In the plastic injection process, the existence of channels through which the injection of the plastic into the mould is performed is usual; however, in this case, the plastic is injected directly into the mould cavity of the part, excluding the need to remove the gypsum from the parts, which usually leaves the mould still attached to the part.

The injection machine is responsible for the injection process of the plastic into the mould and controls all parameters (pressure, temperature, and time, among others) necessary to ensure the quality of the parts' injection. If any defect occurs in the injection process of the parts, the machine immediately gives the order to reject the parts through the use of sensors.

2.1. *Analysis of the Previous Concept*

Prior to this project, the production line consisted of an injection machine, a conveyor belt where the parts were deposited after injection, two operators, and two peripheral equipment whose function was to perform the insertions of the clips and validate the necessary tasks. Figure 3 provides an outline of the current workplace layout and the flow of parts from the mould to the initial storage station, which are then stored at the dispatch station.

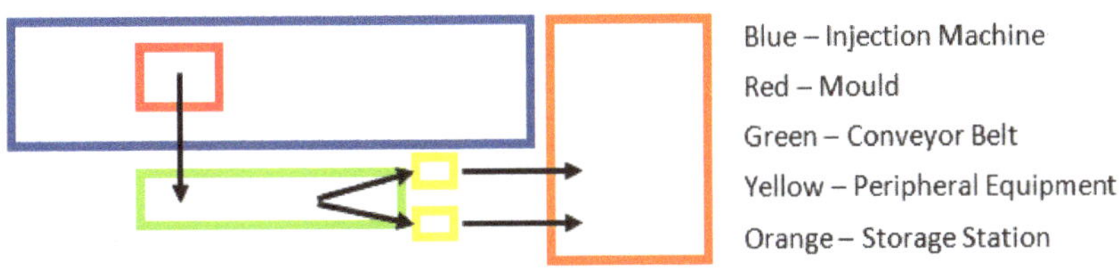

Figure 3. Current workplace layout.

Currently, the control of the parts' quality is performed by the peripheral equipment. When the part reaches the end of the conveyor belt, it is taken by the operators and placed in the peripheral equipment (Figure 4).

Below, it is possible to see which parameters are controlled by the equipment, as well as its verification methods:

- Checking the correct insertion of the springs—Sensor;
- Checking for 'incompletes'—Sensor;
- Version of the part—Vision system;
- Checking for appearance defects—Operator.

If the defined parameters are met, another very important task is performed in the equipment. The parts are pierced two times in order to verify afterwards if they have really passed through the equipment and if they have been properly validated, respectively.

Figure 4. Peripheral equipment.

The passing peak may vary from one equipment to another. Considering that this peak has to happen at the moment when the part is dropped in the equipment, it can be of static or dynamic in nature. It is called static when it is embedded in some surface of the equipment nests, which is in contact with the technical area of the part, and dynamic when it is not. On the other hand, the validation peak, usually designated the OK peak, must be dynamic, which only happens if all of the conditions for the part validation are met. This peak is performed by a pneumatic actuator, which leaves a slight mark on the part in order to show that it has passed the validation process.

2.1.1. Process Cycle

The process starts on the injection moulding machine. If all of the necessary conditions are met, the part is injected, and the machine opens the mould. The task of taking the part from the mould to the conveyor belt is carried out by a three-axis robot. Subsequently, the conveyor belt moves the parts forward to the two operators, where they are handled by each one, thus carrying out the visual check for possible defects, as well as the insertion of the clips. At this moment, if everything is in conformity, the parts are validated by the equipment and stored in the respective station to later be taken to the warehouse.

The three-axis robot responsible for carrying the part from the mould to the conveyor belt has a gripper attached to it, as shown in Figure 5. Taking into account that this project constitutes an improvement of the production process associated with the MO.9163 mould, there is already a valid gripper to do the extraction.

2.1.2. Problem Awareness

With the incessant demand from the automotive industry for new and better products, having resources at a standstill symbolises a loss of profitability, jeopardising any company, regardless of its dimension. Taking into account the high evolution of the automotive industry, what today is a great novelty, within one or two years may already be completely obsolete, which leads to the rapid outdating of the produced equipment. Thus, the problems are:

- Assembly process of components in parts is performed manually, occupying several labour hours;

- Obsolete robotic cell;
- Limited production capacity;
- High costs in the insertion of clips on injected parts.

Figure 5. Three-axis robot responsible by taking the injected parts from the mould.

2.2. *Defined Objectives/Requirements*

The main objective was the improvement of an assembly process of parts for the automotive sector, taking advantage of out-of-service devices already used in previous applications. To accomplish these two goals, a reuse of existing equipment at the work development site was adopted whenever possible. For that, an analysis was performed on the existing stock of material, and a selection of what could still be used was performed. This made the whole project more profitable, more environmentally friendly, and could immediately increase profits. The objectives to achieve this goal were:

- To develop an equipment capable of working autonomously, using the existing three-axis robot, with one type of components and two references;
- To implement a system to feed the components, which are randomly deposited in the equipment;
- To develop an auxiliary tool for the robot, capable of transporting the necessary components for a work cycle;
- To create positioning systems for the parts in order to fix them during the assembly of the components;
- To apply a component-detection system;
- To apply a vision system for component-version detection;
- To carry out the maximum reuse of existing materials.

With the implementation of the equipment in the production lines, the following results were expected to be achieved: reallocating the operators responsible for this assembly to other operations, increasing production capacity, and reducing production costs.

2.3. *Proposed Concept*

With the insertion of a robotic cell in the process, changes in the production must necessarily take place. Thus, the process needs to have a station between the injection machine and the conveyor belt: the cell, which becomes responsible for the tasks previously performed in the peripheral equipment. Consequently, the layout of the production line will change, and a proposal for this concept is presented in Figure 6.

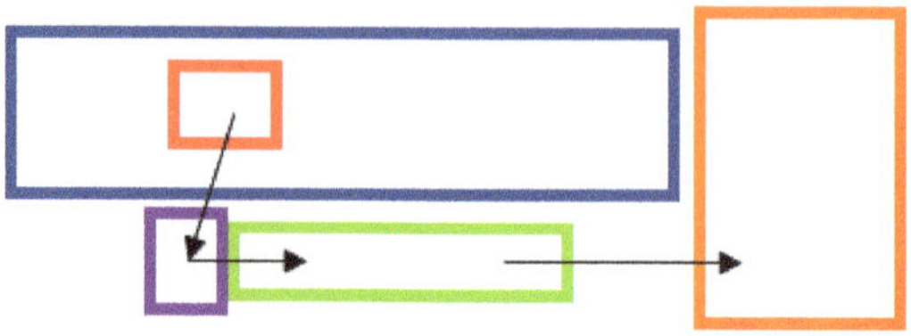

Figure 6. Proposed future workplace layout.

With the insertion of the cell, the order in which the process validation stage was carried out also changed. This validation used to be conducted by the peripheral equipment, and was now be executed by the robotic cell. All functions previously performed by the peripheral equipment were transferred to the cell, and the operator no longer has the function of inserting the clips, with the robot performing this task. Nevertheless, the operator still has to make a visual inspection of the part, but can immediately pack it for dispatch.

The hypothesis of reducing the machine cycle time was analysed, with the possibility of lowering it from 42 s to 38 s, since this was the cycle time of the three-axis robot. However, based on previous tests, this was impossible, as reducing the cycle time would deteriorate the quality of the product, causing the more-frequent appearance of defective parts in the injection.

3. Results

This section intends to present the final developed concept, thoroughly explaining the most relevant components of the system, as well as their functioning. This includes the development of the designed equipment, the new production process, and the results of the implementation, namely the cycle time reduction and the production rate increment.

3.1. Equipment Presentation

The main focus of this work was the production cell; nonetheless, in an initial moment it was necessary to consider several points that would define the unfolding of the whole project. The main parts of the cell and their functions are described below.

3.1.1. Gripper

The gripper used in this project, in contrast with the robot, had to be rethought based on the clips that would be inserted in the parts. Each part housed two clips, so the gripper had to be able to take the necessary number of clips to fulfil an insertion cycle, as shown in Figure 7.

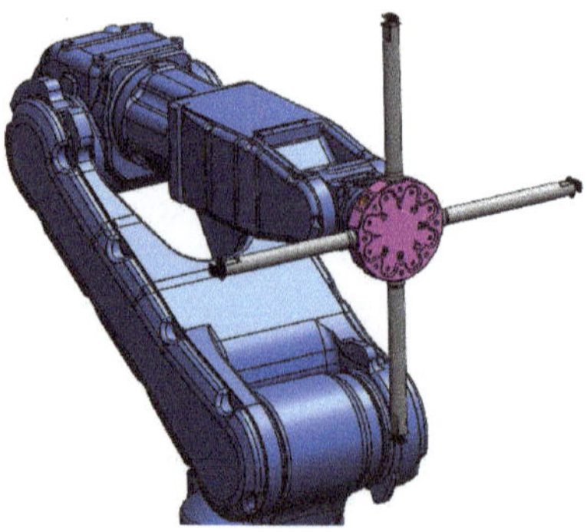

Figure 7. Gripper developed for the robot reused in this project.

3.1.2. Feeding and Rejection Station

The feeder of this project must have cadence to deliver to the robot the required number of clips to perform the insertion cycle. Associated to the feeder, there must be a station where the robot is able to perform the tip-cleaning manoeuvre in case there is an error in some stage of the process. All of these systems must have a set of sensors associated to it that allow the robot to receive the necessary information to execute each operation.

3.1.3. Cell Base

The base of the cell is the robot's jig, i.e., this is where the parts for the robot to insert the clips will be housed. In order to allow the robot to perform this task when entering the base, the parts must be housed in blocks, ensuring the correct fixation of the part. These blocks, highlighted in green in Figure 8, are usually referred to as nests.

Figure 8. Base of the equipment and nests for housing the components.

When the part is dropped into the nests and the three-axis robot gives the order to start the cycle, a pneumatic system is activated which moves forward some stops that fix the part permanently, commonly known as presser foots, in order to reduce gaps in its fixing. To ensure the clamping in the most important points, there was at least one presser foot per clip, which was placed on the upper area of the part, in the clips' insertion zone. In this place, the part had some cavities designated clip holders. In the base there were also several types of sensors that controlled the entire cycle and all actions to be executed by the robot.

3.2. Mechanical Project

In this project, there was no need for calculations or structural dimensioning, since the whole structure was already completed and tested, as the cell had already worked before. The main concerns were essentially related to the part the cell was going to work with. Even so, it was necessary to carry out a check-up on the security and means of access to the cell.

3.2.1. Feeding and Rejection System

The feeding system had the function of supplying clips to the gripper in order to allow the robot to execute its function. In addition, Figure 9 also shows the rejection system, with a basket at the right side of the feeder.

This project's main concern was to ensure the best location for the feeder, in order to reduce the number of movements needed by the robot to perform the feeding. After a careful analysis, a final decision was made, positioning it immediately below the location of the part's nests. The rejection system must be right next to the feeder in order to store the clips which showed an anomaly in some stage of the process. In addition, there must be a window to safely allow operators to supply components to the feeder.

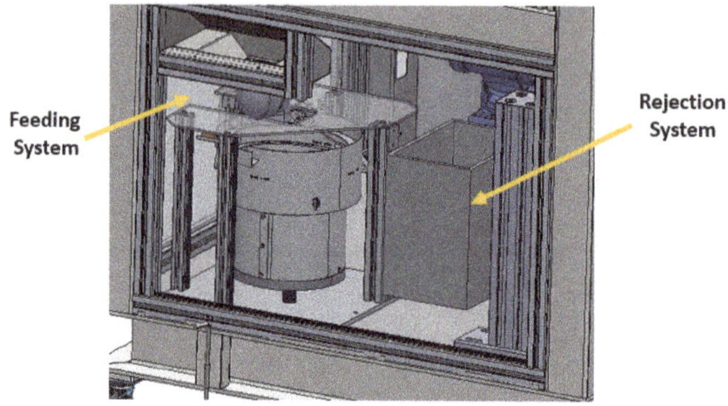

Figure 9. Robotic cell's feeding and rejection system.

3.2.2. Gripper

Some parameters were defined so that the gripper design would meet the necessary requirements, namely (a) it needed to be able to load the needed number of clips for at least one insertion cycle, and (b) the gripper's tips had to be at least 15 mm long to be able to perform the insertion. The gripper was the tool that the robot used to perform the required function, and consisted of three fundamental parts: the core, a protective disc for the robot, and the tips, shown in Figure 10.

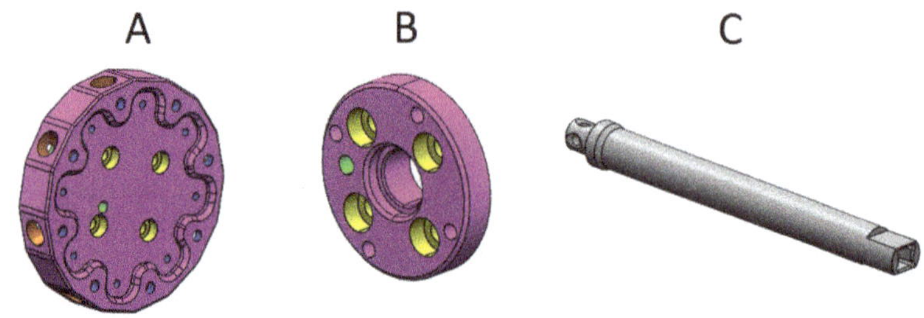

Figure 10. Gripper core (**A**), protection disc (**B**), and tips (**C**).

The core had the function of coupling all of the components, and the tips were the part of the gripper used to carry the clips to the insertion. The protection disk had the function of protecting the robot so that, if any accident happened, the damage was not focused on the robot. It also had a hole for the insertion of a pin, so that the gripper only fit in the robot in a single position.

3.2.3. Base

The base was located at the top of the cell and was removable, as demonstrated in Figure 11. Several systems were coupled to the base which were controlled by the robot in order to perform its function, having to obey to various parameters to be applicable in any cell.

The following are descriptions of the main requirements considered in the execution of the base:

- Removable -> Each cell must be able to work with several moulds. Thus, the base should only be connected to the cell by screws and have lugs to facilitate its transport, for example with a crane.
- Pneumatic supply -> Various mechanisms controlled by compressed air are coupled to the base. Therefore, the base must have a supply that is easy to disassemble, in order to simplify its operation;

- Organisation -> In order to ease the work at the base of the cell, all electrical cables of the pneumatic hoses must be inside the rails. There must be a valve block where all pneumatic controls are performed. There must be an interface box where all the electrical connections between the base and the cell's electrical panel are performed.

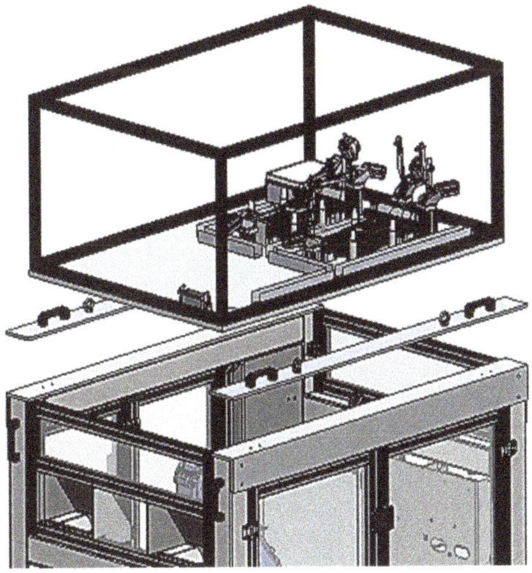

Figure 11. Attachment of the base to the robotic cell.

Nests

The nests are blocks of the base where the parts of the last injection will settle. The main concern with this design element was to ensure the best location and orientation of the blocks to facilitate the cycle of the three-axis robot. For this, a series of specifications were developed to which the nests (Figure 12) needed to obey, allowing for the realization of the project: (1) the part, when pressed, does not allow any displacement; (2) the places where the clips are inserted must have a proper cavity, so that these zones do not have any degree of freedom; (3) the material of the nests must not damage the part; (4) they must be fixed with M5 screws, and there must be boxes for embedding them; and (5) the clip insertion zone shall have an area in which a sensor can be fixed in order to verify the insertion of the clip.

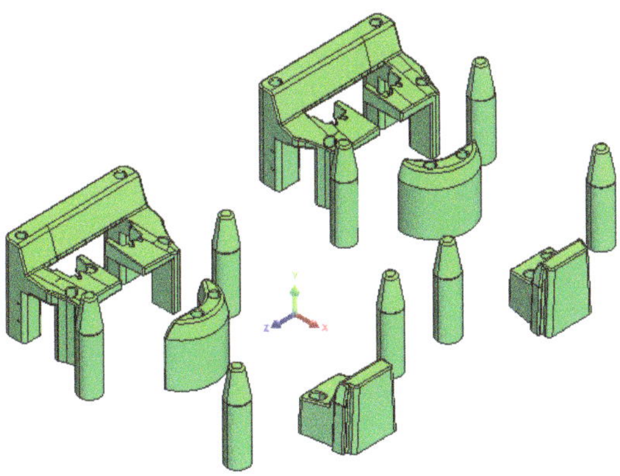

Figure 12. Nests.

The more rotations the three-axis robot needs to make, the longer its cycle time will be. As the parts come out of the mould in a vertical position, it was mandatory that at least one rotation of the three-axis robot occurred before dropping the parts into the nests. Thus, taking into account that the cell would be in a perpendicular position to the conveyor belt, it was decided that the parts would settle at the base, in the same position as they leave the mould, but rotated to the horizontal.

The nests had a cavity that housed the parts' clip holder. Hence, the place where the clips would be inserted was covered by the edges of the part, as can be seen in Figure 13.

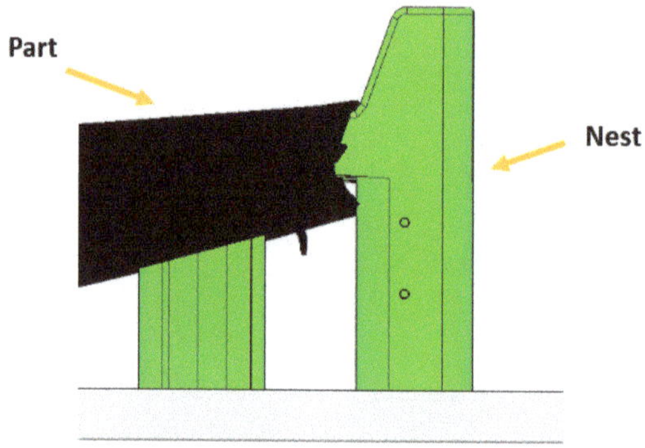

Figure 13. Location of the clip holders.

The detection of the clip in another place was not reliable because the detection must be made in the direction perpendicular to the insertion. Therefore, two threaded holes were drilled in the nests in order to embed a support for a sensor to be placed there, as represented in Figure 14.

Figure 14. Built-in sensor in the nest.

Footing System

When the part fell into the nests and was detected by the sensors, there was a group of pneumatic controllers that received the advance order to fix the part. These pneumatics symbolise a great reuse of existing material. The nests, due to the trajectory performed, imply that these pneumatics need to be fixed in a higher plane than the parts. For that, some bases were accomplished making the fixation of the pneumatic in a profile possible. Then, this profile was coupled to a base with adjustable fixing, making it possible to adjust the height of the clamp from the length of the profile. It was also necessary to build a rod in order to increase the range of the pneumatic. In Figure 15, it is possible to observe the complete assembly.

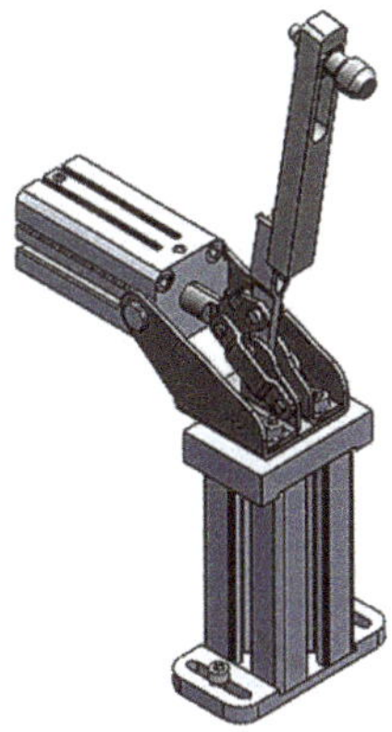

Figure 15. Pneumatic clamp holder with presser foot.

The definition of the points where the part would be pressed was one of the hot topics in this task. When the robot performed the insertion manoeuvre, the presser foot always applied some force on the part, but this could not move. Therefore, it was mandatory to press the part in the areas where there was a clip, thus defining at least two pressing points, corresponding to the location of the clip holders, represented in Figure 16.

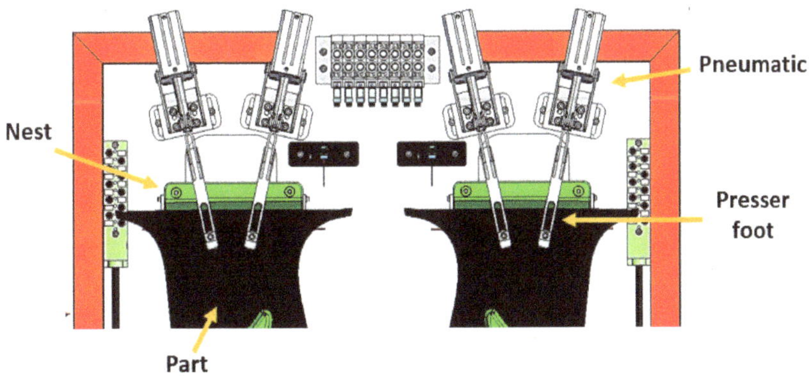

Figure 16. Main footing system.

In this case, however, a third point was defined in order to balance the forces that occurred at the opposite end of the part in relation to the other two. This point was also used to make the passing peak of the part. This way, a threaded hole was drilled in the nest where a pin with a sharp point was housed.

OK Peak

At the end of the robot insertion cycle, if everything complies with the quality criteria, the part is pierced in a specific place. A hexagon must be marked at the place where the validation peak is defined and, to do this, it is necessary to punch the mould in the area where the peak needs to be made, as not all the parts of the mould allow for the easy execution of this marking, such as the areas involving movements of some blocks in the mould.

In order to perform the OK peak, a small pneumatic was used and a support was created to ensure its fixation, the assembly of which is depicted in Figure 17.

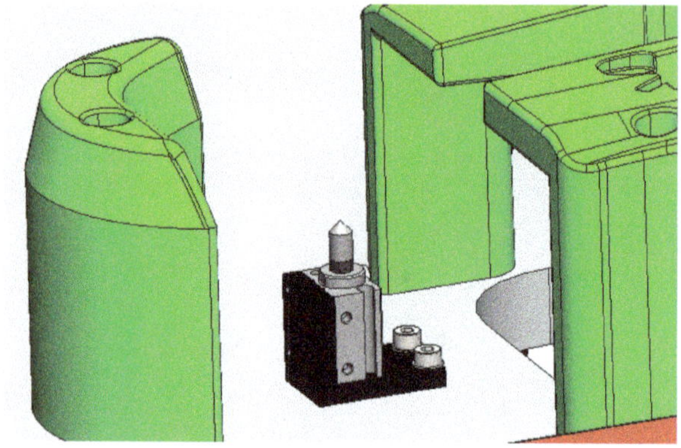

Figure 17. Pneumatic to perform the OK Peak.

Sensor Holders

Taking into account the locations where it was essential to apply the sensors, it was necessary to create supports according to the existing needs.

Incomplete Sensor

The 'incomplete' sensor is a sensor that identifies defective parts. A part is said to be incomplete when the injection process has not happened in the correct way and the part has not reached the final shape, defects which can be controlled using fine detection sensors. In order to overcome this situation, it was necessary to choose a support that would allow for some handling over the positioning of the sensor. In Figure 18 it is possible to see the support and the sensor used.

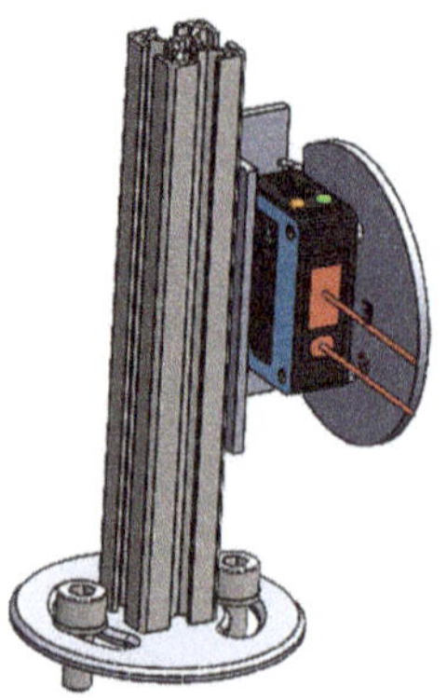

Figure 18. 'Incomplete' sensor holder.

Part Presence Sensor

When the part falls into the nest, its presence must be detected. For this, a sensor whose detection field is larger needs to be used, since the position in which the part falls into the nest may not always be exactly the same. The support used for this sensor can be seen in Figure 19.

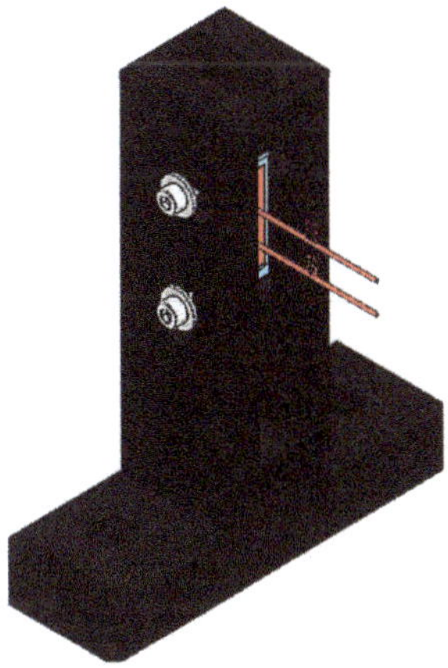

Figure 19. Piece presence sensor holder.

Clip Presence Sensor

The clip presence sensor for this project needed to be designed specifically for this purpose. As it was said before, the clip presence sensor would need to be embedded in the nests. Therefore, a specific support was designed in order to allow for manoeuvring in the sensor positioning. In Figure 20 the sensor with the holder assembly can be seen.

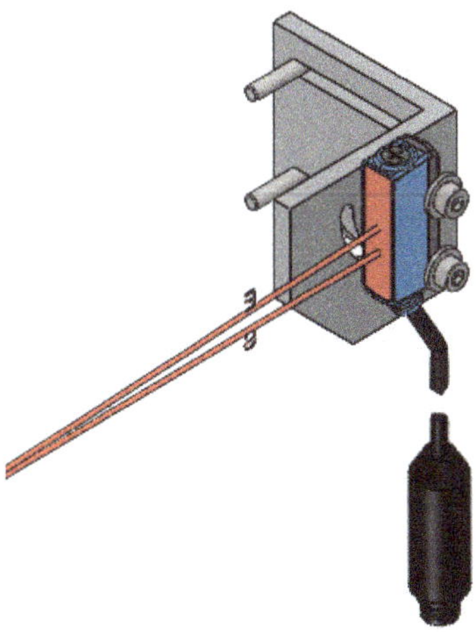

Figure 20. Clip presence sensor holder assembly.

Version Control System

The system used in this project for the version control is the same as was previously used in the peripheral equipment. This was installed under the nests where the operator places the part. At the time the part is placed, the version is automatically validated.

In this project, it was necessary to position the camera close to the nests of the part, and obligatorily close to the place where the version verification was carried out. Thus, a support was made, allowing for the positioning of the camera close to the place where the version is checked (Figure 21).

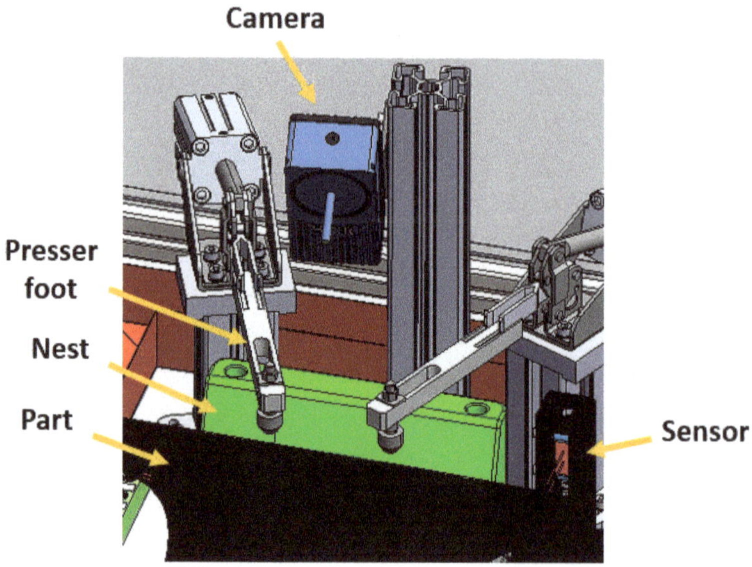

Figure 21. Version control system.

3.3. Cycle Definition

The cycle starts when it is given the starting command. When the start button is triggered, the robot needs to check which is the base coupled to the cell (as mentioned before, a cell has the possibility to work with several bases) and, if it is the base corresponding to the mould working at that moment, the robot advances to the work routine of that base. The first task that the robot needs to perform is the cleaning of the gripper's tips, otherwise a risk of collisions could occur. In other words, the robot needs to ensure the initial conditions so that it can start the work.

Once the condition that the gripper has no clips on its tips is guaranteed, the robot performs the feeding manoeuvre for each tip and then, with the gripper being ready to insert clips, the robot waits for the order to start the insertion operation. When the robot is in this state, it gives the machine ready signal to the three-axis robot, which tells it that the cell is ready to receive parts. Below, the conditions required for the machine ready signal to occur are listed:

- Gripper loaded;
- Part presence sensor turned off;
- Incomplete sensor turned off;
- Clip presence sensor turned off;
- OK peak retracted.

With the machine ready signal ON, the three-axis robot receives permission to drop the part from the last injection onto the base. The moment the part drops into the nests, the part presence sensors become active and the part is pressed, as represented in Figure 22.

Once the part has been pressed, an initial status scan is carried out and, at the same time, it is assured that the part corresponds to the correct version. At this point, the robot verifies whether the parts are incomplete, whether they have clips already inserted, and whether the version being produced matches that of the production order. If any of these conditions are not met, the robot gives the end-of-cycle signal and the parts are rejected. In cases where all conditions are OK, the robot starts the insertion. For this to happen, the presser feet and the OK peak must be advanced, and the part presence sensor, the incomplete sensor, and the clip presence sensor must be turned ON.

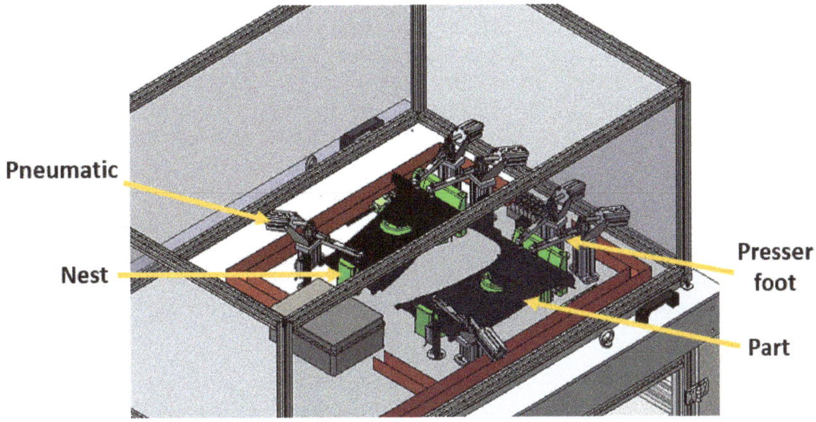

Figure 22. Presser foot closure.

When the robot finishes the insertion operation, a verification of the individual status of the clip presence sensors is performed and, if they are all turned ON, a signal is delivered which gives the information to the 3-axis robot that the parts are OK (Figure 23). If the parts are not OK, this signal is not given. Subsequently, the robot emits the end-of-cycle signal.

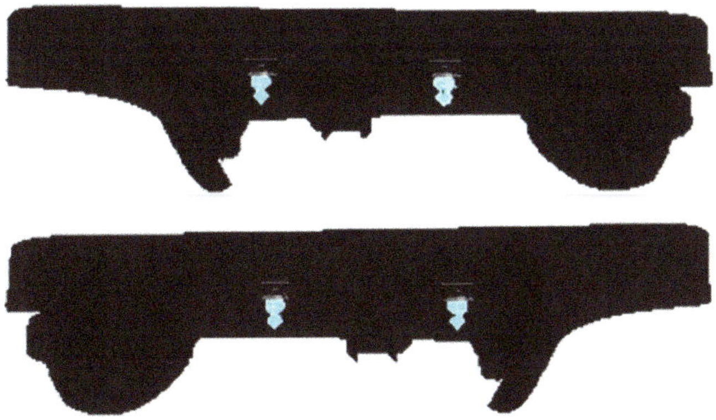

Figure 23. Set of pieces OK.

With the end-of-cycle signal turned ON, the three-axis robot withdraws the parts from the base of the cell and, depending on the previous state of the part-OK signals, it performs the operation of depositing the parts on the conveyor belt or, in the absence of the part-OK signal (therefore only the end-of-cycle order exists), the robot rejects the parts. The moment the part presence signal turns OFF, the part-OK and end-of-cycle signals also turn OFF, and the cycle starts again.

3.3.1. Feeding Cycle

The main purpose of the robot feeding cycle is to prepare the gripper to allow the process to progress. The feeding cycle is associated with the rejection cycle, as both run in parallel and work together to ensure the proper gripper loading. The robot always runs the feeding cycle before performing an insertion, and the four tips have the same loading operation: the robot brings the tip close to the capture rail and confirms the state of a first sensor above the clip capture zone, which verifies the state of the tip. If the tip is empty, the robot performs the clip capture operation when a second sensor is active, which checks for clips in the rail. After the capture, the gripper returns to the previous level and checks if the procedure has taken place in the correct way.

In case any of these processes fail, the robot goes to the rejection station and cleans the tip. If in the previous cycle any of the parts is NOK due to the insertion failure of any clip, the robot performs the cleaning of the respective tip. In addition to the sensors that regulate the robot cycle, there are two more sensors able to check the status of the feeder, and these activate certain signals that warn the operator in case the cell's work is compromised due to the feeder. These sensors detect the state of the feeder's rail, and another one detects the state of the pan. If the rail is empty for a certain amount of time, the feeder sounds an alarm, as this may mean that the feed has jammed. The other sensor checks the status of the pan and, if this sensor is also active for a certain time, another alarm is issued, alerting the operator to supply clips to the feeder.

3.3.2. Cycle Signals

The robotic cell has two banners: one to alert the operator for the state of the feeder (A) and the other for the state of the machine (B). Based on the state of each of these systems, these banners make it easy for the operator to understand the state of the cell's work. The system presented in Figure 24 is located below the zone where the parts are positioned, the base, which can be seen in Figure 11. The robotic arm acts below the superior module to have more space and to perform its tasks more easily. The banners A and B, also located below the base, are additional devices developed to help the operator with the whole process.

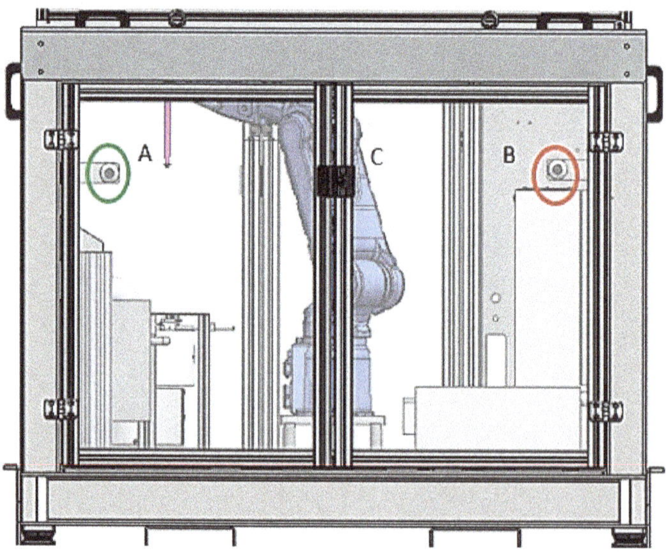

Figure 24. A–Banner of the feeder, B–banner of the robotic cell, C–robotic arm.

Depending on the colour of the cell banner, the operator receives a different alert. Steady yellow indicates machine ready, flashing yellow is the insertion cycle, steady green is part OK, and steady red is part NOK. In the case of the feeder, the same rule applies, with flashing red for missing springs and steady red for jammed feeder.

3.3.3. Insertion Cycle

The insertion cycle is the completion of the entire cell process. Before the three-axis robot drops the part into the cell, it verifies the status of the machine-ready signal. If it is missing, the part cannot be dropped into the cell.

When the necessary conditions exist, the three-axis robot drops the part into the cell. Firstly, the robot scans the state of the 'incomplete' and clip presence sensors. If either of these is turned ON, the parts are immediately rejected, and the end-of-cycle order is given, with the three-axis robot rejecting the parts. If neither of these conditions is ON, the robot advances to the insertion cycle.

At the end of the insertion, the robot rechecks the status of the clip sensors, and if they are all turned ON, the parts-OK signals are activated, and then the cycle ends. These signals are active as long as the part presence signal is turned ON. When the three-axis robot removes the parts from the base, these signals automatically turn OFF and the cycle restarts.

3.4. Equipment Costs

One of the main objectives of this project was to make the maximum reuse of previously used material not needed at the moment. However, even so, it was necessary to resort to suppliers to obtain certain components, the lack of which could not be made up for with the recovered material. The general costs associated with the acquisition of the equipment are present in Table 1, totalizing a value of 16,650 €.

Table 1. General costs associated to the purchase of the equipment.

Acquired Equipment	Cost
Feeding and rejection station	4500 €
Gripper and other accessories	2000 €
Aluminium base plate	150 €
Man labour	10,000 €

3.5. Implementation Results

Some of the gains obtained by inserting the cell in the process are described below:

- There will no longer be stoppages which were mandatory before with the production depending on two people, thus improving the process yield;
- In the short–medium term, it is expected that the quality of the process will improve, thus reducing costs with possible non-conformities in the production;
- Reduction of the probability of non-conformities due to human fatigue;
- With the reallocation of one worker, a cost of 1300 €/month will be saved, and this value can increase if the cell operates for more than one shift per day.

In addition, an overview of all of the reused material is presented: robot and controllers, electric panel and components, protective frame and rims, the aluminium base protection frame, and all of the pneumatic materials (valve block, accessories and 33 pneumatic ramps).

The costs of the equipment, when compared with the reused material and the future gains of having the cell inserted in the process, will be recovered after less than 12 months, if only the gain obtained by the reallocation of an operator is accounted for. It is expected that the quality of the process keeps improving with the robotic cell in production, which will also be reflected in the gains that will be obtained.

It should be noted that the productivity gain ensured by this project has no direct implications for the production cycle, given that the injection cycle is optimized and cannot be reduced. However, just regarding the fact that there are no disturbing factors of the normal production cycle downstream of the injection, i.e., during the inspection and assembly of clips, provides that 16% of the useful production time is not wasted. This is because when the downstream in the inspection and assembly operations decreases, this is implied to stop the injection production cycle, preventing the congestion of the subsequent processes or the use of extra labour to comply with the production schedule. In addition, there is even greater assurance that the product is properly inspected, that is, the inspection is not affected by human fatigue, thus resulting in a more coherent and effective filtering of non-conformities.

4. Discussion

The results obtained show that there is a series of opportunities for improvement in the industry, which must take advantage of them in order to increase economic sustainability, reduce dependence on human labour, and increase quality. On the other hand, the greater sensitivity that exists to environmental issues will certainly influence many decisions, as it is not economically sustainable to replace some equipment prematurely without there being opportunities to reapply it in situations where it can still be useful [41,43]. The work now developed clearly shows that it is possible to reconcile projects to increase productivity with the reuse of devices and components previously used in other projects without any loss of competitiveness in their reuse, thus benefiting the environment, i.e., environmental sustainability.

The concept now presented is in line with other works previously performed, where robotics and conventional automation were used together with a view to increasing productivity and quality assurance. In fact, this solution is perfectly in line with some solutions previously found by other authors to solve problems related to different products, but in which the objective was similar: to increase productivity. Examples of this are the works carried out by Costa et al. [18], Silva et al. [31], and Castro et al. [42]. In fact, Castro et al. [42] also reused a robot that was already out of service, expanding its potential through the possibility of moving this robot on a track, thus increasing its field of action and giving it a new industrial utility where the speed of the execution of tasks was not the main objective, as in the case of the present project. Obviously, the environmental issue is important, but reuse also represents a significant saving in terms of the investment needed to increase the degree of automation of a given process. In the case presented here, and through a relatively rough calculation, the saving in the salary that should be paid to an operator during a year in a work shift is greater than the investment that is necessary to make in this project, which leads to a ROI (return on investment) in less than 12 months. This value fits perfectly with what is usually accepted by industrial companies, which normally consider a maximum of 36 months, in practical terms, to find the project economically viable in view of the estimated life of each product in production. Comparing the ROI period with other works, it is possible to verify that Magalhães et al. [12] achieved an ROI of around 8 months for the automation of a process connected to CNC wire bending equipment, without resorting to any reuse of equipment or components. Furthermore, the complexity of this concept is more modest than that performed in this work. On the other hand, Silva et al. [8] presented a ROI of 5 months for the automation of a process linked to the over-moulding of metallic wires with plastic, which automation involved the orderly arrangement of wires for collection with a gripper, and a dedicated gripper design for feeding and collecting the product. However, the value of the ROI obtained through this work (12 months) is lower than several other works where automation also aimed to increase the productivity and competitiveness of products, although without environmental concerns. Santos et al. [44] developed a conventional automatic system for the textile industry and achieved a ROI of 14.5 months. On the other hand, in the work developed by Silva et al. [31] using the incorporation of a new robot and of all the conventional automation around it, the estimated time for the ROI was 21.5 months, thus showing the difference in the period of ROI that the reuse of an existing equipment can bring in terms of economic advantage. Often, the need to increase productivity is accompanied by the need to guarantee quality levels that are adequate to the requirements imposed by the market. This was also felt in this work, similar to what was also reported by Costa et al. [14]. Some limitations of this work are that the total cycle time can still be improved; however, this depends on the injection cycle time, so there is always a minimum value which cannot be lowered. In addition to this, the position in which the clips are attached can also represent a problem in other works, so the work represented in this article is just an example and has to be adapted when considering other cases.

5. Conclusions

SMEs still have an immense field of progression in terms of productivity gains in component production and assembly operations. Robotics, coupled with conventional automation, is the most suitable way to produce increases in productivity and ensure the increasingly high level of quality demanded by the market. The deactivation of some end-of-life projects leads to the decommissioning of countless devices and components that can still be reused. If projects to increase productivity and quality can take into account devices that are no longer useful in previous projects, the cost of the investment necessary to automate processes and the corresponding increase in productivity and quality can be reduced. Thus, the competitiveness of products benefits in parallel to the environment, as the creation of industrial waste and the consumption of natural resources is minimized.

This work aimed to show that it is possible to create new concepts around the need to increase the productivity and quality of assembly operations underlying the injection of plastic parts for the automotive industry, taking into account devices and components previously used in other projects. The concept is shown through a case study linked to the production of parts for the interior of automobiles in which, after injection, it is necessary to inspect and install clips before the product is packed and sent to the customer. Taking advantage of a robot previously used in another project that is already finished, solutions were created based on conventional automation that complemented the action of the robot, giving rise to a production cell coupled to a plastic injection machine that became much more autonomous in relation to human work, guaranteeing higher levels of production and ensuring a higher level of quality because the control operation is no longer subject to human fatigue.

Given that the new concept completely dispenses one of the operators, as well as almost completely dispensing with the second one, even considering only the dismissal of the first one it is possible to have a return on investment in less than 12 months. This value fits perfectly with the financial performance normally required by industrial companies to consider the investment as profitable, taking into account the estimated life of each manufacturing project. This value is achieved thanks to the use of a robot from a project that has been completed in the meantime, and which has sufficiently attractive characteristics to empower the concept of increased productivity studied here. As previously mentioned, it was not the main objective of the case study used to validate this concept to increase the production rate, since this is imposed by the injection cycle and was already optimized. Nevertheless, the application of this concept prevents the process from having downstream breaks during inspection and assembly operations that were performed manually after the injection cycle. Hence, no productivity gain in terms of cycle is indicated, but taking into account the history of production breaks in manual operations in the past, which congested and forced the injection process to stop, it can be stated that they were around 16 % of the production time.

Therefore, in addition to achieving the desired levels of productivity and quality and reducing the dependence on human labour, it is also possible to take care of the environmental aspect, minimizing the consumption of natural resources and preventing out-of-use equipment from being recycled, thus resulting in energy consumption so that it could be turned into something useful again. This work intended to bring a novel concept of reuse of industrial devices that are still capable of giving a strong contribution to the increase of productivity and quality and minimizing investment costs, i.e. making them more attractive in financial terms, as well as to the environment.

Author Contributions: Formal analysis, R.C. (Rúben Costa), V.S., J.C.S. and A.P.; methodology, R.C. (Rúben Costa) and F.J.G.S.; visualization, R.C. (Rúben Costa), J.P., V.S., J.C.S. and R.C. (Raul Campilho); writing—draft, R.C. (Rúben Costa); writing—reviewing, F.J.G.S.; writing—draft and reviewing, V.S.; conceptualization, F.J.G.S. and J.P.; work orientation, F.J.G.S.; investigation, F.J.G.S. and J.P.; supervision, F.J.G.S.; data curation, R.C. (Raul Campilho), J.C.S. and A.P.; resources, R.C. (Raul Campilho) and A.P.; main research, J.P.; data collection, J.P. All authors have read and agreed to the published version of the manuscript.

Funding: This research received no external funding, and the authors did not receive any funding regarding the development of this work.

Data Availability Statement: No data are available regarding this work.

Acknowledgments: The authors want to thank Diogo Santos due to the data provided, as well as the availability to supervise M.Sc. students from ISEP to carry out their master's thesis.

Conflicts of Interest: The authors declare that they have no known competing financial interests or personal relationships that could have appeared to influence the work reported in this paper. Moreover, the authors declare no conflict of interest.

References

1. Ribeiro, P.; Sá, J.; Ferreira, L.P.; Silva, F.; Pereira, M.; Santos, G. The Impact of the Application of Lean Tools for Improvement of Process in a Plastic Company: A case study. *Procedia Manuf.* **2019**, *38*, 765–775. [CrossRef]
2. Martinho, R.P.; Silva, F.J.G.; Alexandre, R.J.D.; Baptista, A.P.M. TiB_2 Nanostructured Coating for GFRP Injection Moulds. *J. Nanosci. Nanotechnol.* **2011**, *11*, 5374–5382. [CrossRef]
3. Barbosa, R.; Campilho, R.; Silva, F. Injection mold design for a plastic component with blowing agent. *Procedia Manuf.* **2018**, *17*, 774–782. [CrossRef]
4. Silva, F.J.G.; Martinho, R.P.; Alexandre, R.J.D.; Baptista, A.P.M. Increasing the wear resistance of molds for injection of glass fiber reinforced plastics. *Wear* **2011**, *271*, 2494–2499. [CrossRef]
5. Silva, F.; Martinho, R.; Andrade, M.; Baptista, A.; Alexandre, R. Improving the Wear Resistance of Moulds for the Injection of Glass Fibre–Reinforced Plastics Using PVD Coatings: A Comparative Study. *Coatings* **2017**, *7*, 28. [CrossRef]
6. Rashid, O.; Low, K.; Pittman, J. Mold cooling in thermoplastics injection molding: Effectiveness and energy efficiency. *J. Clean. Prod.* **2020**, *264*, 121375. [CrossRef]
7. Venkatesh, G.; Kumar, Y.R.; Raghavendra, G. Comparison of Straight Line to Conformal Cooling Channel in Injection Molding. *Mater. Today Proc.* **2017**, *4*, 1167–1173. [CrossRef]
8. Silva, F.J.G.; Soares, M.R.; Ferreira, L.P.; Alves, A.C.; Brito, M.; Campilho, R.D.S.G.; Sousa, V.F.C. A Novel Automated System for the Handling of Car Seat Wires on Plastic Over-Injection Molding Machines. *Machines* **2021**, *9*, 141. [CrossRef]
9. Shinde, G.V.; Jadhav, V.S. Ergonomic analysis of an assembly workstation to identify time consuming and fatigue causing factors using application of motion study. *Int. J. Eng. Technol.* **2012**, *4*, 220–227.
10. Thun, J.-H.; Lehr, C.B.; Bierwirth, M. Feel free to feel comfortable—An empirical analysis of ergonomics in the German automotive industry. *Int. J. Prod. Econ.* **2011**, *133*, 551–561. [CrossRef]
11. Seeber, C.; Albus, M.; Fechter, M.; Neb, A.; Yoshida, S.I. Automated 2D Layout Design of Assembly Line Workstations through Physical Principles. *Procedia CIRP* **2021**, *104*, 1197–1202. [CrossRef]
12. Magalhães, A.J.A.; Silva, F.J.G.; Campilho, R.D.S.G. A novel concept of bent wires sorting operation between workstations in the production of automotive parts. *J. Braz. Soc. Mech. Sci. Eng.* **2018**, *41*, 25. [CrossRef]
13. Araújo, W.F.S.; Silva, F.J.G.; Campilho, R.D.S.G.; Matos, J.A. Manufacturing cushions and suspension mats for vehicle seats: A novel cell concept. *Int. J. Adv. Manuf. Technol.* **2016**, *90*, 1539–1545. [CrossRef]
14. Costa, M.J.R.; Gouveia, R.M.; Silva, F.J.G.; Campilho, R.D.S.G. How to solve quality problems by advanced fully-automated manufacturing systems. *Int. J. Adv. Manuf. Technol.* **2017**, *94*, 3041–3063. [CrossRef]
15. Moreira, B.; Gouveia, R.M.; Silva, F.; Campilho, R. A Novel Concept of Production and Assembly Processes Integration. *Procedia Manuf.* **2017**, *11*, 1385–1395. [CrossRef]
16. Martins, N.; Silva, F.; Campilho, R.; Ferreira, L. A novel concept of Bowden cables flexible and full-automated manufacturing process improving quality and productivity. *Procedia Manuf.* **2020**, *51*, 438–445. [CrossRef]
17. Nunes, P.M.S.; Silva, F.J.G. Increasing Flexibility and Productivity in Small Assembly Operations: A Case Study. In *Advances in Sustainable and Competitive Manufacturing Systems*; Lecture Note in Mechanical Engineering; Springer: Berlin/Heidelberg, Germany, 2013; pp. 329–340.
18. Costa, R.J.S.; Silva, F.J.G.; Campilho, R.D.S.G. A novel concept of agile assembly machine for sets applied in the automotive industry. *Int. J. Adv. Manuf. Technol.* **2017**, *91*, 4043–4054. [CrossRef]
19. Vieira, D.; Silva, F.J.G.; Campilho, R.D.S.G.; Sousa, V.F.C.; Ferreira, L.P.; Sá, J.C.; Brito, M. Automating equipment towards Industry 4.0—A new concept for a transfer system of lengthy and low-stiffness products for automobiles. *J. Test. Eval.* **2022**, *50*, 20210721. [CrossRef]
20. Groover, M.P. *Fundamentals of Modern Manufacturing Materials, Processes, and Systems*, 4th ed.; John Wiley & Sons: Hoboken, NJ, USA, 2010; p. 889.
21. Aljinović, A.; Gjeldum, N.; Bilić, B.; Mladineo, M. Optimization of Industry 4.0 Implementation Selection Process towards Enhancement of a Manual Assembly Line. *Energies* **2021**, *15*, 30. [CrossRef]
22. Gadaleta, M.; Berselli, G.; Pellicciari, M. Energy-optimal layout design of robotic work cells: Potential assessment on an industrial case study. *Robot. Comput. Manuf.* **2017**, *47*, 102–111. [CrossRef]

23. Gultekin, H.; Gürel, S.; Taspinar, R. Bicriteria scheduling of a material handling robot in an -machine cell to minimize the energy consumption of the robot and the cycle time. *Robot. Comput. Integr. Manuf.* **2021**, *72*, 102207. [CrossRef]
24. Millington, J.; Monfared, R.P.; Vera, D. Innovative mechanism to identify robot alignment in an automation system. *Robot. Auton. Syst.* **2019**, *114*, 144–154. [CrossRef]
25. Mutti, S.; Nicola, G.; Beschi, M.; Pedrocchi, N.; Tosatti, L.M. Towards optimal task positioning in multi-robot cells, using nested meta-heuristic swarm algorithms. *Robot. Comput. Integr. Manuf.* **2021**, *71*, 102131. [CrossRef]
26. Honarpardaz, M.; Ölvander, J.; Tarkian, M. Fast finger design automation for industrial robots. *Robot. Auton. Syst.* **2019**, *113*, 120–131. [CrossRef]
27. Xiao, D.; Song, M.; Ghosh, B.K.; Xi, N.; Tarn, T.J.; Yu, Z. Real-time integration of sensing, planning and control in robotic work-cells. *Control Eng. Pract.* **2004**, *12*, 653–663. [CrossRef]
28. Séguin, S.; Tremblay, H.; Benkalaï, I.; Perron-Chouinard, D.-E.; Lebeuf, X. Minimizing the number of robots required for a Robotic Process Automation (RPA) problem. *Procedia Comput. Sci.* **2021**, *192*, 2689–2698. [CrossRef]
29. Chemweno, P.; Tor, R.-J. Innovative robotization of manual manufacturing processes. *Procedia CIRP* **2022**, *106*, 96–101. [CrossRef]
30. Matenga, A.; Murena, E.; Kanyemba, G.; Mhlanga, S. A novel approach for developing a flexible automation system for rewinding an induction motor stator using robotic arm. *Procedia Manuf.* **2019**, *33*, 296–303. [CrossRef]
31. Silva, F.; Swertvaegher, G.; Campilho, R.; Ferreira, L.; Sá, J. Robotized solution for handling complex automotive parts in inspection and packing. *Procedia Manuf.* **2020**, *51*, 156–163. [CrossRef]
32. Ruiz, R.C.; Ramírez, A.J.; Cuaresma, M.J.E.; Enríquez, J.G. Hybridizing humans and robots: An RPA horizon envisaged from the trenches. *Comput Ind.* **2022**, *138*, 103615. [CrossRef]
33. Johansen, K.; Rao, S.; Ashourpour, M. The Role of Automation in Complexities of High-Mix in Low-Volume Production—A Literature Review. *Procedia CIRP* **2021**, *104*, 1452–1457. [CrossRef]
34. Andronas, D.; Argyrou, A.; Fourtakas, K.; Paraskevopoulos, P.; Makris, S. Design of Human Robot Collaboration workstations—Two automotive case studies. *Procedia Manuf.* **2020**, *52*, 283–288. [CrossRef]
35. Berx, N.; Decré, W.; Pintelon, L. Examining the Role of Safety in the Low Adoption Rate of Collaborative Robots. *Procedia CIRP* **2022**, *106*, 51–57. [CrossRef]
36. Xu, W.; Cui, J.; Liu, B.; Liu, J.; Yao, B.; Zhou, Z. Human-robot collaborative disassembly line balancing considering the safe strategy in remanufacturing. *J. Clean. Prod.* **2021**, *324*, 129158. [CrossRef]
37. Castano-Cano, D.; Grossard, M.; Hubert, A. Resonant sensors for multi-axis force and torque estimation in collaborative robotics. *Mechatronics* **2021**, *82*, 102703. [CrossRef]
38. Gualtieri, L.; Rauch, E.; Vidoni, R. Development and validation of guidelines for safety in human-robot collaborative assembly systems. *Comput. Ind. Eng.* **2021**, *163*, 107801. [CrossRef]
39. Chemweno, P.; Torn, R.-J. Innovative safety zoning for collaborative robots utilizing Kinect and LiDAR sensory approaches. *Procedia CIRP* **2022**, *106*, 209–214. [CrossRef]
40. Raza, M.; Malik, A.A.; Bilberg, A. PDCA integrated simulations enable effective deployment of collaborative robots: Case of a manufacturing SME. *Procedia CIRP* **2021**, *104*, 1518–1522. [CrossRef]
41. Silva, F.J.G.; Gouveia, R.M. *Cleaner Production—Toward a Better Future*; Springer Nature: Cham, Switzertland, 2020; ISBN 978-3-030-23164-4.
42. Castro, A.F.; Silva, M.; Silva, F. Designing a Robotic Welding Cell for Bus Body Frame Using a Sustainable Way. *Procedia Manuf.* **2017**, *11*, 207–214. [CrossRef]
43. Teixeira, P.; Sá, J.; Silva, F.; Ferreira, L.; Santos, G.; Fontoura, P. Connecting lean and green with sustainability towards a conceptual model. *J. Clean. Prod.* **2021**, *322*, 129047. [CrossRef]
44. Santos, P.; Campilho, R.; Silva, F. A new concept of full-automated equipment for the manufacture of shirt collars and cuffs. *Robot. Comput. Manuf.* **2020**, *67*, 102023. [CrossRef]

sustainability

MDPI

Article

Lean Six Sigma with Value Stream Mapping in Industry 4.0 for Human-Centered Workstation Design

Fu-Kwun Wang [1], Benedictus Rahardjo [1,*] and Pol Rifa Rovira [2]

[1] Department of Industrial Management, National Taiwan University of Science and Technology, Taipei City 106335, Taiwan
[2] Thermo King Ireland Limited, D02 Dh60 Dublin, Ireland
* Correspondence: d10901816@mail.ntust.edu.tw

Abstract: Many industries have successfully implemented the define-measure-analyze-improve-control (DMAIC) method of Lean Six Sigma to improve their production activities. Value stream mapping (VSM) for Industry 4.0 allows us to understand the current state of operations in order to plan future improvements. In this study, we propose an improvement model based on DMAIC with VSM 4.0 for a truck cooler manufacturer to improve the picking workstation design with a human-centered approach. We use the DMAIC method to analyze the project step by step. After identifying the root cause, we identified countermeasures to improve the productivity. To reduce human error, the project team adopted a human-centered approach and applied lean tools, such as visual management, error prevention, and waste analysis. As a result of this case study, the yield rate was improved from 98% to 100%, and the direct savings of the project amounted to EUR 3180, mainly due to the freeing up of space. A human-centered LSS framework is also presented as a novel contribution of this study.

Keywords: value stream mapping 4.0; Lean Six Sigma; picking workstation design; human-centered

Citation: Wang, F.-K.; Rahardjo, B.; Rovira, P.R. Lean Six Sigma with Value Stream Mapping in Industry 4.0 for Human-Centered Workstation Design. *Sustainability* **2022**, *14*, 11020. https://doi.org/10.3390/su141711020

Academic Editors: Francisco J. G. Silva, Gilberto Santos, Luís Pinto Ferreira, José Carlos Sá, Manuel Pereira Lopes, Giada La Scalia and Mosè Gallo

Received: 22 June 2022
Accepted: 30 August 2022
Published: 3 September 2022

1. Introduction

Smart workstation design is critical for creating a healthy, comfortable, and productive workplace [1]. In response to disruptions in the global supply chains, automated workstations have been designed to replace labor and shorten working hours [2]. Lean Six Sigma (LSS) is a common strategy for improving operational excellence in manufacturing [3,4]. Lean emphasizes speed and waste, often referred to as efficiency, while Six Sigma emphasizes process variation reduction, often referred to as effectiveness [5]. LSS uses data and statistical analysis to identify the causes of fluctuations that lead to insufficient process outputs [6]. There are many examples of the LSS method's implementation in the manufacturing industry, including the roles of key personnel in all areas of business, such as production, finance, marketing, and even personal management. Chiarini [7] surveyed five European motorcycle part manufacturing companies using various lean tools and found that lean methods reduced waste and had positive environmental impacts. Singh and Rathi [8] conducted a comprehensive analysis of the barriers to LSS's implementation in small and medium-sized enterprises. Lean practices have not been implemented in various sectors in India [9]. However, some businesses have implemented and observed positive benefits of this method, mainly in terms of waste reduction and quality improvement. According to Jasti and Kodali [10], researchers should focus on building and testing novel ideas, including case studies and surveys that analyze academic findings from research on lean practices. Recently, it was found that 63.82% of empirical studies in the LSS literature use real data, and the process can be regulated and validated [11]. The majority of LSS applications in the industry are concentrated in manufacturing, healthcare, automotives, and electronics. Additionally, several articles focus on testing tools, obtaining benefits, and understanding success factors [12,13]. Several authors have discussed the analysis and exploration of key success factors of LSS in order to achieve effective quality management, organizational goals, objectives, and performance in industry [14–16]. Define-Measure-Analyze-Improve-Control (DMAIC) is the problem-solving approach that drives LSS. It is a five-phase method for addressing existing process problems based on a scientific method. In any work environment, continuous improvement can be correlated with the individual performing the activities in order to adapt the improvements to their needs [17]. It has recently been suggested that a human-centered approach can be integrated with the LSS principles [18].

Psomas [19] conducted a systematic literature review of lean manufacturing using a total of 214 articles and discussed a plethora of future research methods. Lobo Mesquita et al. [20] identified forms of integration between lean, Industry 4.0, and environmental sustainability, and showed how these three structures relate to the building of a framework that can help to manage industrial production processes. VSM is often regarded as the starting point for lean projects in the search for opportunities to optimize value and eliminate waste in the manufacturing process. A study by Jasti and Sharma [21] showed that the implementation of VSM had positive impacts in a lean manufacturing environment. VSM 4.0 is a recently developed, collaborative value stream tool for lean management in the context of Industry 4.0. Mapping, design and validate value streams are used to digitally plan improvement actions in order to reduce coordination times within the company. VSM 4.0 is a tool that enables company executives to engage in a series of structured discussions and executive decisions that enable continuous improvement. Numerous authors have provided particular strategies for certain businesses, despite the fact that this strategy is wide and versatile, spanning many industries [22,23].

In this article, we present a method combining the DMAIC approach in LSS and VSM 4.0 to design a human-centered workstation to support workers with intellectual disabilities. A case study is used to illustrate the application of the proposed method, which aims to enable the design of human-centered manufacturing workstations. Section 2 describes a review of VSM 4.0. Section 3 presents the LSS improvement model using VSM 4.0. Section 4 describes the current and future states of operational excellence in the case study company. The last section provides conclusions.

2. Literature Review

VSM creates product lines, records current states, envisions future states, integrates project pipelines, and transforms industrial activities. The map also allows users to focus on materials and information flows [24]. Traditional VSM is mainly used in a series of production lines that produce a single product or a single series of products. It now faces challenges in the big data environments that perform real-time data processing, visualization, and decision assistance. Therefore, it is not suitable for production sites with many products, unstable batches, and frequent line changes [25].

Recently, VSM was combined with Industry 4.0 technology to improve the traditional VSM. Both VSM 4.0 and Dynamic VSM (DVSM) can demonstrate the manufacturing performance of the entire process in real time through a data acquisition system. Meudt et al. [26] applied VSM 4.0 to detect waste and loss in information flows. Hartmann et al. [27] provided steps for designing VSM 4.0 and a new understanding of the information in the value stream. DVSM is based on the integration of traditional VSM and Industry 4.0 technology, connecting production factors such as equipment, workers, and materials. Ramadan et al. [28] combined RFID technology with VSM and proposed a DVSM implementation framework to track and visualize the development cost of a single product. Huang et al. [29] proposed a multi-layer DVSM that combines traditional VSM with a multi-agent system based on cyber-physical system (CPS) technology. An agent can automatically collect data near the site in order to display changes in key metrics related to materials and information flows in a multi-stage production process, either in real time or near-real time. Balaji et al. [25] introduced an integrated model of the internet of things (IoT) and VSM to collect shop floor data and monitor on-site status in real time, helping managers to quickly carry out improvement activities based on experience. Schoeman et al. [30] applied VSM to visualize and analyze waste flows of an iron and steel company in order to achieve an environmentally responsible zero-waste environment. Qin and Liu [31] used VSM to improve the entire e-commerce supply chain process for an amazon retailer.

Guo et al. [32] proposed a hybrid approach combining VSM and DMAIC to overcome the practical implementation of a VSM diagnostic tools and verify the performance throughout an air conditioner assembly line. Salah and Rahim [33] identified waste in the current VSM in the DMAIC program and improved and developed it into the future VSM. DMAIC is a framework suitable for integrating lean practices because it is a problem-structuring tool, providing empirical results that can be used to eliminate waste in structured and semi-structured cases [34]. According to Gupta and Jain [35], every lean tool plays a unique role in specific problems. We must use each tool to understand when, where, why, and how to produce a successful DMAIC process. Most lean tools belong to the improvement and control phases of DMAIC, but once an opportunity has been identified, there is no reason not to use them immediately [36]. Thus, more studies based on the integration of LSS and VSM 4.0 are required.

The potential to improve productivity by implementing LSS and VSM 4.0 using a human-centered approach is worth exploring. An essential aspect of continuous improvement is the inclusion of a human-centered design, as traditional LSS interventions, while attempting to maximize productivity by minimizing resources, often ignore human constraints and needs. When humans are integrated into continuous improvement activities, they provide a whole new dimension in the improvement process and a new perspective on its design [37]. Therefore, to ensure increased productivity and improved working conditions, a continuous improvement process should be implemented whilst simultaneously using a human-centered LSS approach. To achieve this, a comprehensive methodological framework is required.

3. Research Method

Lean and Six Sigma are two compatible problem-solving methods [38]. Lean typically starts by developing an understanding of the customer's added value and then uses VSM to examine the process in detail. When VSM identifies criticalities in process steps [22], Six Sigma provides a data-driven analytical approach to define and quantify the error types. LSS specifies an improvement process called DMAIC. Figure 1 depicts the LSS improvement model using VSM 4.0.

In the DMAIC process, VSM 4.0 is used in the define and improve phases. First, in the define phase, team members write a project charter detailing the project goals and problem statement and use VSM 4.0 to map the process components and process boundaries. In addition, a project management team is assigned to determine the roles and responsibilities and develop the project timeline. Next, in the measure phase, the project team members assess the current situation in order to gather information. Before entering the analyze phase, we must measure the performance metrics of the process.

During the analyze phase, the project team members identify potential causes and evaluate them through hypothesis testing and other explanatory analyses. Some ideas are also generated and must be validated before they can be used in the improve phase. For the next phase of improve, team members plan how they will use VSM 4.0 for a full implementation, select and use tools to eliminate waste, and update the performance metrics of the process as needed. Finally, in the control phase, the team members develop a plan for monitoring and transferring responsibilities to ensure continued success. Additionally, the team finalizes the business case using risk and financial analyses. At each of these phases, tools are used to ensure the correct execution of the picking workstation design, as shown in Table 1.

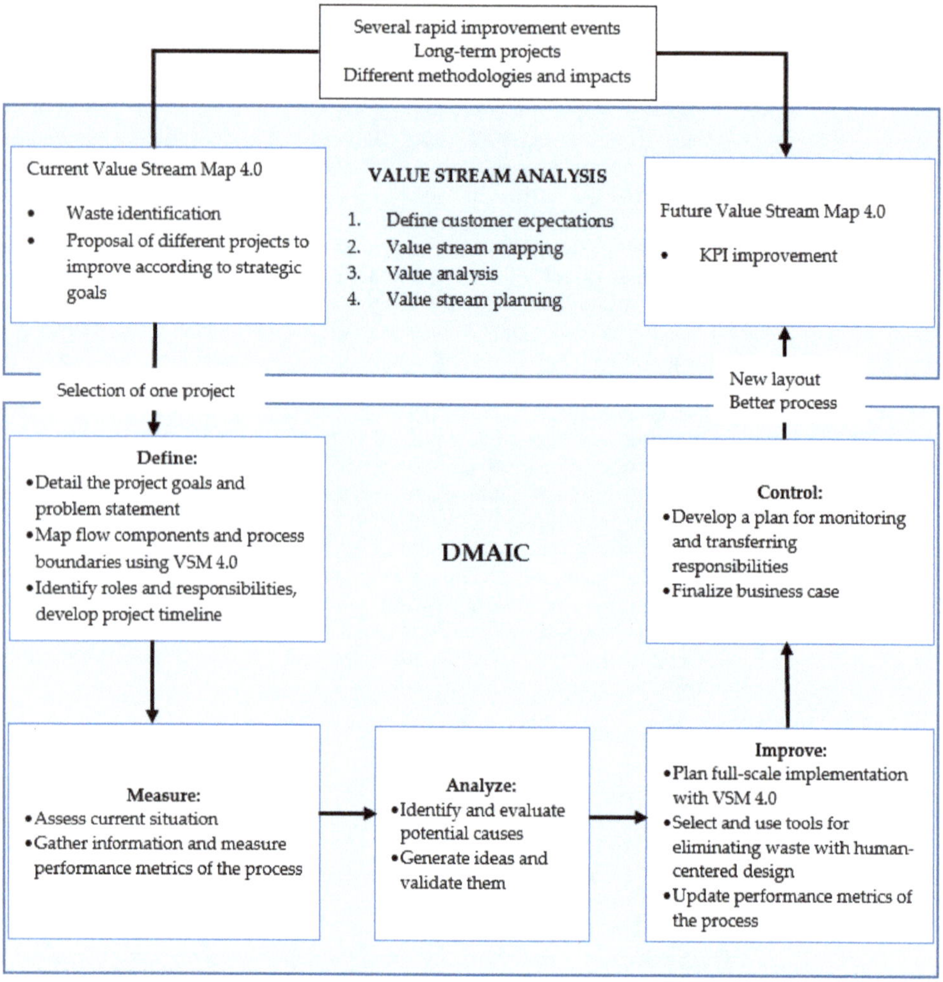

Figure 1. An improvement model of LSS with VSM 4.0.

Table 1. DMAIC approach for the picking workstation design.

Phase	Tools	Activities
Define	VSM 4.0 (current state) Project charter SIPOC diagram Gemba	1. Establish team 2. Assign specific role 3. Create project timeline
Measure	Key performance indicator (KPI) Data visualization	1. Describe process 2. Collect data
Analyze	5 Whys	Identify problem and solution
Improve	Kaizen events VSM 4.0 (future state)	1. Relocate workstation 2. Codify parts and kits

Table 1. *Cont.*

Phase	Tools	Activities
		3. Design identification cards, templates, and trays
		4. Create job breakdown sheets
Control	Brainstorming the risks and countermeasures	1. Risk analysis
		2. Financial analysis

SIPOC stands for suppliers, inputs, process, outputs, customers.

4. Case Study

The case company, called Company T (TC), is located in Spain. It is a world leader with operations on all continents and authorized distribution in 75 countries. TC manufactures truck coolers to ensure world-class quality and promote a diverse culture. The company's managers found, using the annual value stream analysis (VSA), that they needed to respond faster in order to deliver higher quality products to customers, and they identified certain gaps in the manufacturing process. Among the three parallel picking stations at the ends of the three assembly lines, one of the gaps was related to the preparation of bags with distinct sets of electrical components in the style of a typical picking station. Managers found that there was a lot of waste in the process, a lack of inventory management, and frequent customer complaints. Thus, they devised the idea of combining three workstations into one so that they could hire a full-time employee to prepare the small bags with the electronics. Furthermore, because of their strong dedication to diversity, the company seized the opportunity to integrate people with intellectual disabilities into the new workstations. Therefore, the entire workstation had to be designed with great attention to detail, visual cues, and intuitive operation.

The VSA developed certain management agreements, including specific goals and KPIs for the next 12 months, in order to improve the process. These metrics are linked to the six performance indicators listed in Table 2 (safety, quality, delivery, inventory, employee engagement, and growth), as well as statuses and their projects. After reviewing the VSM, the managers identified several contingencies for identifying opportunities for the improvement of the process. A striking highlight was the kit preparation workstation. The cooling unit has three assembly lines, each with a picking station. Each unit takes less than a minute to run. The workers at this workstation pack electronic components into small plastic bags. These components can be used to install cooling units in the customers' trucks. The workstation is subject to the following issues, including the triple inventory points. Since the amount of work required is less than one cycle time, workers must constantly switch between workstations, and they perform other work at the same time. Some customers complained that the components were inefficient for installing the equipment and that the process was inefficient due to inappropriate workstations that are poorly designed for the material flow, ergonomics, and 5S systems. It was determined that combining three workstations into one would provide benefits such as providing a sufficient workload for the 4–5 h shifts, reducing the total inventory, because there is only one point of operation, freeing up more space in the production line, increasing production, and ensuring a calmer environment. At the beginning of the line, to optimize the flow, and because there is no need to change the workstations to the same cell in the line, the overall performance of the line workers was improved.

After the VSA, several project ideas can emerge. Using the DMAIC process is useful for running select projects, as it provides a clear roadmap. This case study explains the DMAIC problem-solving process and all the important aspects of designing this unique workstation. The selected project included a person with an intellectual disability because TC's culture emphasizes community and sustainable development. The company has negotiated an arrangement with AUR, a non-profit organization that helps people with mild intellectual disabilities to find work. Now is the perfect time to build a new picking station, including people with intellectual disabilities, because the labor is simple, safe, and not positioned in the assembly line. As a result, a prospective employee was invited in, a contract with TC was signed, and work began on a new workstation design. This work has to be based on a human-centered strategy based on lean principles; therefore, it starts with attempts to gain a good understanding of people's abilities, speeds of work, and aspirations. Since the prospective employee was due to start work in 21 days, this became the deadline for the project. The DMAIC flow in this case study is shown below.

Table 2. Performance metric review.

Indicator	Safety	Quality	Delivery	Inventories	Employee Engagement	Growth
Current state	No accident in the past 10 years	There is no systematic categorizing of problems	A total of 2–3 days in Southern Europe, 1 week in Northern Europe, 3 weeks in the Middle East, 4 weeks in America	More than 75% of the references had a delivery time of less than three days	Recently, there has been a decrease in employee engagement	Around 30% of the weeks show operation at full capacity
KPI	Number of workdays missed due to working accidents	The number of warrantee requests received within the first six months after delivery	Percentage of units delivered on time	Percentage of the cost of inventories reduced from the current level	Employees' average satisfaction ratings	Maximum capacity
Projects	Keep track of minor bolt and screw inventories	Create an effective diagnosis and classification system	Improve assembly line capability and design a workstation efficiently	Unify workstations and sites along the line where materials are held	Improve line balance and the ability to react and adjust production levels quickly	Restructure the lines and unify some identical workstations in each line

4.1. Define

At the end of each assembly line, there are three picking stations for the preparation of bags with the electronic components (kits). These stations take up a lot of space and have many stock outages. Moreover, workers underperformed due to having to fact that they had to change between activities and the picking workstation waste, which only required 1/10 of their cycle time. Defects related to misidentification occurred frequently, and the missing parts could lead to unforeseen downtime. Figure 2 depicts the current VSM 4.0. If the space and waste of the three production lines were used more efficiently, the factory would save time and space, improve the quality, and reduce the inventory. This time and space could be used to add more workers to the assembly line and create more units. The project timeline was 21 days and included all stages of the DMAIC process (see Figure 3). The project began when the company chose to start the DMAIC process to build a new workstation and ended when the workers started work. The control phase lasted significantly longer, at least three months, to ensure that the quality standards were met. Furthermore, the learning process had to be tracked according to the self-interests of the workers.

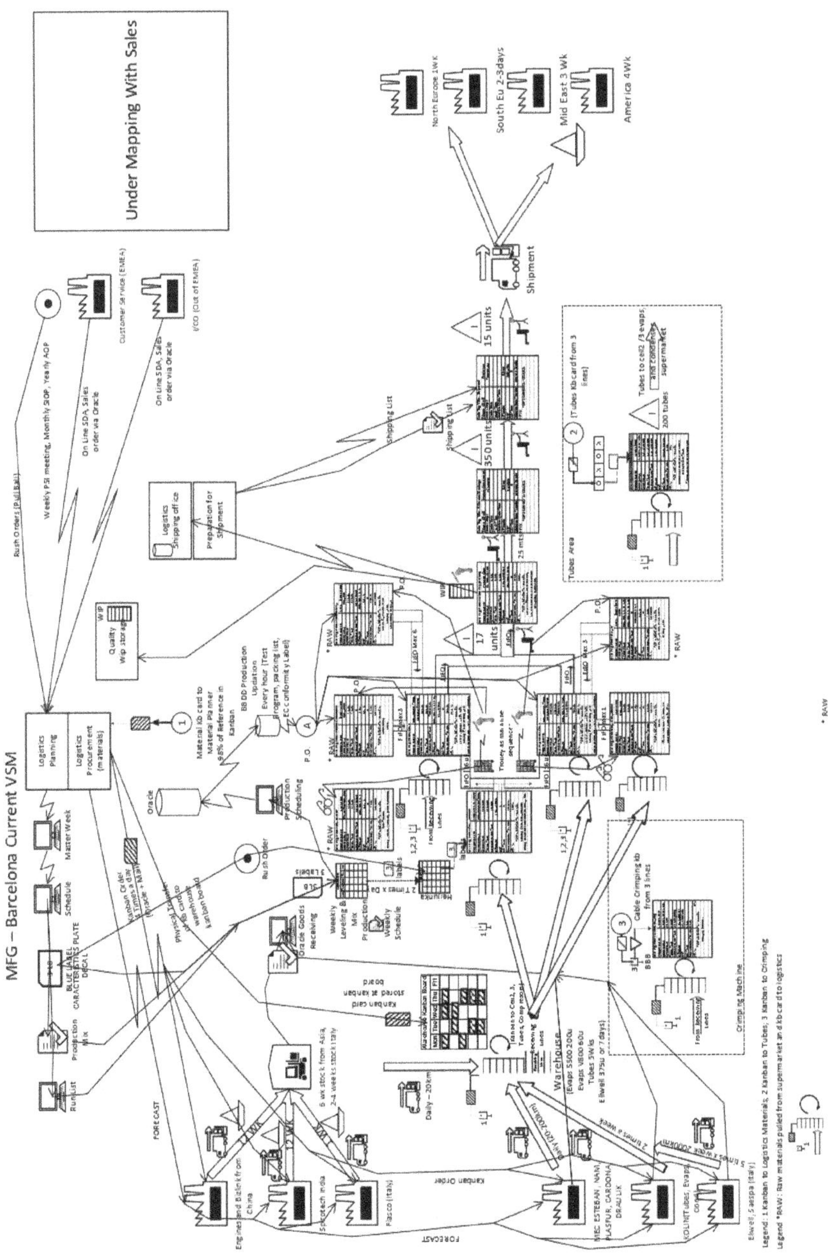

Figure 2. Current VSM 4.0.

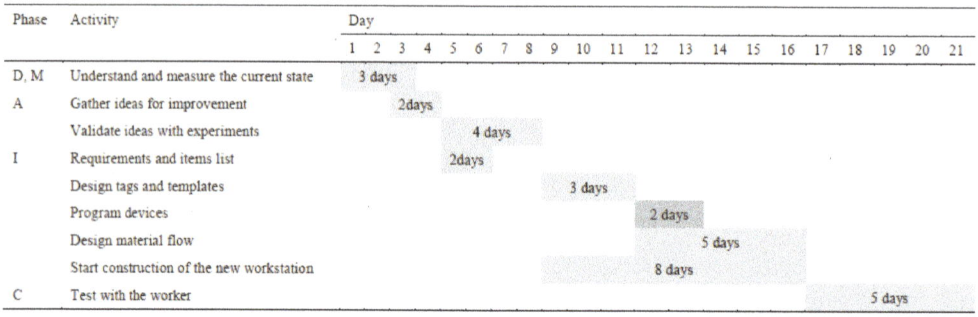

Figure 3. Project timeline.

4.2. Measure

During this phase, the current situation was assessed, and waste was identified and defined. The team first determined what information needed to be gathered and realized that defects were not a quality issue, because the installer had the relevant references; thus, they were fixed within less than a week. However, defects have an impact on delivery, as the customer is not able to utilize the equipment until the installation is complete. Therefore, any defect in the kit had to be considered a delivery issue. The team obtained data by analyzing the last 3000 shipments and found that 97.8% of the defective kits occurred due to missing parts (76%), misidentification (18%), and defective parts (6%). Each production line was assigned a worker who was responsible for the kit preparation. Certain jobs were assigned by this worker to units five or six meters away from the picking station. After completing the mission, the process started with the verification of the unit name and extended to the time when the kit was brought to the unit. Table 3 describes the process and the averages for the types and times of four different workers. Specifically, the process was affected by the following four types of waste:

(1) The worker uses hand movements (1,7) to travel from their current workplace to the picking workstation and back. However, because the worker is continually changing jobs, this mobility wastes time and reduces productivity and concentration.
(2) Picking up (4) parts is inefficient, since workers are continually comparing the panel to the materials.
(3) Mislabeling (6) is common due to the presence of tags from several kits on the table. As a result, the worker must pay close attention to picking up the correct one and comparing it to the unit.
(4) Movement due to an insufficient inventory (8) occurs with every 200 kits, or every two production days. The worker has to leave the line for about two minutes to gather extra material for the kits when the boxes containing parts run out of material.

Table 3. Details of the process description.

No.	Action Name	Type	Rounded Time (s)
1	Going to the workstation after leaving the unit	Movement (5 m)	10
2	Verifying the name of the unit that is being prepared	Visual inspection	2
3	Finding barcodes	Hand movement	4
4	Picking an empty bag	Hand movement	2
5	Finding appropriate objects	Hand movement × 10	3 × 10
6	Tagging the bag	Hand movement	5
7	Return to the unit with the kit	Movement (5 m)	10

Table 3. *Cont.*

No.	Action Name	Type	Rounded Time (s)
		TOTAL	63
8 *	Going to a different place to search for missing components	Movement	120

* is not a part of the process; it occurs when there is a shortage of materials.

4.3. Analyze

After considering the measurements and data gathered, the team proceeded to investigate each defect and determine the root cause. At this stage, a hypothesis is developed to identify the sources of different error types, bottlenecks, and sources of variation. To evaluate the hypothesis, we used five "why" analyses and a quick test. The five main problems observed during the measure phase and their five "why" analyses are summarized, together with the suggested solutions, in Table 4.

Table 4. Five whys analysis of the observed problems.

Waste	Movement to and from the Picking Workstation	Slow Pick-Up Process	Mislabeling	Missing Parts	The Movement Caused by the Inventory Break
Problem	Changing workstations wastes a lot of time for the worker	Picking up materials is a time-consuming and wasteful activity	The barcode stickers on the kits do not identify them well	Many of the kits are incomplete	The worker must leave the line in order to obtain more material
1st Why	A worker must prepare a kit every time a machine is due to be packaged	The worker must keep an eye on the panel at all times	A worker inserts an incorrect barcode	They were not placed inside the bag by the worker	One of the components was depleted
2nd Why	The main assembly process is serially integrated with the kits	There is no easy way to tell what is in each kit	There are many barcodes on the table, and they all look the same	The worker forgot to bring a specific item	The material levels are insufficient for allowing for frequent replenishment
3rd Why	One-piece flow is the production method used	In the boxes, there are no visual aids	Workers have already printed many of them	The worker has no means of knowing if the object was already inside	Every part has three distinct points, and there is not enough room in the workstation for an additional secondary box with parts
4th Why		The boxes come from the resupply, which is not visually altered	The printer is a long distance away, and employees rarely visit it	The worker arranges the items one by one	
5th Why				They have no place to be visualized	
Solution	Remove the process off the mainline and work in batches. The process will become more agile with repetition	Visual identifications on boxes should be visible without having to look at the panel	Tags with various barcodes should not coexist on the table. The printer should be placed close to the table, and only the necessary documents should be printed	Before placing items in the bag, they should all be arranged on a tray to ensure that none are missing	Combine the three inventory points and make room for more component boxes next to the current ones

4.4. Improve

This phase allows us to test the implementation of the idea and the realization of the project. Because it covers all aspects of the workstation design, installation, and worker participation, the improvement phase is the most thorough. Since LSS is a human-centered approach, it aims to reduce human error by involving workers in the process [39]. Therefore, considering their intellectual disability, it was necessary to focus on the workers' needs and test the ideas for improvement in participation with them. In this study, an improvement phase using VSM 4.0 was implemented for the following reasons:

(1) The ability of the worker. Project team members concluded that, if the process was well-structured, simplified, and predictable, workers would be able to complete the work. Therefore, the team established a path that is fully integrated with the workers. The path is designed to last for two months, starting and ending every day for 3 and 6 h successively.

(2) The workstation location. In addition to safety managers and work instructors, the team opted to move the workstations further from the production line, simplifying the material flow. Four types of waste identified in the measure phase were taken into consideration when rearranging the location. The location is at the beginning of the line and the kit is to remain with the structure. The new locations marked with w in the factory layout (Figure 4) illustrate how the workers can be placed at the beginning and center of the production lines in order to distribute the kits to the ends of the three assembly lines. The black arrows in Figure 4 show the incoming and outgoing material flows.

(3) Part and kit codes. To speed up the picking process, all parts and kits required simple coding. The team chose to use symbols to code the parts (Figure 5) and colors to code the kits (Figure 6). Since some parts are already easily identifiable, they do not have any associated symbols. A symbol with an alternate reference number is placed next to the part image. The kits are color-coded and have letters to name them. With this simplification of the parts and kits, templates and tags could be designed effectively.

(4) Templates and trays. The template in Figure 7 depicts the actual dimensions used to identify the surrounding colors of the kit and provide an image of the part, so that workers can easily view the part on the top of the picture. There are tapes behind the template for fixing the metal tray. Figure 8 shows a tray with a funnel placed below it at a low height to prevent parts from falling out of the tray.

(5) Identification card. The appearance of the card allows for quick viewing for the selection of parts. It was deemed that the cards should have a visible picture of the parts and symbols and include colors and quantities to indicate to the workers which kit they correspond to. Figure 9 illustrates how the templates and cards work together.

(6) Box location and size. All parts are divided into three sizes of box: small, medium, and large. The locations of the boxes are sorted according to the frequency of their replenishment: very frequent, common, and rare. Other important considerations are that all boxes must be within reach of the worker (no walking required) and that boxes containing more frequently used items should be in the center. A complete image of the workstation is shown in Figure 10.

(7) Kanban and first-in, first-out (FIFO) order tracking. The team decided to implement a FIFO rail, where the production manager hung Kanban cards according to the production needs. In addition, barcodes on the Kanban cards are to be scanned to identify the bags and boxes where the kits are to be stored in the production line.

(8) Scan and print barcode tags on site. After the worker picked up the Kanban card, they scanned the barcode on the Kanban card to obtain the tag for the automatic printer. The purpose of this process is to avoid mislabeling, as Poka-Yoke prevents workers from making mistakes. Workers fill the bag with the appropriate pieces until all tags are used up, after the tags are created and placed next to the tray in the preparation kit. When all the tags in a batch are used up, workers place a box of kits into the outbox, remove the next Kanban card, and start a new batch.

(9) Outbox. Workers can find boxes with batches of kits on the right side of the map. Replenishers deliver full boxes to the line every hour. Inside the box, there is a level surface and a ramp to place the finished boxes on, and there is space for storing the next batch of empty boxes.
(10) Work breakdown sheet. The team created a work breakdown sheet detailing all the functions of the workstation, as shown in Figures 11 and 12. These two sheets refer to the kit preparation and material replenishment processes, which are required by the picking station.

4.5. Control

In this phase, the team assessed the workstation's effectiveness, as well as the workers' learning paths and progress. Three templates had errors identified within the first week. A control plan enables quality control and ensures that defect-free kits reach the customers. Workers can complete batches on the same day because the templates are easily recreated. As shown in Table 5, some risks and countermeasures for their avoidance were observed during the first month of implementation. The team developed an adaptation plan before the workers started work, but the plan was adjusted based on performance. Workers worked three hours per day and three days per week for the first two weeks, which was not enough to supply all he factories with a kit. After evaluating the efficiency, the company expanded the working hours to 4 h per day and 4 days per week in order to complete all the required kits. Table 6 shows the adjusted planned and actual production kit quantities. In the first two weeks, the kit only had defects related to external issues, such as labeling confusion, misplaced items, and insufficient materials. Once all the issues are corrected, no errors were logged for the next four months, which means that the first pass yield (FPY) of the improved workstation increased from 98% to 100%. In the end, the team decided to hire full-time staff (6 h per day, 5 days per week) and schedule another future project to increase the workload of the workstations and other similar tasks.

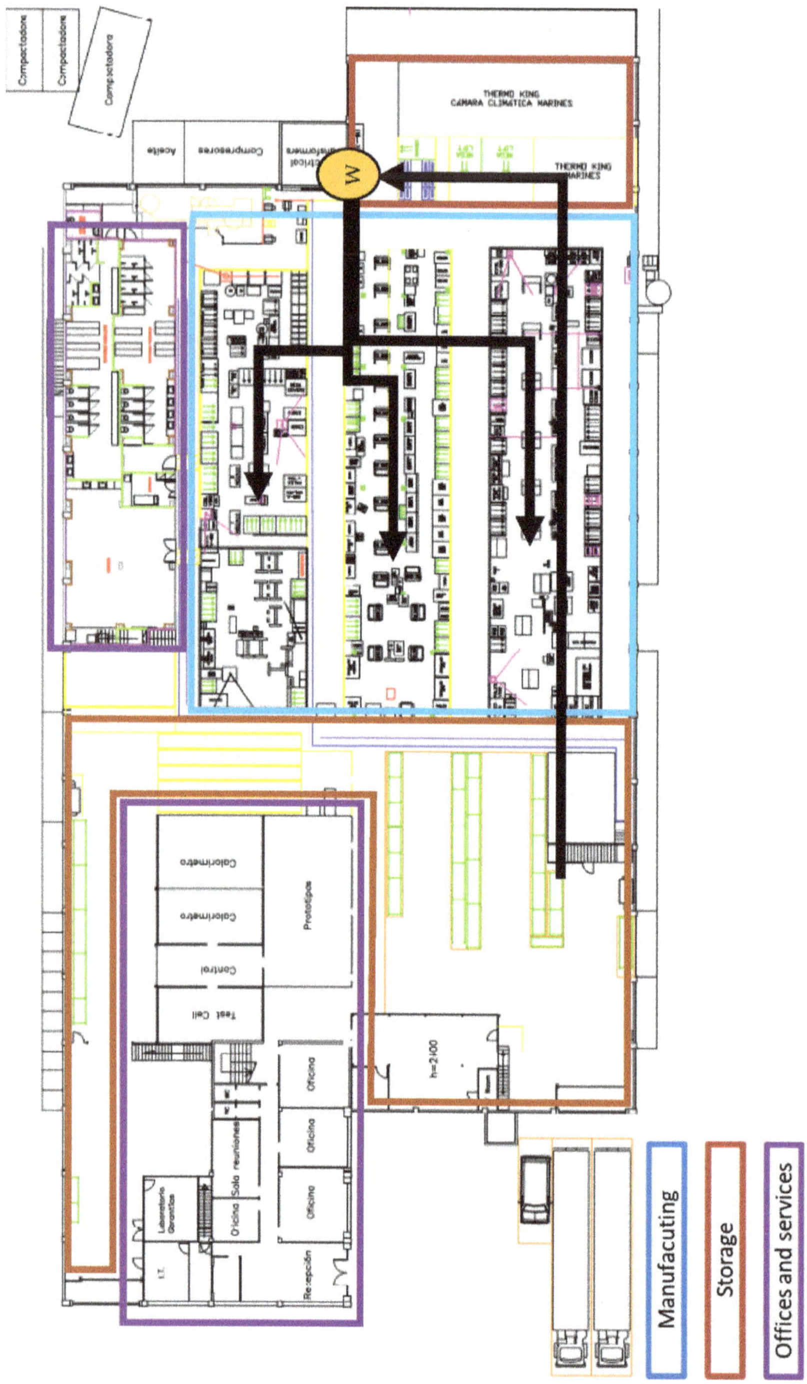

Figure 4. New location for the kit preparation workstation.

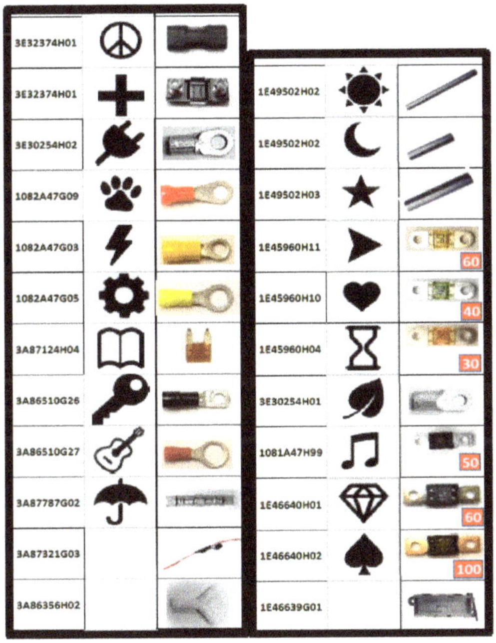

Figure 5. Relation of references, symbols, and parts.

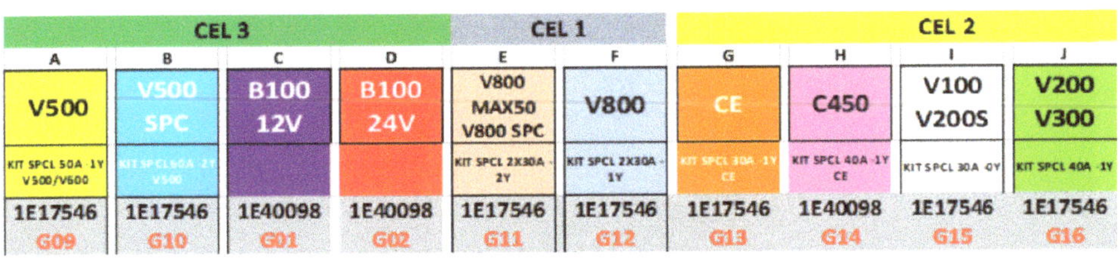

Figure 6. Relationship between references, colors, and kits.

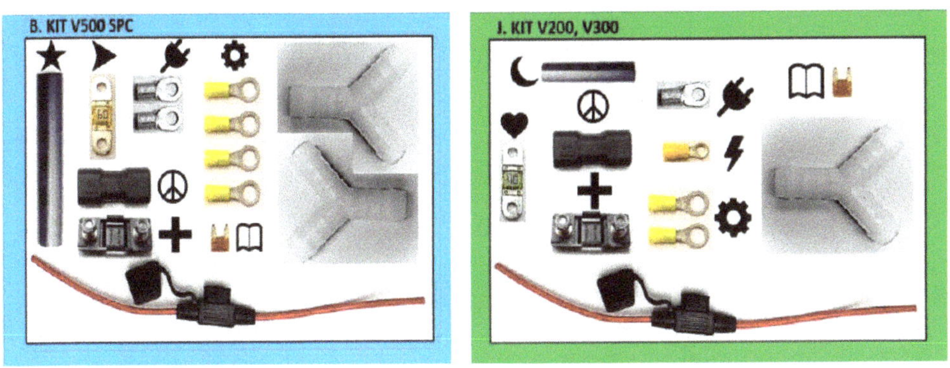

Figure 7. Templates with colors, objects, and symbols.

Figure 8. Place the magnetic template on a metal tray.

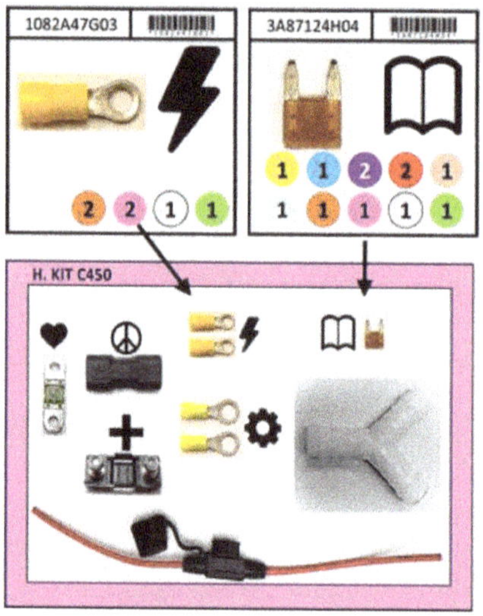

Figure 9. Correspondence between the identification cards and templates.

Figure 10. Picking workstation.

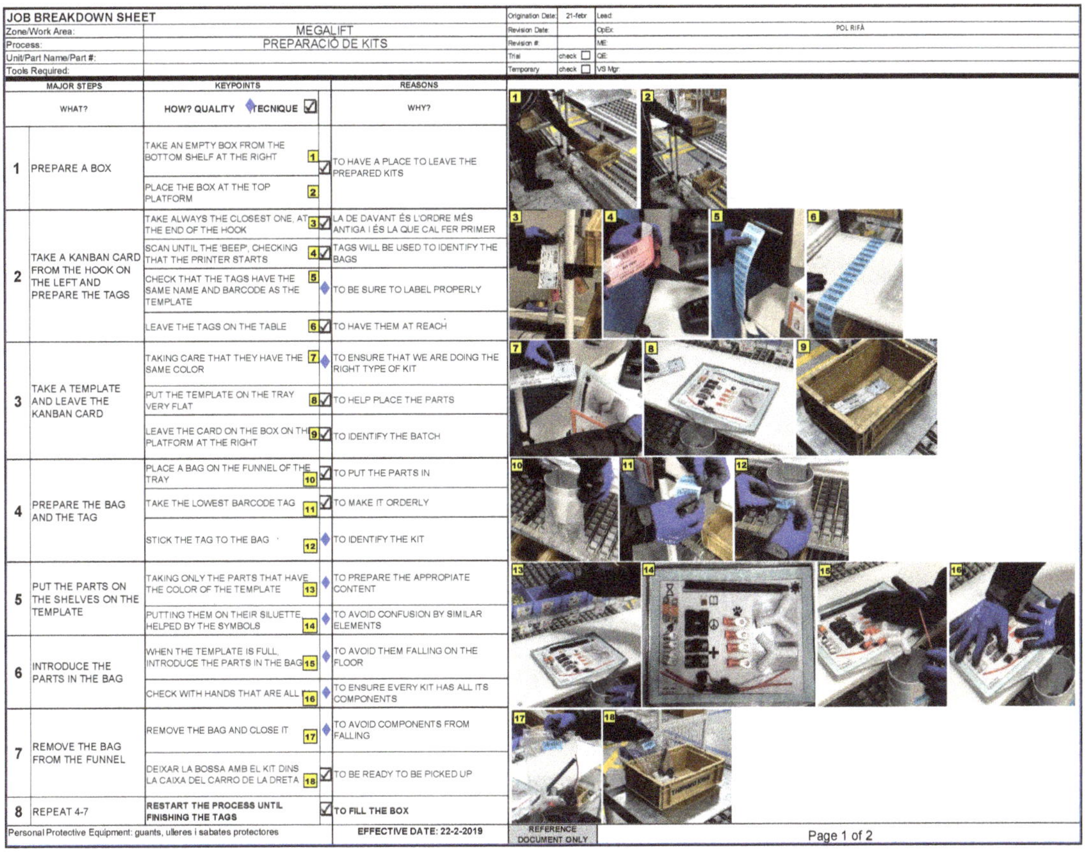

Figure 11. Work breakdown sheet of the kit preparation.

Table 5. Risks and countermeasures in the first month of implementation.

Risk	Countermeasure
The workstation does not include all of the factory's kits	Other picking workstations continue to operate as usual
The template and cards are most likely incorrect	During the first week, small quantities of all the different kits are used and another worker double-checks the quality of the kits
Any discomfort experienced by the worker as a result of the workstation improvements	The worker's motions are studied in real time by the safety department in order to detect any weariness or potentially harmful movements
Replenishers are not assigned to the new material locations	A regular reminder to monitor the workstation is issued over the first two weeks
The worker is not able to deliver satisfactory results	During the first month, the worker keeps track of the number and type of kits created

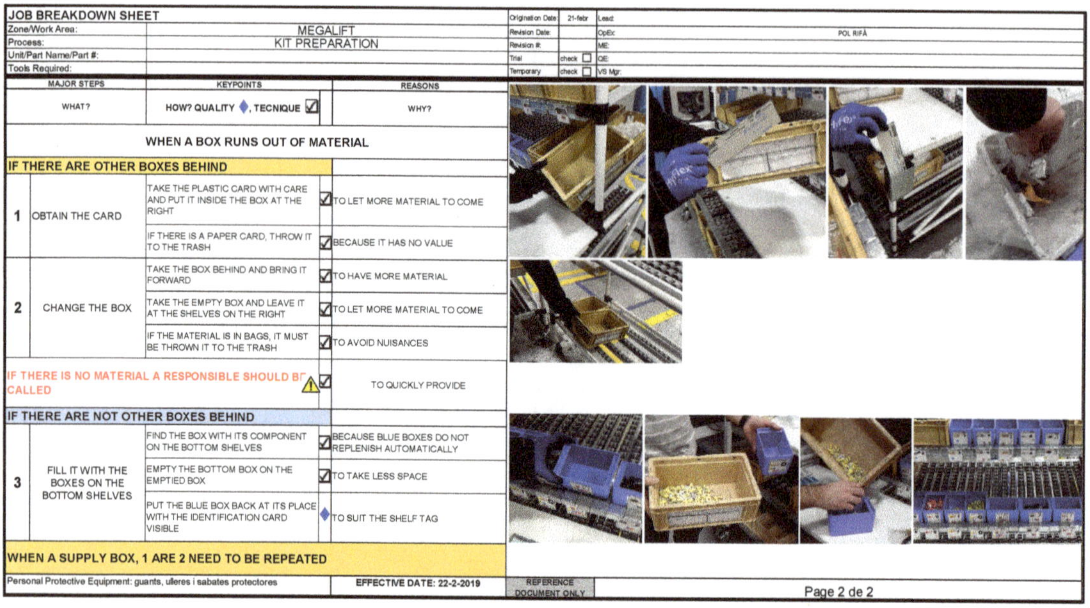

Figure 12. Work breakdown sheet of the material replenishment.

Table 6. Plan for the first month of adaptation.

State	Hours/Day	Week Day	1 1	2	3	2 4	5	6	3 7	8	9	10	4 11	12	13	14	5 15	16	17	18
Taught by instructor	3		56	75																
Surveillance at work	3				70	83	95	90												
Working alone, quality checks	4								133	145	140	135	152	142	135	137				
Normal workload	4																160	153	158	140
		Total		201			268			553				566				611		
Kits/hour			19	25	23	28	32	30	33	36	35	34	38	36	34	34	40	38	40	35

4.6. Financial Effect

The cost analysis of the project is detailed in Table 7. The most relevant suggestion is the addition of another person to the workforce, but the new incorporation of the technology is cheaper than another salaried employee. As the goal was not a cost reduction, this project made some resources available to increase the production, mainly in terms of space and time.

Table 7. Cost analysis of the project.

Description	Item	Cost (EUR)
Cost of the materials for the workstation design	Bench, computer, printer	4.000
Salaried employees cost of project execution	60 h (EUR 30/h)	1.800
Cost of additional employee (within the inclusion plan)	Part-time contract	1.200/month

The impacts of the process can be seen by observing the following indicators:

1. Space: free up 8 m^2, transitioning from 11 m^2 in three different locations to 3 m^2 in one location.
2. Productivity: a 63 s bottleneck for three production lines, enabling increased production.
3. Quality: increase the yield rate from 98% to 100%.
4. Inventory: transition from three places to one place, so that the volume is reduced by 40%. Initial cost: EUR 2200.

The direct savings of the project amounted to EUR 3180, mainly due to the freeing up of space (see Table 8). In the long term, the time saved (at 2.000 units/month) and quality claims (6 claims/month) were estimated to save EUR 100/month. It is difficult to estimate these costs, but the most relevant impacts of the project relate to inclusiveness, space, and predictability.

Table 8. Benefits and savings of the project.

Impact	Item	Savings (EUR)
Space	8 m^2 freed	8.000
Assembly time in the line	63 s/unit	0.4/unit
Inventory reduction	40% of EUR 2200	880
Quality claims by missing parts	5 claims/month	500/month

4.7. Human-Centered LSS Framework

For organizations today, LSS is the most popular methodology for streamlining operations and improving customer satisfaction. Although LSS has many expected benefits, it has one major unintended consequence. Due to its focus on process simplification, employees often feel marginalized. By combining the powerful tools and techniques of LSS with those available using a human-centered approach, as shown in Figure 13, the raised issues can be addressed. This integration enables organizations to engage all different types of employees, partners, and customers in a more effective and productive manner. Integrating LSS tools and technology with a human-centered design and management involves: (1) utilizing VSM to create a more powerful and insightful tool for better understanding current frictions with customers, business partners, or employees; (2) using the customer voice to develop the customer mental model; and (3) developing an employee mental model for each key employee role in the re-engineering business process. This will ensure that the process guides them on a journey that leads to a new business process. In doing so, the employees will be ready, willing, and able to serve customers and business partners as needed. LSS adopts the perspective that any effort to improve systems, processes, products, or services must begin with the customer and the people performing the work. The tools and mindset of a human-centered approach can support performance improvements and take them to the next level in order to understand ambiguous challenges and develop innovative solutions to address them.

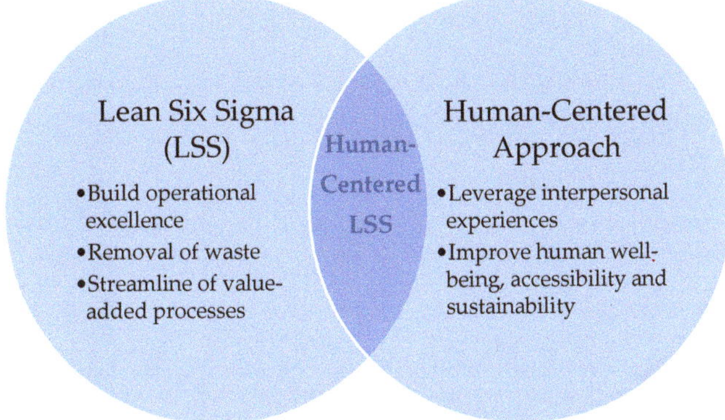

Figure 13. Human-centered LSS framework.

5. Conclusions

Adopting a complementary LSS approach expands the toolbox and offers the prospect of continuous improvement for many industries that have already implemented lean principles. Using VSM 4.0 as a tool, managers can focus on value and develop appropriately, and changes in the production processes can be quickly detected and planned. LSS's DMAIC strategy is a holistic approach that ensures a thorough understanding of potential problems and continuous improvement. We proposed an improvement model based on DMAIC and VSM 4.0 for a truck cooler manufacturer to improve the picking workstation design with a human-centered approach.

The case study provided a step-by-step process framework for, and in-depth analysis of, the implementation of our proposed improvement model. This model allows the company to generate available space on the assembly lines, thereby enhancing the ability to assemble more units and, of course, productivity. In addition, the electrical kit used for the installation achieved 100% FPY, successfully enabling the participation of mentally handicapped workers and raising the company's standards to provide better workstations. The direct savings of the project amounted to EUR 3180, mainly due to the freeing up of space. After the project was completed, not only was the financial performance satisfactory, but the planning capabilities and employee engagement were also improved, and the quality complaints, costs, and deliveries were reduced.

The presented paper is based on a single case study, and the results are limited to this company only. However, the approach of this paper offers a generic learning perspective. Through the effective application of LSS quality initiatives, the means by which the direct savings of the project could be obtained have been demonstrated. Thus, this paper could serve as a unique roadmap for practitioners and academics to improve the material productivity in terms of both product and process.

Author Contributions: Conceptualization, F.-K.W.; formal analysis, P.R.R.; investigation, P.R.R.; methodology, F.-K.W. and B.R.; project administration, B.R.; resources, P.R.R.; supervision, F.-K.W.; validation, F.-K.W. and P.R.R.; visualization, B.R.; writing—original draft, B.R.; writing—review & editing, F.-K.W. All authors have read and agreed to the published version of the manuscript.

Funding: This research received no external funding.

Institutional Review Board Statement: Not applicable.

Informed Consent Statement: Not applicable.

Data Availability Statement: No data are available.

Conflicts of Interest: The authors declare no conflict of interest.

References

1. Turk, M.; Pipan, M.; Simic, M.; Herakovic, N. A Smart Algorithm for Personalizing the Workstation in the Assembly Process. *Appl. Sci.* **2020**, *10*, 8624. [CrossRef]
2. Piccinini, P.; Gamberini, R.; Prati, A.; Rimini, B.; Cucchiara, R. An Automated Picking Workstation for Healthcare Applications. *Comput. Ind. Eng.* **2013**, *64*, 653–668. [CrossRef]
3. Albliwi, S.A.; Antony, J.; Lim, S.A.H. A Systematic Review of Lean Six Sigma for the Manufacturing Industry. *Bus. Process Manag. J.* **2015**, *21*, 665–691. [CrossRef]
4. Timans, W.; Ahaus, K.; van Solingen, R.; Kumar, M.; Antony, J. Implementation of Continuous Improvement Based on Lean Six Sigma in Small- and Medium-Sized Enterprises. *Total Qual. Manag. Bus. Excell.* **2016**, *27*, 309–324. [CrossRef]
5. Raval, S.J.; Kant, R. Study on Lean Six Sigma Frameworks: A Critical Literature Review. *Int. J. Lean Six Sigma* **2017**, *8*, 275–334. [CrossRef]
6. Gupta, V.; Acharya, P.; Patwardhan, M. A Strategic and Operational Approach to Assess the Lean Performance in Radial Tyre Manufacturing in India. *Int. J. Product. Perform. Manag.* **2013**, *62*, 634–651. [CrossRef]
7. Chiarini, A. Sustainable Manufacturing-Greening Processes Using Specific Lean Production Tools: An Empirical Observation from European Motorcycle Component Manufacturers. *J. Clean. Prod.* **2014**, *85*, 226–233. [CrossRef]
8. Singh, M.; Rathi, R. Investigation and Modeling of Lean Six Sigma Barriers in Small and Medium-Sized Industries Using Hybrid ISM-SEM Approach. *Int. J. Lean Six Sigma* **2021**, *12*, 1115–1145. [CrossRef]
9. Panwar, A.; Jain, R.; Rathore, A.P.S. Lean Implementation in Indian Process Industries—Some Empirical Evidence. *J. Manuf. Technol. Manag.* **2015**, *26*, 131–160. [CrossRef]
10. Jasti, N.V.K.; Kodali, R. Lean Production: Literature Review and Trends. *Int. J. Prod. Res.* **2015**, *53*, 867–885. [CrossRef]

11. Thomas, A.J.; Ringwald, K.; Parfitt, S.; Davies, A.; John, E. An Empirical Analysis of Lean Six Sigma Implementation in SMEs—A Migratory Perspective. *Int. J. Qual. Reliab. Manag.* **2014**, *31*, 888–905. [CrossRef]
12. Shokri, A. Quantitative Analysis of Six Sigma, Lean and Lean Six Sigma Research Publications in Last Two Decades. *Int. J. Qual. Reliab. Manag.* **2017**, *34*, 598–625. [CrossRef]
13. Swarnakar, V.; Singh, A.R.; Antony, J.; Tiwari, A.K.; Cudney, E.; Furterer, S. A Multiple Integrated Approach for Modelling Critical Success Factors in Sustainable LSS Implementation. *Comput. Ind. Eng.* **2020**, *150*, 106865. [CrossRef]
14. Habidin, N.F.; Mohd Yusof, S. Critical Success Factors of Lean Six Sigma for the Malaysian Automotive Industry. *Int. J. Lean Six Sigma* **2013**, *4*, 60–82. [CrossRef]
15. Kornfeld, B.; Kara, S. Selection of Lean and Six Sigma Projects in Industry. *Int. J. Lean Six Sigma* **2013**, *4*, 4–16. [CrossRef]
16. Laureani, A.; Antony, J. Critical Success Factors for the Effective Implementation of Lean Sigma: Results from an Empirical Study and Agenda for Future Research. *Int. J. Lean Six Sigma* **2012**, *3*, 274–283. [CrossRef]
17. Dos Santos, Z.G.; Vieira, L.; Balbinotti, G. Lean Manufacturing and Ergonomic Working Conditions in the Automotive Industry. *Procedia Manuf.* **2015**, *3*, 5947–5954. [CrossRef]
18. Nunes, I.L. Integration of Ergonomics and Lean Six Sigma: A Model Proposal. *Procedia Manuf.* **2015**, *3*, 890–897. [CrossRef]
19. Psomas, E. Future Research Methodologies of Lean Manufacturing: A Systematic Literature Review. *Int. J. Lean Six Sigma* **2021**, *12*, 1146–1183. [CrossRef]
20. Lobo Mesquita, L.; Lizarelli, F.L.; Duarte, S.; Oprime, P.C. Exploring Relationships for Integrating Lean, Environmental Sustainability and Industry 4.0. *Int. J. Lean Six Sigma* **2022**, *13*, 863–896. [CrossRef]
21. Jasti, N.V.K.; Sharma, A. Lean Manufacturing Implementation Using Value Stream Mapping as a Tool: A Case Study from Auto Components Industry. *Int. J. Lean Six Sigma* **2014**, *5*, 89–116. [CrossRef]
22. Cavone, G.; Epicoco, N.; Dotoli, M. Process Re-engineering Based on Colored Petri Nets: The Case of an Italian Textile Company. In Proceedings of the 2020 28th Mediterranean Conference on Control and Automation (MED), Saint-Raphael, France, 15–18 September 2020; pp. 856–861.
23. Tresca, G.; Cavone, G.; Carli, R.; Cerviotti, A.; Dotoli, M. Automated Bin Packing: A Layer Building Matheuristics for Cost Effective Logistics. *IEEE Trans. Autom. Sci. Eng.* **2022**, *19*, 1599–1613. [CrossRef]
24. Pinto, C.M.A.; Mendonça, J.; Babo, L.; Silva, F.J.G.; Fernandes, J.L.R. Analyzing the Implementation of Lean Methodologies and Practices in the Portuguese Industry: A Survey. *Sustainability* **2022**, *14*, 1929. [CrossRef]
25. Balaji, V.; Venkumar, P.; Sabitha, M.S.; Amuthaguka, D. DVSMS: Dynamic Value Stream Mapping Solution by Applying IIoT. *Sādhanā* **2020**, *45*, 38. [CrossRef]
26. Meudt, T.; Metternich, J.; Abele, E. Value Stream Mapping 4.0: Holistic Examination of Value Stream and Information Logistics in Production. *CIRP Ann. Manuf. Technol.* **2017**, *66*, 413–416. [CrossRef]
27. Hartmann, L.; Meudt, T.; Seifermann, S. Value Stream Method 4.0: Holistic Method to Analyse and Design Value Streams in the Digital Age. *Procedia CIRP* **2018**, *78*, 249–254. [CrossRef]
28. Ramadan, M.; Salah, B.; Othman, M. Industry 4.0-Based Real-Time Scheduling and Dispatching in Lean Manufacturing Systems. *Sustainability* **2020**, *12*, 2272. [CrossRef]
29. Huang, Z.Y.; Kim, J.; Sadri, A. Industry 4.0: Development of a Multi-Agent System for Dynamic Value Stream Mapping in SMEs. *J. Manuf. Syst.* **2019**, *52*, 1–12. [CrossRef]
30. Schoeman, Y.; Oberholster, P.; Somerset, V. Value Stream Mapping as a Support Management Tool to Identify the Flow of Industrial Waste: A Case Study. *Sustainability* **2021**, *13*, 91. [CrossRef]
31. Qin, Y.; Liu, H. Application of Value Stream Mapping in E-Commerce: A Case Study on an Amazon Retailer. *Sustainability* **2022**, *14*, 713. [CrossRef]
32. Guo, W.; Jiang, P.; Xu, L.; Peng, G. Integration of Value Stream Mapping with DMAIC for Concurrent Lean-Kaizen: A Case Study on an Air-Conditioner Assembly Line. *Adv. Mech. Eng.* **2019**, *11*, 1–17. [CrossRef]
33. Salah, S.; Rahim, A. *An Integrated Company-Wide Management System*; Springer: Berlin/Heidelberg, Germany, 2018.
34. De Mast, J.; Lokkerbol, J. An Analysis of the Six Sigma DMAIC Method from the Perspective of Problem Solving. *Int. J. Prod. Econ.* **2012**, *139*, 604–614. [CrossRef]
35. Gupta, S.; Jain, S.K. A Literature Review of Lean Manufacturing. *Int. J. Manag. Sci. Eng. Manag.* **2013**, *8*, 241–249. [CrossRef]
36. Marques, P.A.; Carvalho, A.M.; Santos, J.O. Improving Operational and Sustainability Performance in a Retail Fresh Food Market Using Lean: A Portuguese Case Study. *Sustainability* **2022**, *14*, 403. [CrossRef]
37. Monroe, K.; Fick, F.; Joshi, M. Successful Integration of Ergonomics into Continuous Improvement Initiatives. *Work* **2012**, *41*, 1622–1624. [CrossRef]
38. Montgomery, D.C. Lean Six Sigma and Promoting Innovation. *Qual. Reliab. Eng. Int.* **2014**, *30*, 1. [CrossRef]
39. Hannola, L.; Richter, A.; Richter, S.; Stocker, A. Empowering Production Workers with Digitally Facilitated Knowledge Processes—A Conceptual Framework. *Int. J. Prod. Res.* **2018**, *56*, 4729–4743. [CrossRef]

sustainability

Article

The Development of an Excellence Model Integrating the Shingo Model and Sustainability

José Carlos Sá [1,2,*], Manuel Reis [1], José Dinis-Carvalho [3], Francisco J. G. Silva [1,2], Gilberto Santos [4], Luis P. Ferreira [1,2] and Vanda Lima [5,6]

1 School of Engineering, Polytechnic of Porto (ISEP), 4200-072 Porto, Portugal; 1171526@isep.ipp.pt (M.R.); fgs@isep.ipp.pt (F.J.G.S.); lpf@isep.ipp.pt (L.P.F.)
2 Institute of Science and Innovation in Mechanical and Industrial Engineering (INEGI), 4200-465 Porto, Portugal
3 Production and Systems Department, University of Minho, 4800-058 Guimarães, Portugal; dinis@dps.uminho.pt
4 Design School, Polytechnic Institute Cavado and Ave, 4750-810 Barcelos, Portugal; gsantos@ipca.pt
5 Polytechnic Institute of Porto, 4200-465 Porto, Portugal; vlima@estg.ipp.pt
6 CIICESI, Escola Superior de Tecnologia e Gestão, Politécnico do Porto, 4610-156 Felgueiras, Portugal
* Correspondence: cvs@isep.ipp.pt

Abstract: Companies are continuously looking to improve their production systems using excellence models, with lean thinking, the Shingo model, six sigma and lean six sigma being the most comprehensive and applied. It is expected that the initial focus for the survival of companies is their economic profitability, but when economic needs are met, the next step is to achieve operational excellence. For this, in addition to economic objectives, it is necessary to include social and environmental objectives, i.e., the other two pillars of sustainability. This study aims to propose a conceptual model identifying the tools that can help achieve the desired results in the three pillars of sustainability aligned with operational excellence. The design of the conceptual model was based on a bibliometric analysis of the literature that relates the concepts of lean thinking, six sigma, lean six sigma and the Shingo model. The Web of Science was the platform selected for the collection of data, and the timeframe considered was 2010 to 2021. A total of 125 articles were analyzed using the VosViewer software, through which it was possible to analyze different topics of study related to the literature. The bibliometric analysis allowed for the identification of the temporal distribution of publications, the categorization of topics, different areas of application and the importance of the tools used in different practical cases. This study points out that companies have at their disposal several tools to achieve economic objectives. On the other hand, there is a set of more restricted tools that are used to meet the objectives of the social and environmental pillars. Future research should focus on identifying tools that meet social and environmental goals in order to strengthen these pillars that are essential for operational excellence and for the sustainability of companies.

Keywords: sustainability; Shingo model; lean six sigma; lean thinking; six sigma; DMAIC

Citation: Sá, J.C.; Reis, M.; Dinis-Carvalho, J.; Silva, F.J.G.; Santos, G.; Ferreira, L.P.; Lima, V. The Development of an Excellence Model Integrating the Shingo Model and Sustainability. *Sustainability* **2022**, *14*, 9472. https://doi.org/10.3390/su14159472

Academic Editor: Jose Arturo Garza-Reyes

Received: 15 June 2022
Accepted: 25 July 2022
Published: 2 August 2022

1. Introduction

Existing global competition has led to significant changes in the market, with a considerable increase in the variety of increasingly complex products with ever shorter and uncertain life cycles [1]. The current problem for manufacturers is how to deliver their products or materials quickly at a low cost and with good quality [2].

In recent years, continuous improvement and operational excellence programs have played an important role in the success and sustainability of organizations, allowing companies to achieve high grades of competitiveness [3]. Lean thinking, the Shingo model, the Toyota way, six sigma and lean six sigma are some examples of applied models. These models guide companies to maintain consistent practices with sustainable development strategies, and, as such, they control waste, reduce production variability and implement practices that are environmentally and socially responsible [4,5].

Organizations have been pressured by stakeholders to adopt not only economic but also social and environmental objectives [6,7]. The growth of ecological consumerism, climate change and the latest government legislation are some of the reasons why companies have improved their performance in all components of sustainable development [8–10]. This growing trend of companies in responding to the challenge of sustainability has resulted in the emergence of the concept of corporate social responsibility, which focuses on creating value that promotes harmony between economic development, social progress, equity and respect for the environment [11].

Lean six sigma (LSS) is one of the excellence models that helps companies improve operational efficiency, reduce costs and increase effectiveness, especially when they need to operate in a highly competitive globalized market [12]. It allows companies to increase the speed of waste elimination, reducing the processes variation as well by integrating the best methodologies and tools associated with lean thinking and six sigma [2,5].

LSS, like other excellence models, has undergone evolutions towards sustainability. First, there was integration between LSS and Green Technology, resulting in GLS (green lean six sigma) methodology that incorporates two dimensions of sustainability (economic and environmental). Recently, sustainable green lean six sigma has emerged, which incorporates all the dimensions of sustainability (economic, environmental and social) and an inclusive approach that makes industries more resilient and promotes a healthier society, meeting Corporate Social Responsibility [13].

The lean six sigma philosophy promotes process and product innovation and has the potential to influence radical innovation in the industry worldwide. To achieve long-term success, organizations need to focus on continuous improvement and problem-solving approaches by changing the organizational culture. Thus, it is possible to adopt these new philosophies, including sustainability, Industry 4.0 and the Circular Economy [14,15].

An important component of this work is to research and categorize the different operational excellence tools used on DMAIC methodology applications, allowing for readers to easily understand the best ones to act in each pillar of the conceptual model under development. In fact, their suitability to act economically, environmentally and socially are different and promotes different effects in achieving operational excellence. The model developed allows for easy perception about what tools can be used to improve each pillar. This categorization of the tools regarding each pillar toward operational excellence is the main novelty presented by this paper, representing an important contribution to the sustainability of companies regarding different pillars. Indeed, if companies improve their performance considering the DMAIC methodology and excellence tools, their sustainability also increases, leading to a better economic, environmental and social performance.

The investigation was based on a literature review in the areas of lean thinking, six sigma, lean six sigma and the Shingo model, through which a bibliometric analysis was performed. The main contribution of this research is the development of a conceptual model that relates excellence tools with the different pillars of sustainable development, in which the DMAIC plays the role of structured methodology, allowing for a strategic direction for improvement. This article is divided into the following sections. First, a theoretical review of the lean thinking literature, six sigma, lean six sigma and the Shingo model is presented. Second, the methodology and research criteria that enabled the treatment of information are presented. Third, the results of the bibliometric analysis are presented. Fourth, the information of the analyzed publications is discussed. Finally, the main findings of the research are revealed.

2. Literature Review

2.1. Lean Philosophy

Production systems have undergone several evolutions throughout history. First, production was essentially handmade, and then it was replaced by mass production with assembly lines. From the middle of the 20th century, the lean approach has increasingly replaced mass production until today [16]. Lean philosophy, or lean thinking, inspired by the Toyota Production System, can be considered as a philosophy that helps in the identification and elimination of waste. In practice, lean philosophy helps maximize product value by minimizing waste [17–20].

Taiichi Ohno, the pioneer of the Toyota Production System (TPS), described the following: "All we are doing is looking at the timeline from the moment a customer gives us an order, to the point when we receive the money. In addition, we are reducing the timeline by reducing waste activities that do not add value" [18].

The objectives of the lean philosophy are: to reduce waste in human efforts, stock, the time of marketing and production space to become agile-looking for the customer while producing quality products more efficiently and economically [21]. This production approach consists of determining the value of each activity and distinguishing the activities that add value from activities that do not add value. The latter has been worked on in order to reduce or even eliminate them [22,23]. Lean philosophy is "a way to do more with less and less human effort, less equipment, less time and less space as it comes closer to providing customers with what they want" [24].

Lean thinking is a critical concept in which value creation and waste reduction are interrelated with each other. Value creation is defined as meeting customer requirements by understanding requirements, mapping the process, promoting value flow to and from stakeholders, extracting customer value and driving toward perfection [25]. In order to reduce or even eliminate waste in the production system, organizations need to follow five key principles, as follows: Define "Value" from the customer's point of view, Identify the "Value Stream", Create a "Continuous Flow", Create "Pull Flow" and "Seek Perfection" [26]. The lean philosophy suggests that, in order to create capacity in the workplace, it is better to identify areas where there is waste and to work in these areas in order to eliminate them from processes [27].

Lean thinking practitioners traditionally focus on what they refer to as the seven forms of waste: Overproduction, Waiting Time, Transportation, Improper Processing, Inventory, Defects and Movements [4,15,28]. Recently, the authors have suggested two more types of waste that should be considered: the Underutilization of People's Creativity and Environmental Waste [29]. The departure of individuals from companies with a certain set of skills and the non-use of talent results in losses of competitiveness and productivity for companies [15].

2.2. Six Sigma

The six sigma philosophy embarks on organized and systematic methodologies that do not have, as their sole purpose, the improvement of a strategic process, but it also allows for the design of new products and services. It achieves significant reductions in defect rates and frequently uses statistical and scientific methods [30]. Essentially, an approach that improves productivity, quality and reduces operating costs is traditionally used to measure and reduce process variability [31]. It is a philosophy of continuous improvement aimed at increasing efficiency and customer satisfaction and reducing operating costs [29]. In general, it is a business improvement strategy, focused on processes to improve the profit yield by eliminating waste, reducing the cost of non-quality and improving the effectiveness and efficiency of the entire system in order to achieve or even exceed the stakeholders' needs and expectations [32].

The six sigma philosophy has two methodological branches: DMAIC (Define, Measure, Analyze, Improve, and Control) and DMADV (Define, Measure, Analyze, Design, Verify). The first focuses on existing products and processes. The second focuses on new products and processes [33,34]. The main differences between the DMAIC and DMADV methodologies are in the objectives and results of the project. DMADV has a more tangible result; however, both methods have better quality, efficiency, more production, profits and greater customer satisfaction as their common goals [35].

The DMAIC methodology provides organizations with a rigorous and disciplined structure [36]. It is used for process improvement and, more recently, for project management. It enables the finding of solutions and opportunities based on decisions supported by data and management standards, and it also provides formal procedures for the implementation of solutions [37,38].

The path of the different DMAIC phases is then described succinctly [39–41]:

- Define: The identification of the problem, the determination of objectives and the appropriation of the relevance of the objectives.
- Measure: The translation of the problem in a measurable way through the observation and research of knowledge.
- Analysis: Making a diagnosis by identifying the factors and causes of influence that determine certain behaviors, and using the aggregated data in the Measurement phase.
- Improve: Implementation of measures that have been designed to improve performance and achieve the desirable state.
- Control: Adjustments to process management and system control; therefore, the results are sustainable.

2.3. Lean Six Sigma

The lean six sigma (LSS) philosophy emerged in the early 2000s and is a hybrid philosophy built from the combination of lean methods and principles with DMAIC [1,42]. Being considered a differentiating philosophy of management and improvement, its acceptance in the generality of the industry has grown rapidly, causing significant improvements in the performance of organizations [43]. These improvements are based on reducing costs, reducing lead time and increasing quality through the elimination of waste and process variation, thus enabling organizations to achieve competitive prices and thereby ensuring customer satisfaction [44–46]. The consistent practice of methodologies such as LSS allows companies to embark on a sustainable development strategy and to be socially and environmentally responsible for their practices, achieving a desirable control that enables them to gain a competitive advantage [47].

The combination of lean thinking and six sigma allows companies to capitalize on the best of both worlds in order to achieve operational excellence [5]. The LSS philosophy "uses tools from both toolboxes to get the best of both philosophies, increasing speed while increasing accuracy" [48]. Lean philosophy, as mentioned above, increases productivity by eliminating waste, such as activities that do not add value, with the goal of achieving the continuous flow of production smoothly and uninterruptedly, producing only what the customer requires. The six sigma philosophy increases productivity by reducing product variations and increases the quality of production processes [49,50]. In addition to that, it also provides methodologies for solving problems in a structured way, which allows a strategic direction for improvement [48].

2.4. The Shingo Model

The Models of Operational Excellence began to emerge around the late 1980s and early 1990s, with the aim of supporting organizations to overcome challenges, to improve the level of performance and to achieve results that are sustainable in the long term [51]. The Models of Operational Excellence, today, are also identified as a potential support for corporate sustainability through the integration of sustainable development criteria in traditional business models. Corporate sustainability is a concept that incorporates areas such as long-term focus, stakeholder needs and the foundations of sustainable development (social responsibility, environmental responsibility and economic responsibility) [52].

In order to achieve operational excellence, there is a panoply of methodologies that can be applied, as is the case of the Shingo model. It was conceived by Japanese industrial engineer Shigeo Shingo, who highlighted the importance of organizational and behavioral culture along with the integration of lean culture; therefore, transformation and development can occur with the desired results [53].

The Shingo model allows for the symbiosis of the multiple strands of a lean management system in a strand of operational excellence, i.e., it is not only focused on the customer (results) and tools but also covers systems, culture and guiding [54]. Organizations design systems with the intention of achieving specific results and select tools to support these systems. However, the tools do not run a business. The behaviors of people in an organization can influence the sustainability of results, and therefore, this model argues that the search for improvement can not only be limited to the application of a new tool or method but should also focus on organizational culture [53]. Organizational culture is the system of principles that members of the organization share, including ways of working, traditions, histories and methods capable of achieving goals [55]. The cultural perspective is important in models of excellence, since it influences the actions and behaviors of people. From the perspective of business excellence, it helps organizations find a motivating context for change and achieve higher performance levels. It is through culture that organizations are able to build systems and processes that are strong and stable to meet the market's needs [56].

The Shingo model presents guiding principles and connects them with systems, tools, results and culture, integrating these five elements around the relationships illustrated in Figure 1.

Figure 1. Shingo model parameters (Shingo Institute, 2022).

The parameters present in Figure 1 correspond to [53]:

- Guiding Principles: Shingo's guiding principles are the basis for a sustainable organizational culture of excellence. They indicate rules that enable the understanding of the consequences of behaviors and, therefore, enable decision making to be conscious and to meet the ideal results;
- Tools: Tools that allow operational execution to be carried out in order to achieve the purpose of the system. Therefore, they must be carefully selected in order for the tools to be aligned with the system;

- Systems: These are sets of processes, persons or procedures that are interconnected and allow the realization of guiding principles;
- Results: The consequence of leadership capacity, behaviors and routines. Ideal results are all those that are sustainable in the long term and that require leadership capacity to create cultures in which behaviors and routines are practiced;
- Culture: The center of the Shingo model, which represents the manifestation of the behaviors and actions of the organization's elements;

The Shingo model presents ten guiding principles that are organized in three dimensions: cultural enablers, continuous improvement and enterprise alignment (Figure 2).

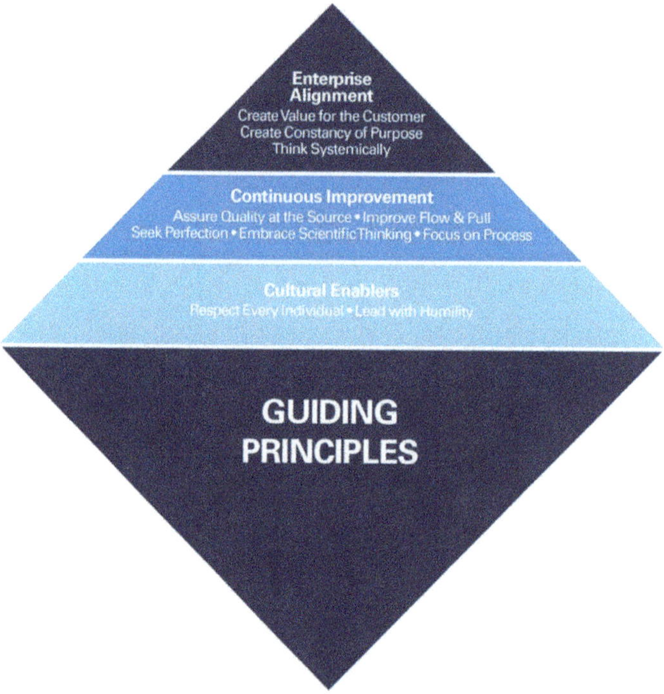

Figure 2. Dimensions of the Shingo model (Shingo Institute, 2022).

Guiding principles are used to align behaviors and actions, a process that is improved by selecting specific tools used to achieve results, where the results then affirm the guiding principles. In the opposite direction, the Shingo model indicates that the guiding principles can generate results in the form of performance indicators and impacts from which the company can obtain interpretations and predictions that can help in the refinement of the tools used to improve the systems that, in turn, direct principles that are anchored in the core values of the company [57].

The principles of the Shingo model are universal and timeless, and therefore, they do not vary in different cultures and eras, although they may manifest themselves differently [57]. Table 1 shows the principles corresponding to the Shingo model's dimensions.

As already referenced, it is known that the Shingo model can help organizations search for optimal results in various dimensions, as long as it is implemented appropriately [57]. Briefly, the model argues that the involvement of all elements of the organization is essential for significant and sustainable improvement and that it is only possible when people's behaviors are aligned with ideal behaviors based on principles.

Table 1. Shingo's model guiding principles [55].

Dimensions	Guiding Principles
Cultural Enablers	Respect each individual
	Lead with humility
Continuous Improvement	Seek perfection
	Adopt scientific thinking
	Focus on processes
	Ensuring quality at the source
	Improve flow and pull
Enterprise Alignment	Thinking systematically
	Create continuity of purpose
	Create value for the customer

3. Methodology

This research was conducted from the Web of Science (WoS) database, and since its scientific quality is internationally recognized, it covers a wide range of publications, a robust citation index as well as other features such as interesting and effective analytical tools. The bibliometric analysis of the referenced articles aims at evaluating the scientific activity of a field of study and performing a meta-analytical evaluation of the literature.

3.1. Database Selection

The Web of Science (WoS) database includes the most important journals in this field of knowledge, such as: Production Planning and Control, Materials Today: Proceedings, the International Journal of Production Research, the International Journal of Productivity and Performance Management, the Journal of Cleaner Production, the International Journal of Advanced Manufacturing Technology, Total Quality Management and Business Excellence, the International Journal of Lean Six Sigma, Clean Technologies and Environmental Policy, the International Journal of Sustainable Engineering, and Sustainability, among others. Moreover, due to restricted access to other databases, this has been the main source of information used by many research groups, being one of the most reputed databases as well because it is linked to Clarivate.

3.2. Publication Search Criteria

The criteria selected with the research are ordered chronologically in Figure 3. First, after the selection of the database (Web of Science), documents written in English and peer reviewed were exclusively considered. Second, the restricted time space between 2010 and 2021 was considered; therefore, the matters that are regarded are current and valid. Data collection was finalized in May 2022. Third, four distinct combinations of keywords ("lean methodology" or "lean manufacturing" or "lean tools", "six sigma" or "DMAIC", "lean six sigma", "sustainability" or "Shingo's model" or "excellence models") were used. This research covered the theme, which included the title, summary and keywords of each post. After these first three steps, articles from the database were randomly selected for each combination of keywords. Finally, a manual selection of publications was carried out on the publications that were selected randomly, and they were fully read. This procedure resulted in the selection of 67 publications based on the theme "Lean", 40 publications based on the theme "Lean Six Sigma", 12 publications based on the theme "Six Sigma" and 6 articles based on the theme "Shingo's Model". A total of 125 publications were consulted, for which a quantitative and qualitative analysis was carried out.

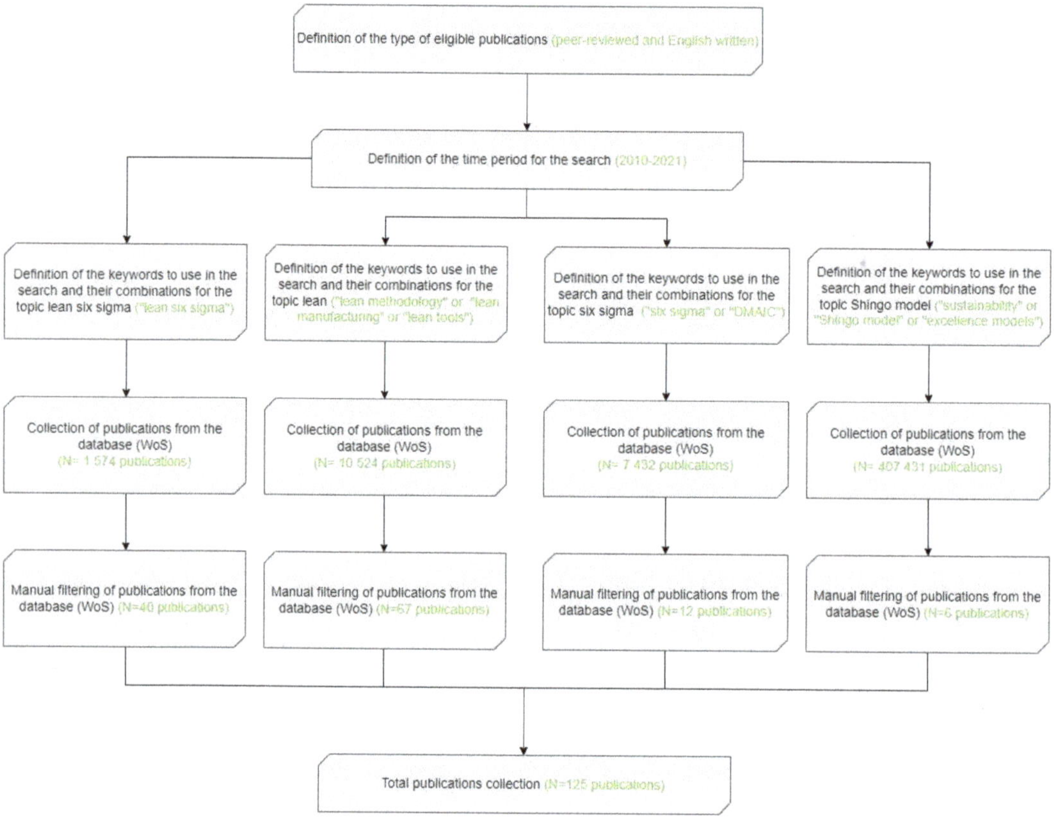

Figure 3. Research strategy steps.

In order to design the respective publication analyses, a database was constructed in an MS Excel® data sheet based on data obtained from the Web of Science, such as the title of the publication, the average number of citations per year and the journal name. Additionally, after reading and analyzing each document, the MS Excel® data sheet was complemented with the area of the application, research objectives, main contributions and suggestions for future research mentioned by the authors of the publications. Finally, these documents were classified into four groups based on research themes, as follows: "Lean", "Six Sigma", "Lean Six Sigma" and "Shingo's Model".

3.3. *Bibliometric Analysis*

The design of the bibliometric analysis was supported by VosViewer software (version 1.6.17) (Universiteit Leiden, Leiden, The Netherlands) through which it was possible to obtain a network visualization map with key concepts of greater relevance of the various publications.

4. Results of the Bibliometric Analysis

4.1. *Distribution of Publications*

With the support of WoS, the publications referenced for several years (Figure 4) were made. It was found that the oldest publications were from the year 2011, with 7 (5.6%) publications, whereas the most recent publications were from 2021, with 14 (11.2%) publications. The years 2017 and 2019 had the most publications, with each having 16 (12.8%)

and 19 (15.2%) publications, respectively. It was noted that there is a growing trend of publications analyzed from more distant years to the most recent.

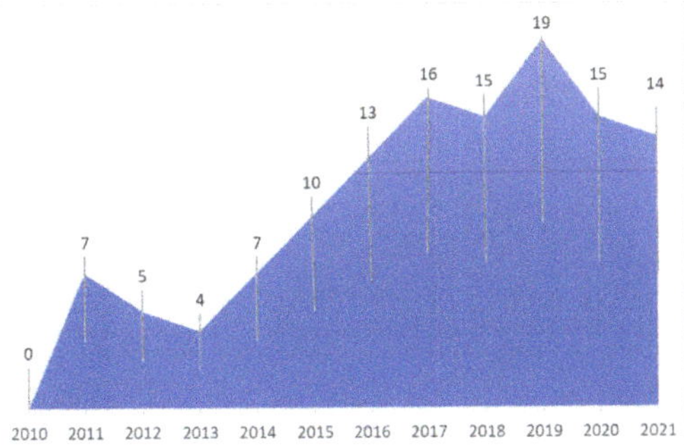

Figure 4. Distribution of publications found under this scope by year of publication.

Regarding the average number of citations per year (Figure 5), there is also an increasing trend. In addition to the upward trend in the number of publications, there is also an upward trend in the number of citations. This shows the increasing interest of the scientific community in studying the concepts lean, six sigma, lean six sigma and the Shingo model in recent years. The publications of the years 2020 and 2021 were the most cited, with each surpassing the barrier of 500 citations.

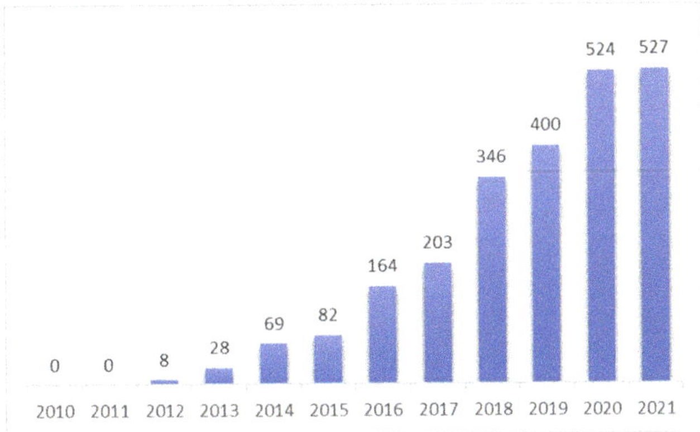

Figure 5. Distribution of citations by year of publication.

In relation to the journals used as sources of information (Figure 6), it was verified that Production Planning and Control and Materials Today: Proceedings were the most common, with a total of 9 and 7 publications, respectively. The following was Total Quality Management and Business Excellence, with 5 publications. The International Journal of Production Research, the International Journal of Productivity and Performance Management, the Journal of Cleaner Production and the International Journal of Advanced Manufacturing Technology completed the group of journals with the most publications associated with, respectively, 4 publications each. Regarding the axis identified as "Other", it represents the journals that had fewer than 4 publications associated with the work. Making a total of 88 journals on the "Other" axis, this group includes journals such as the Central European Journal of Operations Research, the International Journal of Lean Six Sigma, the International Journal of Sustainable Engineering, Quality Engineering and Trends in Food Science and Technology, among others. The high number of associated journals in the distribution of publications based on the concepts "Lean", "Six Sigma",

"Lean Six Sigma" and "Shingo's Model" highlights the scope and diversity of the fields of study related to those subjects.

Figure 6. Distribution of Publications by Scientific Journal.

4.2. Cluster Identification

The construction of the clusters was based on the intersection of articles based on the topics "Lean", "Six Sigma", "Lean Six Sigma" and "Shingo's Model". The first cluster consisted of 17 terms. The second cluster consisted of 14 terms. The third and fourth clusters were composed of 11 terms. The fifth cluster consisted of 10 terms. The sixth and seventh clusters were composed of 8 terms. The eighth cluster consisted of 7 terms. The ninth cluster consisted of 6 terms. The tenth and eleventh clusters were composed of 4 terms, summing up a set of 100 study topics.

By observing the map presented in Figure 7, it seems that the ten most mentioned topics are: "manufacturing, technology, lean manufacturing, productivity, management, lean, efficiency, work, DMAIC and sustainability". The ten topics less mentioned are: "responsibility, leadership, SMED, workstation, rework, variation, automation, energy, VSM and flow".

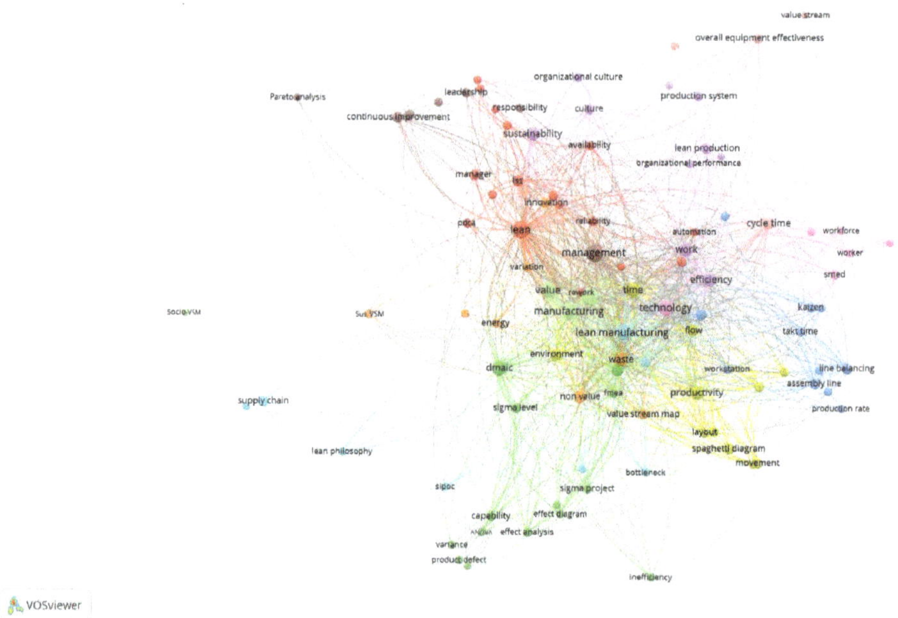

Figure 7. Most important study areas for lean, six sigma, lean six sigma and the Shingo model.

By analyzing Figure 7, it is possible to observe that the most dispersed areas encompass topics such as: "value chain, production rate, movements, bottleneck, sigma design, capability, logistics chain, Sus-VSM and OEE". Later, some of these topics are studied in greater detail.

Following the network visualization map (Figure 7), the constitution of each of the eleven clusters is presented in Table 2.

Table 2. Most important clusters on lean, six sigma, lean six sigma and the Shingo model.

Clusters	Items
Cluster 1 (17 items)	Automation, availability, competitive advantage, green lean, infrastructure, lean, lean six sigma, maintenance, manager, SMEs, operational excellence, PDCA, production cost, reliability, rework, commitment of top management, training;
Cluster 2 (14 items)	ANOVA, "Capability", cost, DMAIC, effect analysis, effect diagram, FMEA, inefficiencies, product defect, product quality control, sigma level, sigma method, sigma design, variance;
Cluster 3 (11 items)	Production line, kaizen, kanban, lead time, line balancing, production rate, takt time, waste reduction techniques, labor standardization, workstation, yamazumi;
Cluster 4 (11 items)	Environment, flow, inventory, layout, lean principle, lean thinking, movements, productivity, spaghetti, staff, time;
Cluster 5 (10 items)	Culture, efficiency, environmental management, lean production, lean production system, organizational culture, organizational performance, production system, sustainability, work;
Cluster 6 (8 items)	Bottleneck, customer satisfaction, inventory management, lean manufacturing, lean philosophy, OEE, SIPOC, logistics chain;
Cluster 7 (8 items)	Energy, innovation, non-value, Sus-VSM, sustainable manufacturing, VSM, variation, waste;
Cluster 8 (7 items)	Business excellence, continuous improvement, leadership, management, Pareto analysis, quality management, responsibility;
Cluster 9 (6 items)	Time set-up, human factor, SMED, technology, worker, workforce;
Cluster 10 (4 items)	Cycle time, OEE, project management, value chain;
Cluster 11 (4 items)	Industry, manufacturing, Socio-VSM, value;

As shown in Table 2, it is possible to identify 11 clusters in relation to the intersection of concepts among lean, six sigma, lean six sigma and the Shingo model. In cluster 1, it shows the correlation between different philosophies, strategic methodologies and critical success factors for the improvement of key indicators. Cluster 2 focuses on the weight of certain tools and methodologies in controlling waste and defects. Cluster 3 shows tools capable of generally improving the operation of production lines. Cluster 4 depicts the connection of some classic waste scum of lean thinking and tools capable of diminishing its effects on organizations and their eventual connections to the environment. Cluster 5 lists concepts such as sustainability, lean and the relevance of culture to the success of these concepts. Cluster 6 describes productivity and flow-related tools that can help meet customer needs. Cluster 7 describes the need to reduce the environmental impact of organizations, more critical areas related to sustainability and some tools capable of helping in this direction. Cluster 8 indicates some critical success factors and support areas for achieving excellence. Cluster 9 contains factors and techniques capable of reducing non-productive times for organizations. Cluster 10 shows classic production and tools capable of increasing the efficiency of production systems. Cluster 11 essentially focuses on the importance that social factors should have in organizations for the production of widespread value.

5. Structural Analysis

5.1. Distribution of Publications by Search Topic and Application Area

Through this analysis, the publications were categorized according to four research topics, which were: lean, six sigma, lean six sigma and the Shingo model (Figure 8). Through this categorization, it was possible to observe that the lean topic was the one with the most associated publications, with a total of 67 (54%). Next, the lean six sigma topic had about 40 (32%) associated publications. The six sigma topic had 12 (9%) associated publications. Finally, the Shingo model topic had 6 (5%) associated publications. This discrepancy in the number of six sigma publications in relation to lean and lean six sigma is explained by the fact that the focus was on the research of methodical structures and not six sigma tools. These data reveal that lean has had greater prominence and importance in studies in the area of industry and whose tendency is to increase the relevance of lean six sigma by being able to adopt concepts and methodologies of both lean and six sigma areas. The Shingo model allows us to relate these concepts to operational excellence.

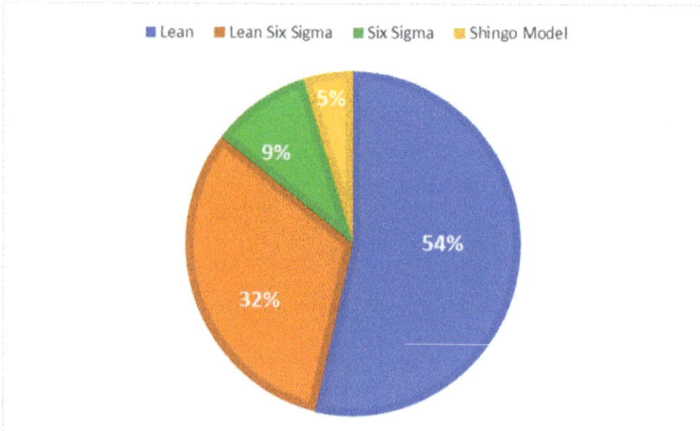

Figure 8. Categorization of publications for each topic.

Table 3 shows a categorization of the papers consulted, taking into account the topic and quantifying the number of papers consulted under each topic.

Table 3. Categorization of the analyzed publications.

Topic	Publications	Total
Lean	[2,4,16–18,21–23,26–28,58–112]	67; 54%
Six Sigma	[3,30–33,35,40,41,113–116]	12; 9%
Lean Six Sigma	[1,5,12,14,15,24,25,29,34,36–39,42–50,117–134]	40; 32%
Shingo Model	[51,52,54–57]	6; 5%

Additionally, a graph (Figure 9) is displayed from which publications are distributed throughout their years and search topics. From this chart, it is possible to verify that publications related to the lean topic were in primacy from the year 2016, with a total of 48 publications until 2021. This corresponds to about 71% of the total publications with the lean topic. Publications with the topic lean six sigma began to gain prominence from the year 2017, with a total of 29 publications by 2021. This is about 72% of the total publications with the topic lean six sigma. Publications with the topic six sigma were dispersed in a constant way over the years while never surpassing the two associated publications per year. The years 2013 and 2021 were the exceptions that did not present any publications associated with the theme six sigma. Publications with the Shingo model topic were concentrated in the years 2015, 2017, 2018 and 2019. The year 2018 contained half of the publications under this topic, with about 3 publications.

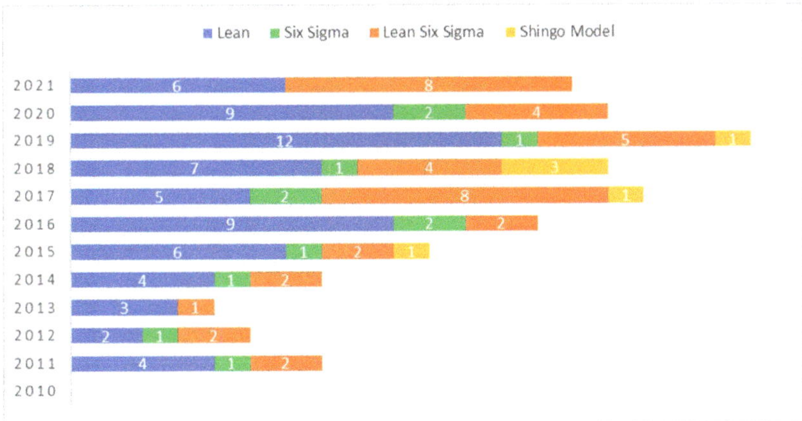

Figure 9. Distribution of publications analyzed by topic.

The publications were also distributed depending on the area of application (Figure 10). From this distribution, it was possible to notice that the most referenced areas were theoretical, automotive and metallurgical, with a total of 30, 17 and 13 publications, respectively. Unsurprisingly, lean dominated the automotive area, whereas lean six sigma began to gain prominence in the theoretical and metallurgical area. The areas that completed the most outstanding ranges were health, multisector, food and textile, with a total of 10, 7, 6 and 5 publications, respectively. In the area classified as "Others", there were documents with areas of application presenting fewer than 5 references, such as services, agriculture, aerospace, electronics, pharmaceuticals and logistics, among others.

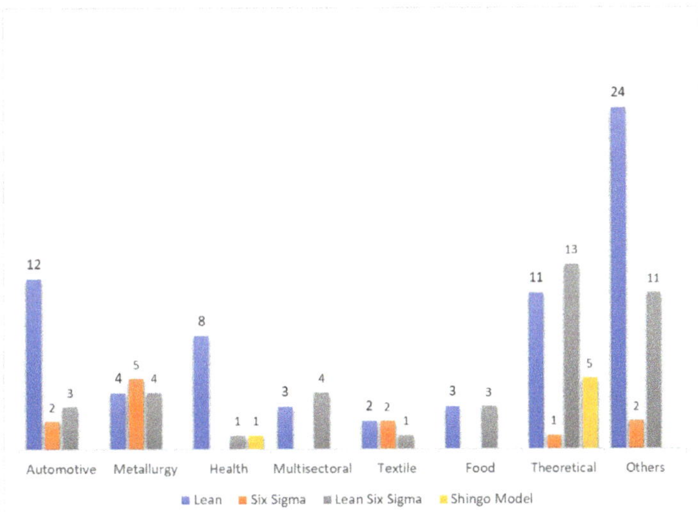

Figure 10. Classification of publications by application area.

5.2. Presentation of Case Studies

Some case studies are presented from the publications analyzed in this research (see Table 4).

Table 4. Description of case studies of the publications analyzed.

Authors	Application Area	Description
[66]	Manufacturing	In this work, the focus was on the implementation of the Work Standardization tool on a production line. The implementation of the tool resulted in a 37.5% decrease in work at one of the workstations and an increase of 304.7% for the daily income at the workstation that was the bottleneck of the production line.
[67]	Car industry	Labor Standardization was applied in several operations of the production line, resulting in a decrease in cycle time by 350 s and reducing setup time by 1500 s. Working procedures were built and placed on all machines in the line; therefore, they were easily accessible to all operators.
[70]	Manufacturing	In this work, the focus was on the Standardization of the Work of a manufacturing company in India, resulting in a decrease of 31.6 s in product cycle time and a production increase to 58 parts that initially floated around 45–50 parts in a 7 h shift. It should be noted that production improvements were achieved without the need for investment in machines or tools by the company.
[71]	Printing Shop	In this work, the focus was on optimizing resources in a small printing shop in order to be able to satisfy requests. To this end, a study was conducted on the times of the processes, the movements of the operators and the respective layout of the production process, identifying sources of waste and opportunities for improvement. Finally, Work Standardization occurred with the help of Process Charts and visual help of the execution of operations, allowing for a reduction in unnecessary movements by 66%, and the standard time on the workstation decreased from 244 to 199 s for each product, increasing the production rate by 63.2% as well.
[72]	Health	In this work, the focus was on improving the processing efficiency of patients in a hospital's emergency department. In order to reduce patient waiting times, two tools were developed that allowed for the standardization of the admission processes of patients in the hospital. After the implementation of the tools, the admission time decreased from 154 to 144 min, although it did not significantly improve the time of admission, but it improved the patients' transfer times (from 30.5 to 21.7 min) and the time required to receive or deliver medical reports (from 3.8 to 2.8 min). Thus, there was a significant improvement in the flow of patients.
[68]	Manufacturing	In this work, the focus was on the implementation of a methodology developed by the author for micro and small enterprises that allowed for the improvement of the OEE index of an operation. After an initial analysis, a set of lean tools was recommended to correct inefficiencies. The practical case presented demonstrated that, after observing the data, availability was the main cause of overall efficiency loss. When applying the Pareto diagram, it was observed that the change in tools was the highest stopping time. The application of SMED was suggested, and, after its implementation, a significant increase in the availability of the machine was observed, increasing the OEE from a range between 70% to 75% to an interval between 81% to 85%.
[69]	Plastic Extrusion	In this work, the focus was to increase the efficiency of an extrusion process of a company in the plastic industry. After the collection and analysis of the process data, it was found that the main cause of the low OEE indicator was the long downtime without producing. To this end, an innovative proposal was developed that involved the use of SMED and preventive maintenance techniques. For the validation of this model, arena simulation software was used to analyze what results this model may lead to, showing that the proposals could result in a reduction in non-productive times by 36.67% and an improvement in the OEE by 9.02%.
[85]	Semiconductor industry	In this work, the author developed an innovative model that allows for a systematic approach to improve the OEE of a system, a framework that incorporates the advantages of OEE with conventional improvement models. It was validated with a 38-week study case by a semiconductor company in Malaysia. The production system had a low OEE of 73.4%, caused by the loss of availability (76.5%). The application of the model developed by the author resulted in an improvement of the OEE and availability to 76.5% and 80%, respectively. In financial terms, it resulted in savings of about 565,000 USD for each conversion of the production line. It should be noted that this model was subject to a single case study. Therefore, it is necessary to test it in different scenarios to prove its robustness.
[86]	Manufacturing	The authors developed a method capable of combining SMED with preset systems (tool anticipation or device adjustments) in order to improve the productivity of production systems. It was validated with a study case in the industry that allowed the following results to be recorded: an 87% reduction in the setup times of the bottleneck resource (labelling machine), a 33.8% increase in the OEE of the labelling machine, a 17% increase in the OEE of the production line and a 45% reduction in lot size.
[87]	Machining	In this work, the objective was to improve the OEE of the CNC machines of a company of the mold industry in Portugal. Initially, they had an OEE value of about 50%. Through the application of tools such as 5S, SMED, Visual Management and Work Standardization, it was possible to improve the performance of the machines. Overall, the OEE was improved by about 20%.

Table 4. *Cont.*

Authors	Application Area	Description
[88]	Pipe Manufacturing	In this work, the authors highlighted the importance of a new indicator capable of combining OEE and Sustainability. This new indicator was designated as Overall Environmental Equipment Effectiveness (OEEE). In order to assess its impact, the authors conducted a case study at a pipe manufacturing company. By analyzing the application of this indicator on a production line, it concluded that the raw material of the product analyzed would have to be replaced by another one in order to comply with the sustainability requirements. Other modifications were also performed, such as the elimination of excessive stocks of raw materials and increased flexibility of the production line (from 2301 to 685 s for the production time of the first part in the line) in order to respond in a timely manner to the customer's needs. With these changes, the cost of production decreased by 6.2%, highlighting the company's increased competitiveness with respect to price, flexibility and sustainability.
[89]	Cellulose	The authors developed an integrated model based on six sigma and TPM in order to improve the performance indicators of the production systems. The model was applied in a manufacturing cell consisting of two machines from a pulp company. The goal of the study was achieved with the application of multiple tools, such as Brainstorming Sessions, Pareto Diagram, Ishikawa Diagram, histograms, FMEA and control charts, among others. At the end of the project, there was a significant improvement in the OEE, which was in a range between 50% and 54% for a range between 76% to 83%, also reducing rework from 22% to 10% and the defect rate from 24.82% to 5%, leading to financial savings of around 2 million USD per year.
[90]	Car industry	In this work, the focus was on the integration of the production teams linked to an assembly system with the measuring system performance indicators in real time. The validation of this system took place in a factory of the automotive industry in Spain. The goal of this system was to focus the organization on value chains and to improve their performance based on the OEE. The system provided real-time OEE values from value chains and allowed production teams to use meetings to discuss and improve the provided indicators. Thus, greater relevance was directed to key points that had a critical impact on production volume. The solution presented allowed for a significant improvement of OEE values in general between 5 and 10% from January 2009 to January 2012 in the case study. It should be noted that this model has been validated with only one practical case. It is necessary to test it in different scenarios to prove its veracity.
[100]	Food industry	In this work, there was a lean implementation in a food company with the help of two tools (SMED and VSM). VSM allowed for the identification of the different wastes associated with the production line (84% of the total production). Then, with the help of SMED, it was possible to reduce setup times by 34% and to promote an increase in line productivity by 11%, allowing the company to avoid the use of temporary workers in periods of high demand.
[101]	Food industry	In this work, there was a lean implementation in a set of companies in the wine sector. VSM was used as the main tool for the determination of waste and points of improvement for the production process. From the VSM information, some tools were selected that were capable of solving the problems found. In the end, it was possible to reduce the lead time between 50% and 65% and to induce a reduction in raw materials between 8% and 16%.
[102]	Logistics	In this work, the focus was on improving the logistics processes of a military unit. VSM and VSD (Value Stream Design) served as the basic tools for improving the processing of item orders, allowing for eliminating or reducing activities that did not add value and improved order processing procedures. The results that were obtained in a simulation showed that the "future state map" could allow for increases in activities that add value, from 44% to 70%, and a reduction in lead time of 69.6% was achieved.
[103]	Car industry	In this work, the focus was on improving the productivity of a production line of an automobile company. For this, VSM and the simulation approach were adopted to validate the improvements. It was observed that it was possible to reduce cycle time by 87.59%, to reduce WIP inventories by 76.47%, to reduce lead time by 95.41%, to reduce the number of operators by 57.14% and to reduce setup times by 70.67%.
[104]	Services	In this work, the focus was on improving the efficiency of services provided by a maintenance team and repairing buildings of a public university. VSM was used as a basis for the identification of waste and improvement points. Then, the simulation was used to validate the proposals. The changes made, based on the simulation, allowed for a reduction in lead time by 26.8% and a reduction in waiting times by 33.6%. This case highlights the increased difficulties of implementing lean concepts in the area of services. However, they can be implemented and provide very good results.
[36]	Theory	In this work, a VSM model focused on sustainability was used, designating it as *Sus-VSM*. It emerged with the emergence of sustainable manufacturing systems. The particularity of this model is that, in addition to evaluating the classic VSM metrics such as cycle time, lead time and material cost, it also analyzes a set of metrics related to the environment and the social environment. For example, it analyzes chemical consumption, water consumption, energy consumption, noise level and the ergonomics of jobs. Thus, the *future state map* is not only aimed at improving the product and process flow, but it also improves environmental performance and work conditions.

Table 4. *Cont.*

Authors	Application Area	Description
[128]	Theory	In this work, a VSM model focused on energy was developed. The authors' model was designated as Lean Energy Six Sigma Value Stream Mapping (LESSVSM) and was based on Energy Value Stream Mapping (EVSM). It has emerged to combat waste existing in the energy area. It evaluates consumption along the manufacturing system chain and identifies foci where consumption can be eliminated or managed more efficiently. The design of the future state map from this model aims to achieve sustainable manufacturing by reducing energy and waste costs, increasing efficiency in their jobs.
[16]	Theory	In this work, a model was proposed that integrates the Eco-Function Matrix with VSM. The process starts with the construction of the current state map in which it identifies and categorizes waste. Areas classified as more critical are configured as requirements in the Eco-Function Matrix. Brainstorming sessions are then held building proposals and improvements that can be designated as attributes. Then, the conventional Procedure of QFD (Quality Function Deployment) begins to consider the environmental perspective in the construction of the matrix and its interrelationships. Finally, the waste and proposals for improvements is ordered as a priority; therefore, the construction of the future state map is one of the most important topics.
[18]	Theory	In this work, a model designated as *Variability Source Mapping* (VSMII) is proposed, with a focus on the identification and reduction in variability throughout the production system. The construction steps are similar to the classic VSM model, but it adds some details, such as the mean, standard deviation and coefficient of variation of the cycle time for each job. It also assesses the variability of the flow on arrival between jobs or phases, developing a system variability metric that is able to identify the most critical locations or jobs with variability levels. With the help of lean tools and control production policies, it is possible to build a future state map that allows for the reduction or elimination of the sources of variability, thus decreasing the variability level of the system
[105]	Theory	In this work, the authors provide a new approach capable of incorporating the PDCA (Plan–Do–Check–Act) methodology with environmental VSM (E-VSM). This is an alternative proposal capable of improving the environmental performance of operations using principles, techniques and lean tools. Reductions in energy consumption and waste production are some of the environmentally analyzed wastes with great focus on E-VSM. The PDCA approach allows for a systematic methodology in the elaboration of the current state map and future state map. According to the author, the main focus of this model is for industrialists who intend to achieve environmentally sustainable operations.
[106]	Theory	In this work, the author presents a new tool called Ergo-VSM, based on the VSM methodology and incorporating ergonomic indicators related to the physical and psychological factors of workers. The goal is to improve the ergonomic conditions of workers without compromising the classic productivity indicators. The inclusion of factors related to the physical environment of workers, such as noise, temperature and luminosity, can affect not only the health of workers but also their own performance. Therefore, this tool can also have a positive effect on productivity.
[107]	Construction	In this work, the authors propose a new model based on the integration of LCA (Life Cycle Assessment) and VSM, called LCA-VSM, a tool capable of prioritizing measures that improve economic and environmental performance, encouraging a continuous process improvement approach. The model was validated by a construction company that produces paint materials and tools, providing a reduction between 5% and 15% in environmental impact (across nine environmental categories), reducing lead time from 103.26 to 24.01 days and promoting a reduction in cycle time from 35.7 to 33.75 s.
[108]	Theory	In this work, the authors provide a model capable of modeling VSM from an economic and environmental perspective. The methodology is called E^2-VSM. It is a simulation model that considers the dynamic behavior of a multi-product production system and assesses its environmental impact, such as energy consumption. The model also evaluates the financial and environmental performance of each of the manufactured products. The goal is to optimize the machines, production orders and production parameters that lead to the best solution through energy efficiency and resources for a production system.
[109]	Multisector	This work shows the practicality of Sus-VSM. Three cases of studies demonstrating their applicability in different areas of industry are presented. It enables improvement in the ratio of activities with added value (such as the classic VSM model) as well as the improvement of environmental and social indicators. At the environmental level, for example, reductions in energy consumption, water consumption and raw materials consumption were achieved. At the social level, for example, workers' exposure to ergonomic hazards and risks in the workplace were also reduced.
[111]	Electronics	This paper presents a new methodology called Socio-VSM, the objective of which is to incorporate environmental and social indictors in order to accelerate the transition to sustainable manufacturing. For the validation of this methodology, a case study was carried out by an electronics company in Malaysia. Indicators such as noise level, ergonomic workstation conditions and classical productivity indicators were evaluated. On the basis of these indicators, changes were made to reduce the risk of injury or occupational disease by workers by reducing noise exposure and ergonomic changes in jobs. At the environmental and economic level, no changes were made to the production system.

Table 4. *Cont.*

Authors	Application Area	Description
[112]	Car industry	This paper presents a new methodology called scrap value stream mapping (S-VSM), a tool that allows for the mapping and identification of scrap and its costs throughout the process chain, from which it is possible to act with lean tools in order to attenuate the volume of scrap and generate potential savings. The validation of this tool was achieved with a case study by an automotive company, in which savings of €44,782 were possible in the following years if the volume of plastic waste remained at the same level in 2015.
[79]	Car industry	In this work, the focus was on minimizing the sources of waste resulting from the transport of raw materials in a manufacturing plant. With the spaghetti diagram, inefficiencies were identified with respect to layout and transport, allowing for optimizing the distances traveled from raw materials in the factory from 152 km to 117 km per year. With respect to time consumed in material transports, a decrease from 67 h to 30.1 h per year was achieved.
[80]	Health	In this work, the focus was on improving the satisfaction of patients in a hospital through the standardization of drug logistics processes. By applying the spaghetti diagram, it was possible to optimize the routes of the medicine cart and thus decrease the travel times through the hospital. About 92% of patients reported that, after the intervention in the cart routes, the medications were provided in less time. In addition, the application of the 5S in the cart allowed for reductions in the time spent in the search for drugs from 50.8 s to 30.2 s per unit.
[81]	Feed	In this work, the focus was on the application of lean philosophy in a food industry factory. A spaghetti diagram allowed for the tracking of worker movements and identifying unnecessary movements that could be eliminated, from which it was possible to redesign the layout, and there was a reduction from 6 m to 2 m of distance traveled by the workers for each cycle of time. Other tools such as VSM, OEE and Job Balancing were applied in the project. Overall, productivity increased by around 40%.
[82]	Health	In this work, the focus was on the application of lean six sigma in the medical records department of a hospital. Several tools were used, including a spaghetti diagram, from which it was possible to record the patterns of movement in the work area under study and to identify unnecessary movements between various points of the department. A new layout was designed to minimize movements and delays. The study resulted in a decrease in the processes, on average, from 19 to 8 min. Overall, the project allowed the hospital to save about 20,000 USD in human resources and fixed costs related to the hospital's bureaucratic processes.
[83]	Car industry	In this work, the objective was to improve the productivity of a seating factory in the automotive industry through the lean methodology. Several tools were used in the project. The spaghetti diagram allowed for the study of the layout, cycle times and movements of the workers. This work led to outlining a new layout proposal that allowed for reducing the number of workstations from 6 to 5, the cycle time from 807 to 697 s and the movements of the workers for each seat produced on the line from 15 to 9.6 m.
[84]	Health	In this work, the focus was on improving efficiency and quality in the provision of medical care in the urology department of a hospital with lean methodology. Data such as patient volume, waiting times, cycle times and the movements of physicians were collected with the help of a spaghetti diagram. As result of that work, it was possible to reduce the average cycle time per patient from 46 to 41 min in a period of 90 days. In addition, it allowed for reductions in waiting times and increases in the available contact time between the doctor and the patient from 7.5 to 10.6 min. The time of added value in the patient's visits to the hospital increased, in proportion, from 30.6% to 66.3% at the end of the 90-day period.
[91]	Car industry	Through the Yamazumi tool, it was possible to analyze three jobs and identify sources of waste. Improvements were made with respect to job design and the sequence of operator activities, allowing for the value of the takt time imposed on the assembly line to be fulfilled.
[94]	Manufacturing	In this work, the objective was to improve the efficiency of an assembly line of a refrigerator factory in Turkey. With Yamazumi's help, the situation presented was analyzed, resulting in improved productivity from 118 units to 155 units per shift. The number of operators was reduced from 32 to 28. The average idle time of the operators dropped from 155 s to 70 s.
[95]	Textile	In this work, the focus was on improving the efficiency of a production line of a textile factory. Yamazumi and 5S were the tools selected for process analysis. The improvement proposals were designed and implemented, resulting in a 34% decrease in production cycle time and a 32% decrease in time without added value. The application of the 5S was crucial for the increase in the line efficiency by 12.5%; therefore, human resources were well-used.

Table 4. *Cont.*

Authors	Application Area	Description
[96]	Graphic	In this work, two practical cases related to two departments of a printing press are presented. The first case was carried out on a line of the production department. Standardization of Work and Yamazumi were the tools chosen to increase productivity. Line balancing was achieved with a reduction in the variation of waste throughout the processes, allowing for decreases in the number of workers from 17 to 15. The bottleneck operation of the production line was optimized to increase production from 16.3 copies per hour to 21.3 copies per hour. The second practical case was carried out in the packaging department. Line balancing was achieved by reducing movement times and reducing material flow distances through changes in layout organization. The bottleneck of the production line was also optimized, and some manual operations were eliminated, resulting in increased productivity from 22 boxes per hour to 25 boxes per hour.
[64]	Polymer industry	In this work, the authors describe the adaptation of a *lean production* system for Industry 4.0. The practical case, which took place in a polymer industry organization, consisted in the development of a *Poka-Yoke* tool through a sensor system, which was able to manage the interactions between the physical components and the corresponding manufacturing processes, thus communicating instructions and corrections through block functions. Lean 4.0 tools can interact with the production database and can be updated depending on the scenario, making Lean 4.0 tools dynamic, intelligent and flexible, and ensuring compliance with lean principles, organization objectives, the elimination of waste and increased productivity and value creation for the customer.
[2]	Aerospace	In this work, the focus was on production in an aerospace company. First, the layout was identified, and the operations in the production lines were sequenced. Through observation, it was possible to reconfigure the layout to reduce transport times and to improve the production flow. With the 5S tool application, a more organized system was constituted, reducing movements and waits and also eliminating inventory and unnecessary materials in the production line. A worker was added at one of the stations to decrease wait times. Work standardization was also crucial to increase efficiency in addition to promoting a better work environment. These changes resulted in increased productivity and decreased cycle times of operations.
[65]	Multisector	In this work, the application of lean techniques in small and medium-sized enterprises in India is analyzed. Nine manufacturing units were subjected to the lean program, which lasted about 18 months. Depending on the diagnosis performed in each manufacturing unit, an improvement project was developed using tools such as VSM, 5S, Kaizen and SMED. Overall, these projects achieved savings in the order of 9 million Rupees (official currency of India) and a reduction in the setup time from 135 min to 45 min, and more than 300 *Kaizen* measures were implemented, resulting in material savings, time and improved working conditions. Manufacturing units have managed to create jobs and compete with more attractive prices and higher quality.
[120]	Logistics	In this work, the focus was on optimizing the time spent by the product in logistics processes. With the application of DMAIC and VSM methodologies, and with the perspective of lean thinking, it was possible to identify foci of waste in order to improve the processes. VSM allowed for the understanding of the logic of flow in logistics and for the optimization of the times of the various processes identified by changing the current map state. The development and application of the future state map allowed for the achievement of reductions in lead time and reductions in processing time.
[36]	Manufacturing	In this work, the DMAIC methodology was applied together with the *Sus-VSM* approach, seeking to achieve sustainable manufacturing. Thus, not only was it possible to make improvements in the set of classical metrics of production management, but improvements in a set of environmental and social metrics were also achieved. The significant reduction in energy consumption and chemicals in the practical case that is presented must be highlighted.

5.3. Tool Effects

Based on the publications analyzed, an impact matrix was constructed in order to evaluate the impact of the tools on the economic, social and environmental dimensions. The "+" sign indicates that the tool has a positive impact, whereas the "-" sign points out that the tool has a negative impact on a given dimension (economic, social and environmental). Regarding the number of repetitions of each signal ("+" or "-"), it represents the relevance of an effect that has been reported in the literature (Table 5).

Table 5. Matrix of tools' impact on economic, environmental and social pillars.

Tool	Effect(s)	Impact Dimension		
		Economic	Environmental	Social
Standardized Work	Reduction in the workforce; Improvement in process productivity; Reduction in non-productive times; Reduction in workers' movements;	++		++
SMED	Improvement in process efficiency; Increased energy consumption;	+	-	
5S	Improved organization and safety; Reduction in material consumption, movements, energy and waste;	++++	+++	++
Visual Management	Improvement in process efficiency; Process control;	++		
TPM	Improvement in process management; Reduction in non-productive times, rework and energy consumption; Prevention of mechanical problems;	+++++	+	
VSM	Improvement in process efficiency; Identification and reduction in waste such as: the excessive use of raw materials, energy, waste, non-productive times, stocks and a reduction in th workforce;	+++++	+++	+
Sus-VSM	Reduction in the environmental and social impact of processes; Reduction in consumption of chemicals, water, energy and noise; Evaluation of ergonomic conditions;	+++	++++	+++
Socio-VSM	Assessment of risk factors for workers' health and safety;	+		+
S-VSM	Identification of and reduction in waste and scrap throughout the process chain;	++	++	
Spaghetti Diagram	Reduction in movements, transport and work;	+++		++
Yamazumi	Balancing jobs; Compliance of takt time; Improvement in process productivity; Reduction in the workforce;	++++		
Poka-Yoke	Reduction in process inefficiencies; Reduction in energy consumption;	+	+	
Kaizen	Reduction in energy, waste and material consumption;	+++	+++	
Kanban	Improvement in the operational efficiency of the processes; Reduction in material consumption;	++	+	
LESSVSM	Reduction in waste related to energy consumption;	+	+	
VSM II	Identification and reduction in the variability of the production system;	+	+	
E-VSM	Identification and reduction in energy waste sources;	+	+	
Ergo-VSM	Evaluation of ergonomic conditions related to physical and psychological factors of workers;			++

5.4. Conceptual Model

Based on the analyzed studies, through a bibliometric analysis, a conceptual model (Figure 11) is proposed. This model is able to relate the different tools analyzed with the different pillars (economic, social and environmental); therefore, organizations can choose the best tools for each scope, serving DMAIC as a structural basis for the construction of the projects, with the ultimate goal of achieving operational excellence. The tools in the economic pillar are the ones that best suit economic growth. The tools that are in the social pillar are the ones that best suit the achievement of social responsibility, i.e., the widespread improvement of working conditions. The tools that are in the environmental pillar are the ones that best suit the achievement of the sustainability of operations and environmental goals. This model aims to help the industry clarify which tools are most relevant for meeting different objectives (economic, social and environmental) [135]. It should be noted that this model was not subjected to practical validation, although it was based on empirical studies.

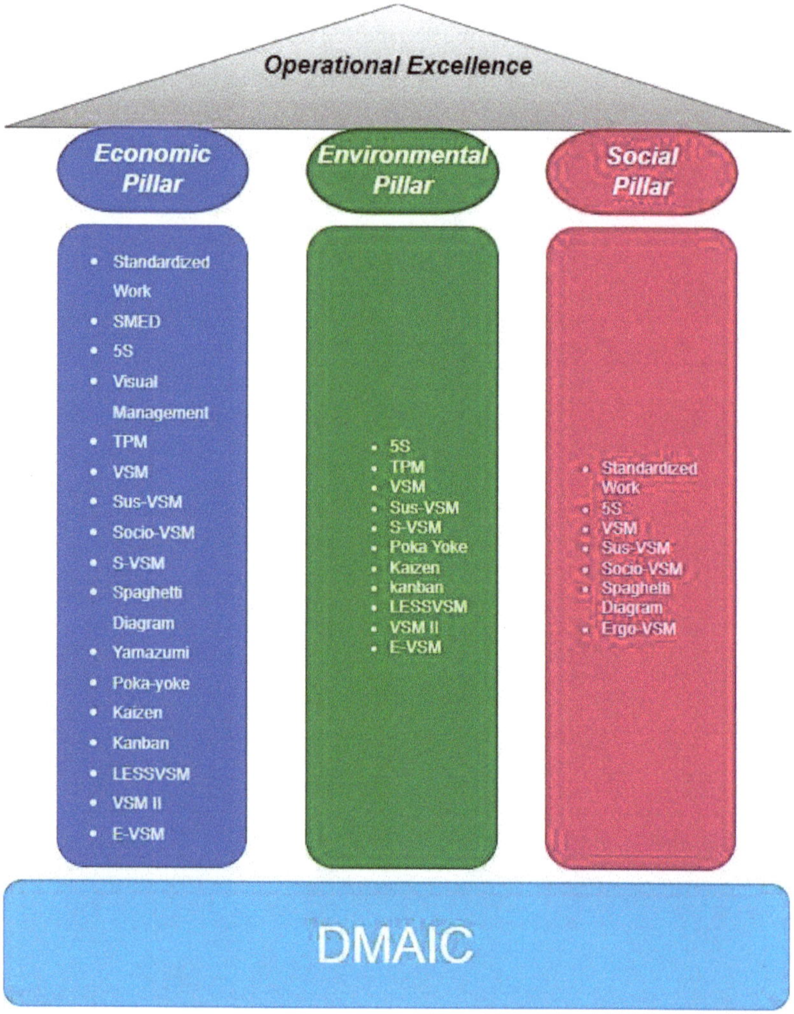

Figure 11. Conceptual Model.

6. Discussion

6.1. Evolution of Publications

The bibliometric analysis developed in this work, based on the Web of Science, allowed for the distribution of publications per year and average annual citations for the period between 2010 and 2021. Through these fields, it was possible to conclude that these terms are receiving increased attention. With respect to publications, the peak was reached in 2019, with about 19 publications. In relation to the average annual number of citations, the peak was reached in 2021, with about 527 citations. The analysis also involved the design of a cluster map based on the link between keywords, illustrating an intellectual structure that allows for the detection of trends and flaws in research on lean, six sigma, lean six sigma and the Shingo model. This map clarifies that the research based on these four topics considered the following: (1) the strong link of lean tools in combating different sources of the waste of organizations; (2) the strong connection of DMAIC to projects to improve organizations; (3) the strong link between operational excellence and improvement philosophies such as lean, lean six sigma, green lean and PDCA, which allow for the construction of a competitive advantage; (4) the weak link between the environmental concept and lean philosophy; (5) the strong link between organizational culture and sustainability and the overall performance of organizations; (6) the strong link between innovation and factors related to the environmental pillar; (7) the weak link between the tools related to the social pillar and the industry area; and (8) the strong connection of the concepts of responsibility and leadership capacity in achieving a level of business excellence.

6.2. Relationship between Lean, Six Sigma, Lean Six Sigma and the Shingo Model

The bibliometric analysis involved the categorization of the selected publications based on the four research topics (lean, six sigma, lean six sigma and the Shingo model), from which a temporal distribution and application area were made. With respect to research topics, lean was the topic that was most addressed, whereas the Shingo model topic was the least addressed. The results obtained show that:

- For the lean topic, the automotive industry is the most debated area, also being compatible with six sigma through lean six sigma;
- For the lean six sigma topic, the theoretical area is the most debated area, despite the growing importance in the automotive and metallurgical industry that relates lean and six sigma concepts;
- For the six sigma topic, the metallurgical area is the most debated, which relates lean concepts through lean six sigma;
- For the Shingo's model topic, the theoretical area is the most debated, demonstrating a weak relationship of this topic with practical areas, with only one publication with this topic in a practical area (health);
- In the "Other" area, the presence of Lean is highlighted in several areas of which there is a tendency for the adoption of these concepts [135], involving areas such as logistics, aerospace, agriculture and pharmaceuticals, among others.

From this set of studies, the description of positive effects on the economic pillar is highlighted with greater emphasis. With less intensity, there were also positive effects on the social and environmental pillars, demonstrating that the environmental and social pillars suffer detriment to the economic pillar. The literature, in general, corroborates this disproportion of positive effects on the different pillars [9]. Lean interventions may result in green benefits, but not all Lean practices are in line with green strategies [136]. Lean tools and interventions tend not to impact the different pillars of sustainability in an integrated way [7].

However, in order to achieve operational excellence, organizations need to gradually redirect their efforts to these issues; therefore, sustainability is effectively achieved. According to Teixeira et al. [137], although the economic and environmental pillars have a stronger positive relationship with the organizational performance of companies, the social pillar remains relevant for achieving a superior competitive advantage, highlighting the need for organizations to improve the performance of sustainability pillars in an integrated way to maximize competitive advantage.

In the impact matrix, it is possible to observe that only 3 tools in 18 of those reported had a positive impact on all sustainability pillars (5S, VSM and Sus-VSM). This analysis allowed for the construction of a conceptual model that incorporates different tools in the three pillars of sustainability in order to help organizations achieve better results and achieve operational excellence.

7. Conclusions

This work presents a bibliometric analysis on lean philosophy, six sigma, lean six sigma and the Shingo model. It involves an evaluation of the selected literature based on the temporal distribution of publications and annual average citations. This analysis allows for the design of a new model, through which it is possible to select the best lean tools to improve the operational performance of a company. In fact, companies can present asymmetries with respect to performance regarding each operational pillar. The categorization achieved through the developed model allows for selecting, firstly, the pillar where the DMAIC needs to be focused on, and, after that, for selecting the best tools that are able to fit this goal. Indeed, the different lean tools are categorized considering each pillar of the DMAIC methodology, helping readers and researchers to easily identify the best tools they need to improve each one of the pillars that is usually considered to obtain operational excellence using the DMAIC methodology. This topic represents the main novelty of this paper, contributing to an easier selection of the tools that can lead a company to the operational excellence in each pillar, since companies can be differently prepared to be sustainable in each of the three pillars that are widely and usually considered in the literature. It can be highlighted that this categorization is only possible based on an extensive literature review, which points out each tool that is used and the main purposes behind its use. The contributions, limitations and recommendations for future research are as follows.

7.1. Contributions

This study contributes to the topic under discussion through the following points:

- The identification of the most and least discussed items related to lean, six sigma, lean six sigma and the Shingo's model;
- The categorization of the literature according to the four topics (lean, six sigma, lean six sigma and the Shingo's model), including its distribution by area of application and reporting its main conclusions;
- Highlighting the main effects of the tools identified in the literature and their relationship with the objectives of sustainable development;
- Proposing a conceptual model based on sustainability in order to achieve operational excellence.

7.2. Limitations

This study has the following limitations:

- Research was restricted to peer-reviewed publications;
- Publications came from scientific journals, discarding other sources of information such as dissertations, reports and theses;
- The Web of Science was the only data collection platform in this research, thus rejecting publications from this field of study that are on other platforms and that may have had a significant impact on data representation;
- The conceptual model was not subjected to empirical validation in a real case study.

7.3. Recommendations for Future Research

In future research, some additional efforts are recommended in the identification of tools and their effects in the social and environmental area because they are the pillars presenting the less volume of tools in the conceptual model, determining a methodology for quantifying the effects of these tools through key performance indicators in order to direct them to more specific objectives. To do so, it is necessary to validate the conceptual model in real case studies, preferably in different sectors of activity, to be able to evaluate the adaptability of the model in different contexts. Moreover, as main future work, researchers can use this model to apply the selected tools in specific sectors and to validate its suitability in achieving the desired results with respect to each pillar.

Author Contributions: The authors confirm their contributions to this paper as follows: Conceptualization, J.C.S., M.R., J.D.-C. and F.J.G.S.; Data curation, J.C.S. and M.R.; Formal analysis, J.C.S., M.R., J.D.-C. and L.P.F.; Investigation, J.C.S., M.R., J.D.-C., F.J.G.S. and G.S.; Methodology, J.C.S., M.R., J.D.-C., G.S. and V.L.; Writing—original draft preparation, J.C.S., M.R. and J.D.-C.; Writing—review and editing, F.J.G.S., G.S., L.P.F. and V.L. All authors have read and agreed to the published version of the manuscript.

Funding: The work of the author Vanda Lima is supported by national funds, through the FCT—Portuguese Foundation for Science and Technology under the project UIDB/04728/2020.

Institutional Review Board Statement: Not applicable.

Informed Consent Statement: Not applicable.

Data Availability Statement: No data are available.

Acknowledgments: The authors would like to thank CIDEM and LAETA/INEGI/CETRIB for all support provided.

Conflicts of Interest: The authors declare no conflict of interest.

References

1. Nandakumar, N.; Saleeshya, P.G.; Harikumar, P. Bottleneck identification and process improvement by lean six sigma DMAIC Methodology. *Mater. Today Proc.* **2020**, *24*, 1217–1224. [CrossRef]
2. Garre, P.; Bharadwaj, V.V.S.N.; Shashank, P.S.; Harish, M.; Dheeraj, M.S. Applying lean in aerospace manufacturing. *Mater. Today Proc.* **2017**, *4*, 8439–8446. [CrossRef]
3. Srinivasan, K.; Muthu, S.; Devadasan, S.R.; Sugumaran, C. Enhancement of sigma level in the manufacturing of furnace nozzle through DMAIC approach of Six Sigma: A case study. *Prod. Plan. Control* **2016**, *27*, 810–822. [CrossRef]
4. Fercoq, A.; Lamouri, S.; Carbone, V. Lean/Green integration focused on waste reduction techniques. *J. Clean. Prod.* **2016**, *137*, 567–578. [CrossRef]
5. Ruben, R.B.; Vinodh, S.; Asokan, P. Lean Six Sigma with environmental focus: Review and framework. *Int. J. Adv. Manuf. Technol.* **2018**, *94*, 4023–4037. [CrossRef]
6. Teixeira, P.; Sá, J.C.; Silva, F.J.G.; Ferreira, L.P.; Santos, G.; Fontoura, P. Connecting lean and green with sustainability towards a conceptual model. *J. Clean. Prod.* **2021**, *322*, 129047. [CrossRef]
7. Teixeira, P.; Sá, J.C.; Silva, F.J.; Santos, G.; Fontoura, P.; Coelho, A. Lean Contribution to the Companies' Sustainability. *IFIP Adv. Inf. Commun. Technol.* **2021**, *610*, 400–408. [CrossRef]
8. Silva, V.; Lima, V.; Sá, J.C.; Fonseca, L.; Santos, G. B Impact assessment as a sustainable tool: Analysis of the Certification Model. *Sustainability* **2022**, *6*, 5590. [CrossRef]
9. Silva, F.J.G.; Sá, J.C.; Ferreira, L.P.; Santos, G.; Nogueira, M.C. The three pillars of sustainability and agile project management: How do they influence each other. *Corp. Soc. Responsib. Environ. Manag.* **2020**, *19*, 1–18. [CrossRef]
10. Silva, F.J.G.; Gouveia, R.M. *Cleaner Production—Toward a Better Future*; Springer Nature Switzerland: Cham, Switzerland, 2020; ISBN 978-3-030-23164-4.
11. Fonseca, L.; Silva, V.; Carlos, S.J.; Lima, V.; Santos, G.; Silva, R. B Corp versus ISO 9001 and 14001 certifications: Aligned, or alternative paths, towards sustainable development? *Corp. Soc. Responsib. Environ. Manag.* **2021**, *29*, 1–13. [CrossRef]
12. Sreedharan, V.R.; Raju, R.; Rajkanth, R.; Nagaraj, M. An empirical assessment of Lean Six Sigma Awareness in manufacturing industries: Construct development and validation. *Total Qual. Manag. Bus. Excell.* **2018**, *29*, 686–703. [CrossRef]
13. Kaswan, M.S.; Rathi, R.; Garza-Reyes, J.A.; Antony, J. Green lean six sigma sustainability—Oriented project selection and implementation framework for manufacturing industry sigma. *Int. J. Lean Six Sigma* **2022**. [CrossRef]
14. Yadav, V.; Gahlot, P.; Rathi, R.; Yadav, G.; Kumar, A.; Kaswan, M.S. Integral measures and framework for green lean six sigma implementation in manufacturing environment. *Int. J. Sustain. Eng.* **2021**, *14*, 1319–1331. [CrossRef]
15. Jordan, E.; Kušar, J.; Rihar, L.; Berlec, T. Portfolio analysis of a Lean Six Sigma production process. *Cent. Eur. J. Oper. Res.* **2019**, *27*, 797–813. [CrossRef]

16. Vinodh, S.; Arvind, K.R.; Somanaathan, M. Tools and techniques for enabling sustainability through lean initiatives. *Clean Technol. Environ. Policy* **2011**, *13*, 469–479. [CrossRef]
17. Azevedo, J.; Sá, J.C.; Ferreira, L.P.; Santos, G.; Cruz, F.M.; Jimenez, G.; Silva, F.J.G. Improvement of production line in the automotive industry through lean philosophy. *Procedia Manuf.* **2019**, *41*, 1023–1030. [CrossRef]
18. Rodrigues, J.; Sá, J.C.V.; Ferreira, L.P.; Silva, F.; Santos, G. Lean MANAGEMENT "Quick-Wins": Results of Implementation. A case study. *Qual. Innov. Prosper.* **2019**, *23*, 3–21. [CrossRef]
19. Dinis-Carvalho, J.; Monteiro, M.; Macedo, H. *Continuous Improvement System: Team Members' Perceptions*; Springer: Berlin/Heidelberg, Germany, 2020; Volume 122.
20. Dinis-Carvalho, J.; Moreira, F.; Bragança, S.; Costa, E.; Alves, A.; Sousa, R. Waste identification diagrams. *Prod. Plan. Control* **2015**, *26*, 1–13. [CrossRef]
21. Arunagiri, P.; Gnanavelbabu, A. Identification of high impact lean production tools in automobile industries using weighted average method. *Procedia Eng.* **2014**, *97*, 2072–2080. [CrossRef]
22. Balamurugan, R.; Kirubagharan, R.; Ramesh, C. Implementation of lean tools and techniques in a connecting rod manufacturing industry. *Mater. Today Proc.* **2020**, *33*, 3108–3113. [CrossRef]
23. Salem, R.; Musharavati, F.; Hamouda, A.M.; Al-Khalifa, K.N. An empirical study on lean awareness and potential for lean implementations in Qatar industries. *Int. J. Adv. Manuf. Technol.* **2016**, *82*, 1607–1625. [CrossRef]
24. Neves, P.; Silva, F.J.G.; Ferreira, L.P.; Pereira, T.; Gouveia, A.; Pimentel, C. Implementing lean tools in the manufacturing process of trimmings products. *Procedia Manuf.* **2018**, *17*, 696–704. [CrossRef]
25. Shokri, A. Quantitative analysis of Six Sigma, Lean and Lean Six Sigma research publications in last two decades. *Int. J. Qual. Reliab. Manag.* **2017**, *34*, 598–625. [CrossRef]
26. Satolo, E.G.; Hiraga, L.E.d.; Goes, G.A.; Lourenzani, W.L. Lean production in agribusiness organizations: Multiple case studies in a developing country. *Int. J. Lean Six Sigma* **2017**, *8*, 335–358. [CrossRef]
27. Gomes, A.M.; Senna, P.; Monteiro, A.; Pinha, D. Study on techniques and tools used in lean healthcare implementation: A literature review. *Braz. J. Oper. Prod. Manag.* **2016**, *13*, 406–420. [CrossRef]
28. Sivaraman, P.; Nithyanandhan, T.; Lakshminarasimhan, S.; Manikandan, S.; Saifudheen, M. Productivity enhancement in engine assembly using lean tools and techniques. *Mater. Today Proc.* **2020**, *33*, 201–207. [CrossRef]
29. Vinodh, S.; Kumar, S.V.; Vimal, K.E.K. Implementing lean sigma in an Indian rotary switches manufacturing organization. *Prod. Plan. Control* **2014**, *25*, 288–302. [CrossRef]
30. Peruchi, R.S.; Junior, P.R.; Brito, T.G.; Paiva, A.P.; Balestrassi, P.P.; Araujo, L.M.M. Integrating multivariate statistical analysis into six sigma DMAIC projects: A case study on AISI 52100 hardened steel turning. *IEEE Access* **2020**, *8*, 34246–34255. [CrossRef]
31. Rahman, A.; Shaju, S.U.C.; Sarkar, S.K.; Hashem, M.Z.; Hasan, S.M.K.; Mandal, R.; Islam, U. A case study of six sigma define-measure-analyze-improve-control (DMAIC) methodology in garment sector. *Indep. J. Manag. Prod.* **2017**, *8*, 1309. [CrossRef]
32. Yu, K.T.; Ueng, R.G. Enhancing teaching effectiveness by using the Six-Sigma DMAIC model. *Assess. Eval. High. Educ.* **2012**, *37*, 949–961. [CrossRef]
33. Pereira, M.T.; Bento, M.I.; Ferreira, L.P.; Sá, J.C.; Silva, F.J.G. Using six sigma to analyse customer satisfaction at the product design and development stage. *Procedia Manuf.* **2019**, *38*, 1608–1614. [CrossRef]
34. Baptista, A.; Silva, F.J.G.; Campilho, R.D.S.G.; Ferreira, S.; Pinto, G. Applying DMADV on the industrialization of updated components in the automotive sector: A case study. *Procedia Manuf.* **2020**, *51*, 1332–1339. [CrossRef]
35. Stanivuk, T.; Gvozdenovic, T.; Mikulicic, J.Z.; Lukovac, V. Application of six sigma model on efficient use of vehicle fleet. *Symmetry* **2020**, *12*, 857. [CrossRef]
36. Costa, T.; Silva, F.J.G.; Ferreira, L.P. Improve the extrusion process in tire production using Six Sigma methodology. *Procedia Manuf.* **2017**, *13*, 1104–1111. [CrossRef]
37. Tenera, A.; Pinto, L.C. A Lean Six Sigma (LSS) project management improvement model. *Sel. Pap. 27th Ipma Int. Proj. Manag. Assoc.* **2014**, *119*, 912–920. [CrossRef]
38. Cheng, C.Y.; Chang, P.Y. Implementation of the Lean Six Sigma framework in non-profit organisations: A case study. *Total Qual. Manag. Bus. Excell.* **2012**, *23*, 431–447. [CrossRef]
39. Cabrita, M.d.R.; Domingues, J.P.; Requeijo, J. Application of Lean Six-Sigma methodology to reducing production costs: Case study of a Portuguese bolts manufacturer. *Int. J. Manag. Sci. Eng. Manag.* **2016**, *11*, 222–230. [CrossRef]
40. de Mast, J.; Lokkerbol, J. An analysis of the Six Sigma DMAIC method from the perspective of problem solving. *Int. J. Prod. Econ.* **2012**, *139*, 604–614. [CrossRef]
41. Girmanova, L.; Solc, M.; Kliment, J.; Divokova, A.; Miklos, V. Application of six sigma using DMAIC methodology in the process of product quality control in metallurgical operation. *ACTA Technol. Agric.* **2017**, *20*, 104–109. [CrossRef]
42. Costa, L.B.M.; Filho, M.G.; Fredendall, L.D.; Paredes, F.J.G. Lean, six sigma and lean six sigma in the food industry: A systematic literature review. *Trends Food Sci. Technol.* **2018**, *82*, 122–133. [CrossRef]
43. Indrawati, S.; Ridwansyah, M. Manufacturing continuous improvement using lean six sigma: An iron ores industry case application. *Procedia Manuf.* **2015**, *4*, 528–534. [CrossRef]
44. Swarnakar, V.; Singh, A.R.; Tiwari, A.K. Evaluating importance of Critical Success Factors in successful implementation of Lean Six Sigma Framework. In Proceedings of the International Conference on Advances in Mechanical Engineering and Nanotechnology (ICAMEN), Manipal, Jaipur, India, 8–9 March 2019; Volume 2148. [CrossRef]

45. da Silva, O.R.; Rosini, A.M.; Guevara, A.J.H.; Palmisano, A.; Venanzi, D. Lean six sigma: Multiple case study. *Risus-J. Innov. Sustain.* **2018**, *9*, 74–84. [CrossRef]
46. Alexander, P.; Antony, J.; Cudney, E. A novel and practical conceptual framework to support Lean Six Sigma deployment in manufacturing SMEs. *Total Qual. Manag. Bus. Excell.* **2021**, 1–31. [CrossRef]
47. Alhuraish, I.; Robledo, C.; Kobi, A. A comparative exploration of lean manufacturing and six sigma in terms of their critical success factors. *J. Clean. Prod.* **2017**, *164*, 325–337. [CrossRef]
48. Venanzi, D.; Faustino, D.L.; Silva, O.R.; Hasegawa, H.L. Lean Six Sigma—Multiple case study. *Rev. Gestão Inov. Tecnol.* **2017**, *7*, 4059–4073. [CrossRef]
49. Gleeson, F.; Coughlan, P.; Goodman, L.; Newell, A.; Hargaden, V. Improving manufacturing productivity by combining cognitive engineering and lean-six sigma methods. *Procedia CIRP* **2019**, *81*, 641–646. [CrossRef]
50. Kumar, P.; Singh, M.; Phull, G.S. Production lessening analysis of manufacturing unit in India: Lean Six Sigma perspective. *J. Proj. Manag.* **2019**, *4*, 281–290. [CrossRef]
51. Toma, S.-G.; Marinescu, P. Business excellence models: A comparison. *Proc. Int. Conf. Bus. Excell.* **2018**, *12*, 966–974. [CrossRef]
52. Wolniak, R. The level of maturity of quality management systems in Poland-Results of empirical research. *Sustainability* **2019**, *11*, 4239. [CrossRef]
53. The Shingo Model—Shingo Institute—Home of the Shingo Prize. Available online: https://shingo.org/ (accessed on 27 June 2022).
54. Kelly, S.; Hines, P. Discreetly embedding the Shingo principles of enterprise excellence at Abbott Diagnostics manufacturing facility in Longford Ireland. *Total Qual. Manag. Bus. Excell.* **2019**, *30*, 1235–1256. [CrossRef]
55. Pakdil, F.; Leonard, K.M. The effect of organizational culture on implementing and sustaining lean processes. *J. Manuf. Technol. Manag.* **2015**, *26*, 725–743. [CrossRef]
56. Edgeman, R. Routinizing peak performance and impacts via virtuous cycles. *Meas. Bus. Excell.* **2017**, *21*, 261–271. [CrossRef]
57. Edgeman, R. Excellence models as complex management systems: An examination of the Shingo operational excellence model. *Bus. Process Manag. J.* **2018**, *24*, 1321–1338. [CrossRef]
58. Leksic, I.; Stefanic, N.; Veza, I. The impact of using different lean manufacturing tools on waste reduction. *Adv. Prod. Eng. Manag.* **2020**, *15*, 81–92. [CrossRef]
59. Palange, A.; Dhatrak, P. Lean manufacturing a vital tool to enhance productivity in manufacturing. *Mater. Today Proc.* **2021**, *46*, 729–736. [CrossRef]
60. Rahani, A.R.; Al-Ashraf, M. Production flow analysis through Value Stream Mapping: A lean manufacturing process case study. In *Procedia Engineering*; Elsevier: Amsterdam, The Netherlands, 2012. [CrossRef]
61. Wahab, A.N.A.; Mukhtar, M.; Sulaiman, R. A conceptual model of lean manufacturing dimensions. *Procedia Technol.* **2013**, *11*, 1292–1298. [CrossRef]
62. Hodge, G.L.; Ross, K.G.; Joines, J.A.; Thoney, K. Adapting lean manufacturing principles to the textile industry. *Prod. Plan. Control* **2011**, 237–247. [CrossRef]
63. Panwar, A.; Nepal, B.P.; Jain, R.; Rathore, A.P.S. On the adoption of lean manufacturing principles in process industries. *Prod. Plan. Control* **2015**, *26*, 564–587. [CrossRef]
64. Dănuţ-Sorin, I.R.; Opran, C.G.; Lamanna, G. Lean 4.0 dynamic tools for polymeric products manufacturing in Industry 4.0. *Macromol. Symp.* **2021**, *396*, 3–5. [CrossRef]
65. Ramakrishnan, V.; Jayaprakash, J.; Elanchezhian, C.; Ramnath, B.V. Implementation of lean manufacturing in Indian SMEs-A case study. *Mater. Today Proc.* **2019**, *16*, 1244–1250. [CrossRef]
66. Lu, J.C.; Yang, T. Implementing lean standard work to solve a low work-in-process buffer problem in a highly automated manufacturing environment. *Int. J. Prod. Res.* **2015**, *53*, 2285–2305. [CrossRef]
67. Nallusamy, S.; Saravanan, V. Enhancement of overall output in a small scale industry through VSM, line balancing and work standardization. *Int. J. Eng. Res. Africa* **2016**, *26*, 176–183. [CrossRef]
68. Dresch, A.; Veit, D.R.; de Lima, P.N.; Lacerda, D.P.; Collatto, D.C. Inducing Brazilian manufacturing SMEs productivity with Lean tools. *Int. J. Product. Perform. Manag.* **2019**, *68*, 69–87. [CrossRef]
69. Poves-Calderno, I.G.; Ramirez-Mendoza, J.A.; Nunez-Ponce, V.H.; Alvarez-Merino, J.C. Application of lean manufacturing techniques in a Peruvian plastic company. In Proceedings of the 2019 IEEE International Conference on Industrial Engineering and Engineering Management (IEEM), Macao, China, 15–18 December 2019; pp. 546–550. [CrossRef]
70. Mor, R.S.; Bhardwaj, A.; Singh, S.; Sachdeva, A. Productivity gains through standardization-of-work in a manufacturing company. *J. Manuf. Technol. Manag.* **2019**, *30*, 899–919. [CrossRef]
71. Realyvásquez-Vargas, A.; Flor-Moltalvo, F.J.; Blanco-Fernández, J.; Sandoval-Quintanilla, J.D.; Jiménez-Macías, E.; García-Alcaraz, J.L. Implementation of production process standardization—A case study of a publishing company from the SMEs Sector. *Processes* **2019**, *7*, 2019.
72. Wolak, E.; Jones, C.; Leeman, J.; Madigan, C. Improving throughput for patients admitted from the emergency department: Implementation of a standardized report process. *J. Nurs. Care Qual.* **2020**, *35*, 380–385. [CrossRef]
73. Cantini, A.; de Carlo, F.; Tucci, M. Towards forklift safety in a warehouse: An approach based on the automatic analysis of resource flows. *Sustainability* **2020**, *12*, 8949. [CrossRef]

74. Gladysz, B.; Santarek, K.; Lysiak, C. Dynamic spaghetti diagrams. A case study of pilot RTLS implementation. *Adv. Intell. Syst. Comput.* **2018**, *637*, 238–248. [CrossRef]
75. Frosini, F.; Miniati, R.; Cecconi, G.; Dori, F.; Iadanza, E.; Vezzosi, S.; Mechi, M.T.; Belardinelli, A. Lean thinking in hospital: Case study at the pathology laboratory. *IFMBE Proc.* **2015**, *45*, 613–616. [CrossRef]
76. Burroni, L.; Bianciardi, C.; Romagnolo, C.; Cottignoli, C.; Palucci, A.; Fringuelli, F.M.; Biscontini, G. Jacopo Guercini Lean approach to improving performance and efficiency in a nuclear medicine department. *Clin. Transl. Imaging* **2021**, *9*, 129–139. [CrossRef]
77. Jessome, R. Improving patient flow in diagnostic imaging: A case report. *J. Med. Imaging Radiat. Sci.* **2020**, *51*, 678–688. [CrossRef] [PubMed]
78. Pitapurapu, R.; Gupta, A.; Maly, K.; Nadeem, T. Dead reckoning with smartphone. *Sensors* **2015**, *1*, 206–217. [CrossRef]
79. Mahajan, M.; Chistopher, K.B.; Harshan; Prasad, H.C.S. Implementation of lean techniques for sustainable workflow process in Indian motor manufacturing unit. *Procedia Manuf.* **2019**, *35*, 1196–1204. [CrossRef]
80. Yang, J.X.; Hunt, T.D.; Ting, H.H.; Henderson, D.; Finkelstein, J.; Davidson, K.W. Improving value-add work and satisfaction in medical residents training: A Resident-Led quality improvement project employing the lean method to improve hospital supply usage. *Postgrad. Med. J.* **2017**, *93*, 193–197. [CrossRef]
81. Tanco, M.; Santos, J.; Rodriguez, J.L.; Reich, J. Applying lean techniques to nougat fabrication: A seasonal case study. *Int. J. Adv. Manuf. Technol.* **2013**, *68*, 1639–1654. [CrossRef]
82. Alfraih, M.M. Productivity and performance improvement in the medical records department of a hospital an application of Lean Six Sigma. *Electron. Libr.* **2016**, *34*, 1–5.
83. Rekha, R.S.; Periyasamy, P.; Nallusamy, S. An optimized model for reduction of cycle time using value stream mapping in a small scale industry. *Int. J. Eng. Res. Africa* **2016**, *27*, 179–189. [CrossRef]
84. Skeldon, S.C.; Simmons, A.; Hersey, K.; Finelli, A.; Jewett, M.A.; Zlotta, A.R.; Fleshner, N.E. Lean methodology improves efficiency in outpatient academic uro-oncology clinics. *Urology* **2014**, *83*, 992–998. [CrossRef]
85. Cheah, C.K.; Prakash, J.; Ong, K.S. An integrated OEE framework for structured productivity improvement in a semiconductor manufacturing facility. *Int. J. Product. Perform. Manag.* **2020**, *69*, 1081–1105. [CrossRef]
86. Pacheco, D.A.D.; Heidrich, G.D.G. Revitalising the setup reduction activities in Operations Management. *Prod. Plan. Control* **2021**. [CrossRef]
87. Pereira, A.M.H.; Silva, M.R.; Domingues, M.A.G.; Sá, J.C. Lean six sigma approach to improve the production process in the mould industry: A case study. *Qual. Innov. Prosper.* **2019**, *23*, 103–121. [CrossRef]
88. Domingo, R.; Aguado, S. Overall environmental equipment effectiveness as a metric of a lean and green manufacturing system. *Sustainability* **2015**, *7*, 9031–9047. [CrossRef]
89. Sharma, R.K.; Sharma, R.G. Integrating six sigma culture and TPM framework to improve manufacturing performance in SMEs. *Qual. Reliab. Eng. Int.* **2014**, *30*, 745–765. [CrossRef]
90. Gregorio, J.O.R. A case study of an integrated manufacturing performance measurement and meeting system. *J. Manuf. Technol. Manag.* **2015**, *26*, 515–535.
91. Sabadka, D.; Molnar, V.; Fedorko, G.; Jachowicz, T. Optimization of production processes using the Yamazumi Method. *Adv. Sci. Technol. Res. J.* **2017**, *11*, 175–182. [CrossRef]
92. Cannas, V.G.; Pero, M.; Pozzi, R.; Rossi, T. Complexity reduction and kaizen events to balance manual assembly lines: An application in the field. *Int. J. Prod. Res.* **2018**, *56*, 3914–3931. [CrossRef]
93. Czifra, G.; Szabó, P.; Mĺkva, M.; Vaňová, J. Lean principles application in the automotive industry. *Acta Polytech. Hung.* **2019**, *16*, 43–62. [CrossRef]
94. Yılmaz, F.; Demirel, F.; Zaim, S.; Sevim, S. Assembly line balancing by using axiomatic design principles: An application from cooler manufacturing industry. *Int. J. Prod. Manag. Eng.* **2020**, *8*, 3143. [CrossRef]
95. Kumar, D.V.; Mohan, G.M.; Mohanasundaram, K.M. Lean tool implementation in the garment industry. *Fibres Text. East. Eur.* **2019**, *27*, 19–23. [CrossRef]
96. Chan, C.O.; Tay, H.L. Combining lean tools application in kaizen: A field study on the printing industry. *Int. J. Product. Perform. Manag.* **2018**, *67*, 45–65. [CrossRef]
97. Salonitis, K.; Tsinopoulos, C. Drivers and barriers of lean implementation in the Greek manufacturing sector. *Procedia CIRP* **2016**, *57*, 189–194. [CrossRef]
98. Kafuku, J.M. Factors for effective implementation of lean manufacturing practice in selected industries in Tanzania. *Procedia Manuf.* **2019**, *33*, 351–358. [CrossRef]
99. Eswaramoorthi, M.; Kathiresan, G.R.; Prasad, P.S.S.; Mohanram, P.V. A survey on lean practices in Indian machine tool industries. *Int. J. Adv. Manuf. Technol.* **2011**, *52*, 1091–1101. [CrossRef]
100. Maalouf, M.M.; Zaduminska, M. A case study of VSM and SMED in the food processing industry. *Manag. Prod. Eng. Rev.* **2019**, *10*, 60–68. [CrossRef]
101. Jiménez, E.; Tejeda, A.; Pérez, M.; Blanco, J.; Martínez, E. Applicability of lean production with VSM to the Rioja wine sector. *Int. J. Prod. Res.* **2012**, *50*, 1890–1904. [CrossRef]
102. Acero, R.; Torralba, M.; Pérez-Moya, R.; Pozo, J.A. Value stream analysis in military logistics: The improvement in order processing procedure. *Appl. Sci.* **2020**, *10*, 106. [CrossRef]

103. Singh, J.; Singh, H. Application of lean manufacturing in automotive manufacturing unit. *Int. J. Lean Six Sigma* **2020**, *11*, 171–210. [CrossRef]
104. Cavdur, F.; Yagmahan, B.; Oguzcan, E.; Arslan, N.; Sahan, N. Lean service system design: A simulation-based VSM case study. *Bus. Process Manag. J.* **2019**, *25*, 1802–1821. [CrossRef]
105. Garza-Reyes, J.A.; Romero, J.T.; Govindan, K.; Cherrafi, A.; Ramanathan, U. A PDCA-based approach to Environmental Value Stream Mapping (E-VSM). *J. Clean. Prod.* **2018**, *180*, 335–348. [CrossRef]
106. Dominguez-Alfaro, D.; Mendoza-Muñoz, I.; Montoya-Reyes, M.I.; Navarro-González, C.R.; Cruz-Sotelo, S.E.; Vargas-Bernal, O.Y. Ergovsm: A new tool that integrates ergonomics and productivity. *J. Ind. Eng. Manag.* **2021**, *14*, 552–569. [CrossRef]
107. Salvador, R.; Barros, M.V.; Santos, G.E.T.D.; van Mierlo, K.G.; Piekarski, C.M.; de Francisco, A.C. Towards a green and fast production system: Integrating life cycle assessment and value stream mapping for decision making. *Environ. Impact Assess. Rev.* **2021**, *87*, 106519. [CrossRef]
108. Alvandi, S.; Li, W.; Schönemann, M.; Kara, S.; Herrmann, C. Economic and environmental value stream map (E2VSM) simulation for multi-product manufacturing systems. *Int. J. Sustain. Eng.* **2016**, *9*, 354–362. [CrossRef]
109. Brown, A.; Amundson, J.; Badurdeen, F. Sustainable value stream mapping (Sus-VSM) in different manufacturing system configurations: Application case studies. *J. Clean. Prod.* **2014**, *85*, 164–179. [CrossRef]
110. Santos, D.L.d.; Giglio, R.; Helleno, A.L.; Campos, L.M.S. Environmental aspects in VSM: A study about barriers and drivers. *Prod. Plan. Control* **2019**, *30*, 1239–1249. [CrossRef]
111. Gholami, H.; Jamil, N.; Zakuan, N.; Saman, M.Z.M.; Sharif, S.; Awang, S.R.; Sulaiman, Z. Social Value Stream Mapping (Socio-VSM): Methodology to Societal Sustainability Visualization and Assessment in the Manufacturing System. *IEEE Access* **2019**, *7*, 131638–131648. [CrossRef]
112. Carmignani, G. Scrap value stream mapping (S-VSM): A new approach to improve the supply scrap management process. *Int. J. Prod. Res.* **2017**, *55*, 3559–3576. [CrossRef]
113. Gupta, A.; Sharma, P.; Malik, S.C.; Agarwal, N.; Jha, P.C. Productivity improvement in the chassis preparation stage of the amplifier production process: A DMAIC six sigma methodology. *Int. J. Reliab. Qual. Saf. Eng.* **2016**, *23*, 1–14. [CrossRef]
114. Gandhi, S.K.; Sachdeva, A.; Gupta, A. Reduction of rejection of cylinder blocks in a casting unit: A six sigma DMAIC perspective. *J. Proj. Manag.* **2019**, *4*, 81–96. [CrossRef]
115. Lucato, W.C.; Vieira, M.; Santos, J.C.D. Eco-Six Sigma: Integration of environmental variables into the Six Sigma technique. *Prod. Plan. Control* **2015**, *26*, 605–616. [CrossRef]
116. Rahman, A.; Shaju, S.U.C.; Sarkar, S.K. Application of Six-Sigma using define measure analyze improve control (DMAIC) methodology in garment sector. *Indep. J. Manag. Prod.* **2018**, *9*, 810–826. [CrossRef]
117. Antony, J.; Scheumann, T.; Sunder, V.M.; Cudney, E.; Rodgers, B.; Grigg, N.P. Using Six Sigma DMAIC for Lean project management in education: A case study in a German kindergarten. *Total Qual. Manag. Bus. Excell.* **2021**. [CrossRef]
118. Yadrifil, A.A.; Septyanti, A.M.; Rus, M. Implementation of lean-DMAIC method for reducing packing defect in a flour company. *AIP Conf. Proc.* **2020**, *2227*, 040014. [CrossRef]
119. Kansal, J.; Singhal, S. Application and validation of DMAIC Six-Sigma tool for enhancing customer satisfaction in a government R&D organization. *Int. J. Qual. Res.* **2017**, *11*, 931–944. [CrossRef]
120. De Pinto, J.; Schuwarten, L.A.; de Junior, G.C.; Novaski, O. Proposal the application of DMAIC tools and Value Stream Mapping under the perspective of the Lean phylosophy for process improvement: A case study. *Braz. J. Oper. Prod. Manag.* **2017**, *14*, 556–566. [CrossRef]
121. Tampubolon, S.; Purba, H.H. Lean six sigma implementation, a systematic literature review. *Int. J. Prod. Manag. Eng.* **2021**, *9*, 125–139. [CrossRef]
122. Chiarini, A.; Bracci, E. Implementing Lean Six Sigma in healthcare: Issues from Italy. *Public Money Manag.* **2013**, *33*, 361–368. [CrossRef]
123. Leite, D.G.; Montesco, R.A.E.; Sakuraba, C.S. Increasing a gas distributor net profit through Lean Six Sigma. *Qual. Eng.* **2018**, *30*, 359–370. [CrossRef]
124. Vinodh, S.; Gautham, S.G.; Ramiya, R.A. Implementing lean sigma framework in an Indian automotive valves manufacturing organisation: A case study. *Prod. Plan. Control* **2011**, *22*, 708–722. [CrossRef]
125. Tran, T.-A.; Luu-Nhan, K.; Ghabour, R.; Daroczi, M. The use of Lean Six-Sigma tools in the improvement of a manufacturing company—Case study. *Prod. Eng. Arch.* **2020**, *26*, 30–35. [CrossRef]
126. Nedra, A.; Nejib, S.; Yassine, C.; Morched, C. A new Lean Six Sigma hybrid method based on the combination of PDCA and the DMAIC to improve process performance: Application to clothing SME. *Ind. Text.* **2019**, *70*, 447–456. [CrossRef]
127. Hardy, D.L.; Kundu, S.; Latif, M. Productivity and process performance in a manual trimming cell exploiting Lean Six Sigma (LSS) DMAIC—A case study in laminated panel production. *Int. J. Qual. Reliab. Manag.* **2021**, *38*, 1861–1879. [CrossRef]
128. Verma, N.; Sharma, V.; Badar, M.A. Entropy-based lean, energy and six sigma approach to achieve sustainability in manufacturing system. *Arab. J. Sci. Eng.* **2021**, *46*, 8105–8117. [CrossRef] [PubMed]
129. Kaswan, M.S.; Rathi, R. Investigating the enablers associated with implementation of Green Lean Six Sigma in manufacturing sector using Best Worst Method. *Clean Technol. Environ. Policy* **2020**, *22*, 865–876. [CrossRef]
130. Alhuraish, I.; Robledo, C.; Kobi, A. Assessment of lean manufacturing and six sigma operation with decision making based on the analytic hierarchy process. *IFAC-PapersOnLine* **2016**, *49*, 59–64. [CrossRef]

131. Alnadi, M.; McLaughlin, P. Critical success factors of Lean Six Sigma from leaders' perspective. *Int. J. Lean Six Sigma* **2021**. [CrossRef]
132. Timans, W.; Antony, J.; Ahaus, K.; van Solingen, R. Implementation of Lean Six Sigma in small- and medium-sized manufacturing enterprises in the Netherlands. *J. Oper. Res. Soc.* **2012**. [CrossRef]
133. Stankalla, R.; Koval, O.; Chromjakova, F. A review of critical success factors for the successful implementation of Lean Six Sigma and Six Sigma in manufacturing small and medium sized enterprises. *Qual. Eng.* **2018**, *30*, 453–468. [CrossRef]
134. Gnanaraj, S.M.; Devadasan, S.R.; Murugesh, R.; Sreenivasa, C.G. Sensitisation of SMEs towards the implementation of Lean Six Sigma-an initialisation in a cylinder frames manufacturing Indian SME. *Prod. Plan. Control* **2012**, *23*, 599–608. [CrossRef]
135. Silva, F.J.G.; Ferreira, L.P. *Lean Manufacturing: Implementation, Opportunities and Challenges*; Nova Science Publishers: New York, NY, USA, 2019; ISBN 978-1-53615-725-3.
136. Silva, S.; Sá, J.C.; Silva, F.J.G.; Ferreira, L.P.; Santos, G. Lean Green—The importance of integrating environment into lean philosophy—A case study. *Lect. Notes Netw. Syst.* **2020**, *122*, 211–219. [CrossRef]
137. Teixeira, P.; Coelho, A.; Fontoura, P.; Sá, J.C.; Silva, F.J.G.; Santos, G.; Ferreira, L.P. Combining lean and green practices to achieve a superior performance: The contribution for a sustainable development and competitiveness—An empirical study on the Portuguese context. *Corp. Soc. Responsib. Environ. Manag.* **2022**, 1–17. [CrossRef]

sustainability MDPI

Article

The Importance of Fab Labs in the Development of New Products toward Mass Customization

Sérgio Carqueijó [1], Delfina Ramos [2], Joaquim Gonçalves [1], Sandro Carvalho [1,3], Federica Murmura [4], Laura Bravi [4], Manuel Doiro [5], Gilberto Santos [6,*] and Kristína Zgodavová [7]

1 School of Technology, Polytechnic Institute Cavado Ave, 4750-810 Barcelos, Portugal; sergiocarqueijo96@gmail.com (S.C.); jgoncalves@ipca.pt (J.G.); scarvalho@ipca.pt (S.C.)
2 School of Engineering of Porto (ISEP), Polytechnic of Porto, 4200-465 Porto, Portugal; dgr@isep.ipp.pt
3 2Ai—Applied Artificial Intelligence Laboratory, 4750-810 Barcelos, Portugal
4 Department of Economics, Society, Politics, University of Urbino, 61029 Urbino, Italy; federica.murmura@uniurb.it (F.M.); laura.bravi@uniurb.it (L.B.)
5 Organizacion de Empresas e Márketing, Escola Ingeniería Industrial, Vigo University, 36208 Vigo, Spain; mdoiro@uvigo.es
6 Design School, Polytechnic Institute Cavado Ave, 4750-810 Barcelos, Portugal
7 Faculty of Materials, Metallurgy and Recycling, Institute of Materials and Quality Engineering, Technical University of Košice, 04200 Košice, Slovakia; kristina.zgodavova@tuke.sk
* Correspondence: gsantos@ipca.pt

Citation: Carqueijó, S.; Ramos, D.; Gonçalves, J.; Carvalho, S.; Murmura, F.; Bravi, L.; Doiro, M.; Santos, G.; Zgodavová, K. The Importance of Fab Labs in the Development of New Products toward Mass Customization. *Sustainability* **2022**, *14*, 8671 . https://doi.org/10.3390/su14148671

Academic Editor: Fernando Almeida

Received: 18 June 2022
Accepted: 13 July 2022
Published: 15 July 2022

Abstract: The idea of developing Fab Labs (Fabrication Laboratories) was originated by Neil Gershenfeld of the Massachusetts Institute of Technology (MIT) Center for Bits and Atoms in 2006, where it signaled the start of a new era that is changing the world economy by breaking down the boundaries between the digital and physical worlds. The Portuguese Fab Labs are analyzed and a subsequent comparison with others European countries and in the USA was made. This comparison is based on aspects of the profile, the knowledge, the services, and the users. The survey was made by questionnaire, where the Portuguese version of it was adapted from another one disseminated at European level and in the USA, created in connection with a doctoral thesis in Italy. There are 25 active Fab Labs in Portugal, of which 16 responses were obtained and considered valid, so they represent the sample of our study. The results show that the Portuguese Fab Labs are in an embryonic phase with few associated or registered users. Portuguese Fab Labs have areas of work and investment capacity in machinery and technology similar to those of other European countries. However, in terms of turnover, there is a big difference between Portugal and some of the European countries, with American Fab Labs having completely different realities from the European ones. This work is relevant because it compares the Fab Labs of developed countries with those of Portugal. To overcome the difference in good practices existing in other countries, the Portuguese Fab Labs need: (1) better publicity, as well as more support for volunteer workers at Fab Labs, so that more ideas will appear and therefore more products; (2) the facilitation of the use of Fab Labs to have more volunteer workers, who must receive experimental courses, in order to make the best use of the available equipment; (3) evolution, from the current subtractive manufacturing to the additive manufacturing looking for innovation; (4) improved quality, ergonomics, and safety in the design of their own products; (5) and on the part of those responsible, therecognition, dissemination, and celebration of the best ideas that have turned into good products, in order to spread good practices.

Keywords: Fab Labs; product development; Mass Customization; Portugal

1. Introduction

Over the past few years, the production paradigm has undergone several changes mostly motivated by technological developments [1,2]. The ability to respond to consumers' needs become a fundamental aspect in the competitiveness of companies [3–5]. In this highly competitive market that requires reduced response times, highly complex products, diversity, and mass production personalization, the reduction of production costs is a problem [6–8]. This requires collaborative spaces, such as Fab labs, which can be defined as a localized space that offers open access to resources, such as machines and prototyping tools [9–11] for industry 4.0 [12–14].

Today, consumers are pushing the technological development of industry, demanding complex, diversified, updated, and even personalized products, as well as very short delivery times. The design and production of "made-to-stock" are changing to "made-to-order"; the Mass Production (MP) is replaced by a production typology based on the need to accommodate new versions and options [15,16]. In various industrial sectors, the globalization of the economy has created a growing need to respond quickly to market demands, which translates into a drastic reduction in the "time to market" new products, making the life cycle of products drastically reduced. The fast and judicious product development becomes a critical factor for the competitiveness and commercial aggressiveness of companies, determining their subsistence capacity [17]. Thus, Mass Customization (MC) began, which is a production system that allows the personification and personalization or individualization of products, as well as services for a value comparable to that of MP [18]. The essence of MC is to transform a customer into a "co-designer", in which the customer is able to gain access and, simultaneously, participate in the design process [15]. The concept of design and product development can be expressed by the requirements or even co-designing the product with the configuration toolkit [19].

Hence, Mass Customization allows a customer to design certain parts or features of a product. With this, the costumer keeps costs closer to that of mass-produced products. In many cases, the components of the product can be modular. This flexibility allows customers to incorporate their ideas in the product [20]. Thus, the customer can mix-and--match options to create a semi-custom final product.

The development process gives rise to essentially interactive and necessarily multidisciplinary activities. These activities allow a large number of methodologies, systems, tools, and solutions developed by professionals and/or companies from different areas, to be shared by all involved. For this integration and unification of views around the product to be developed, the old sheets of paper containing a two-dimensional expression of what was planned to be produced were no longer enough, as they were time-consuming and dubious [21]. One of the most decisive technologies in the renewal of the industry's operation was the introduction of CAD/CAM systems and 3D CAD modelling capabilities. However, although 3D CAD models provide us with a better view of the object under development, they do not offer the tactile sensation or the notion of assembly.

The prototype came to fill this weakness, giving the opportunity to have a better perception of the object under development. According to Jacobs [22], "there is no better way to make sure that a complex piece has all the desired characteristics than to hold it in your hand, rotate it a few times and look at it from all sides", and thus the Rapid Prototyping (RP) and Rapid Tool Manufacturing (RTM) technologies have significantly enhanced the ability to reduce time to market [17]. These prototyping techniques have evolved over the years, and today, they present a much higher execution speed, compared to conventional prototypes. Through the case of the emergence of Fab Labs, a global network of several hundreds of organizations aims to make digital fabrication machines, such as 3D printing, accessible to diverse audiences [23,24]. The transformation of Fab Labs from elite to collective leads many authors to shed light on changes in the governance of innovation processes [25,26].

The aforementioned authors and others focus on the cultural value of making, identifying the methods for sharing the knowledge and the technical skills, particularly, in the context of digital design and fabrication [27–29]. In this case can appear a new trend in engineering education [30–32] able to create new business models with new digital technologies [33–35]. It is very important to create value [36–38] and to protect the intellectual property [39,40]. At the same time, it is also important to protect the environment [41–43] through concerted actions between organizations [44–46] toward sustainability [47,48].

The activity of producing by adding material (AM) instead of removing it, through 3D printers in which objects are generated by stratification and addition of material, is a revolutionary aspect of prototyping techniques. In addition to the possibility of creating more diversified products with different geometries, the possibility of redefining the activities of the production and logistics processes is offered, new professional figures in the area of manufacturing may be created, new "prototyping platforms for exploration, innovation, invention and learning, providing stimulation for entrepreneurship" [49] as is the case with Fab Labs, which despite developing in a social context, are gradually penetrating the industrial context, in companies such as Airbus, Safran, Airliquide, Orange, and above all at Renault [50,51], which has pioneered the implementation of a corporate Fab Lab [52,53].

The idea of developing Fab Labs (Fabrication Laboratories) was originated by Neil Gershenfeld of the Massachusetts Institute of Technology (MIT) Center for Bits and Atoms in 2006, where it signaled the start of a new era that is changing the world economy by breaking down the boundaries between the digital and physical worlds [49]. The Fab Lab project was created from an experimental course at MIT launched by Gershenfeld in 1998 called "How to Make (Almost) Anything", whose intention was to bring together personal and digital fabrication, individual creativity, and group collaboration. The name illustrates the idea that inspired the Fab Labs: the creation of places where information technology serves the productive activity with a good quality [54–56]. Thus, Fab Labs provide people with the right tools, so they can design and build the most extraordinary things [57], where it exists a bridge from the idea to new product development [58]. In other words, new objects are created with digital design interacting with machines that operate on physical materials [59], where new products are developed [60–62] with designers taking into account the rules of quality [63–65] and also the environmental goal [66–68] of sustainability [69–71]. Yet sometimes problems arise with Indoor Air Quality [72], among others.

The Fab Foundation defines the Manufacture of Laboratories (Fab Lab) as "a technical prototyping platform for innovation and invention, providing a stimulus to local entrepreneurship." At the same time, Fab Lab is a platform for learning and innovation, a place to play, create, learn, guide, invent. A Fab Lab means connection to a global community of students, educators, technologists, researchers, manufacturers, and innovators; in practice, it is a knowledge sharing network that spans 30 countries and 24 time zones. Since all Fab Labs share common tools and processes, the program is building a global network, a distributed laboratory for research and invention [73].

Fab Lab is a prototyping platform for learning and innovation that provides important stimuli for local entrepreneurship and is based mainly on four key factors: openness, interdisciplinary collaboration, effectiveness, and transferability. Currently, the Fab Lab concept is not an alternative to mass production in the creation of large-scale products, but it is committed to demonstrating its potential in modifying the manufacturing logic, offering individuals the ability to create bespoke products, for local and personal needs, to be considered economical according to the logic of mass production [59]. It is a space with a marked social character that offers accessible manufacturing tools and, sometimes, it is conceived as an appropriate platform to quickly start prototyping and development processes of any type of object [74]. In addition, Fab Labs can be incubated by already mature companies, which intend to create laboratories with social, educational, research, and dissemination of their products and services, just like Renault, which is a pioneer in the industrial sector in the development of its own Internal Fab Lab [50]. Increasingly, with respect to their service portfolios, many of them appear to be working likewise to other existing concepts of innovation intermediation such as living laboratories, fab laboratories, business incubators, and co-working spaces [75].

Realizing the impact that the availability and use of Fab Labs can have on the economies of countries, this study aims to understand how Fab Labs are used in Portugal and compares them to similar realities in different countries.

Therefore, the research question that the paper investigates is:

RQ1: What are the differences and similarities among Portuguese Fab Labs and the main European and the American realities of Fab Labs?

The objective of this work was to analyze the use of Fab Labs in Portugal and to compare the sociodemographic and economic reality of Portuguese Fab Labs to the Fab Labs of the main European countries (Italy, France, Germany, Netherlands, and Spain) and the USA.

The structure of this work begins with the introduction, followed by the materials and methods. The results are then presented. Data are analyzed and statistical indicators are displayed. A comparative analysis is made between different countries. Finally, the conclusions appear.

2. Material and Methods

A Portuguese version of a questionnaire was developed. It was adapted from another one disseminated at European level and in the USA, created in connection with a doctoral thesis in Italy [59]. There is a total of 25 Fab Labs active in Portugal, found on the Fab Foundation's official website (http://fabfoundation.org/ (accessed on 2 May 2019)); 16 responses were obtained that were considered valid, so they represent the sample of our study. The questionnaire was designed on a digital platform 'Google Forms'. The contacts with the Fab Labs were carried out via telephone, e-mail, or through the Association of Fab Labs Portugal being asked to answer the questionnaire that was available at the indicated link. In all contacts made, it was indicated how the scope of the responses would serve the purpose of the work, the rules of confidentiality and anonymization of data.

In the construction of the questionnaire, three sections were considered:

- The first section investigates the profile of the respective Fab Lab, considering the location, the number of workers, the size of the structure, the revenue, the average number of users, and their investments in machinery and technology;
- The second section describes the skills and competencies of Fab Labs, namely, who are their main customers, what kind of products do they do and what sector are they targeting, what kind of new digital machines do they use most, their main skills, and services that they deliver to customers;
- Section three takes into account the use of digital technologies and the interconnection with the industry. In this section, we try to investigate what percentage of prototypes developed entirely at a Fab Lab actually arrived in the industry, also to understand the level of connection of Fab Labs with the prototyping industries and beyond, as well as to understand if external organizations choose to develop projects to address deficiencies in the Fab Labs incubators. In this part, the contribution of Quality Management Systems (QMSs), namely through metrology, to product innovation within the scope of Fab Labs will also be investigated. At the end of the section, it was also asked, based on experiences, customers' demand for new projects and the experience of the market's evolution and progress, which technologies will be decisive in the near future.

The first contacts took place in July 2019, and the data collection process was completed in September. The data were processed and analyzed with the statistical software SPSS (IBM SPSS Statistics 22, Armonk, NY, USA) and the graphics created with the Microsoft Office Excel.

3. Results

3.1. Data Analysis

In order to study the different competence dimensions of the Fab Labs, we carried out a Principal Component Analysis (PCA), with Oblimin rotation [76,77].

Variables whose factor loadings are less than 0.6 were excluded from the analysis. The scale used for the variables under study was a 5-point Likert scale. All tests were performed with 95% confidence. In all next tables, N represents the number of Fab Labs surveyed, which is 16.

3.2. Statistical Indicators

A descriptive analysis was carried out to highlight the main characteristics of Fab Labs in order to determine their socio-demographic and economic profile. Tables 1–7 show these indicators. In addition, Cronbach's alpha value was measured to evaluate discriminating capacity of each questions group [78]. It was considered that any group with an alpha value greater than or equal to 0.6 have a discriminating capacity. Table 1 shows who are the main Portuguese Fab Labs users. The individual customers (3.44) are the customers with which the laboratories work most. On the other hand, it is manufacturing companies (2.06) that least seek the services offered by Fab Labs.

Table 1. Main Portuguese Fab Labs users ($\alpha = 0.862$).

	N	Minimum	Maximum	Average	Standard Deviation
Manufacturing companies	16	1	4	2.06	0.854
Individual customers	16	2	5	3.44	1.365
Professionals	16	2	5	3.00	1.033
Institutions/Schools	16	1	4	2.81	1.109
Universities	16	1	5	2.38	1.088
Artists	16	1	5	2.81	1.167
Designers	16	1	5	2.81	1.167

Table 2. Sectors with presence in Portuguese FabLabs ($\alpha = 0.523$).

	N	Minimum	Maximum	Average	Standard Deviation
Fashion	16	1	4	2.38	1.025
Wood Industry	16	1	5	2.56	1.209
Mechanic	16	1	4	2.25	0.931
Automotive	16	1	3	1.50	0.730
Food	16	1	4	1.94	0.998
Electronic Technology	16	1	4	2.56	0.814
IoT	16	1	4	2.31	0.946
Software	15	1	5	1.93	1.100

Table 3. Developed product type ($\alpha = 0.770$).

	N	Minimum	Maximum	Average	Standard Deviation
Products—commercialization	16	1	4	2.31	0.946
Products—single customer	16	2	4	2.75	0.856
Prototypes—enterprises	15	1	3	2.07	0.594
Prototypes—single customer	16	1	5	2.25	0.856

Table 4. Frequency of use of equipment in Portuguese Fab Labs ($\alpha = 0.453$).

	N	Minimum	Maximum	Average	Standard Deviation
3D Printer	16	2	5	3.69	1.014
3D Scanner	16	1	6	3.00	1.897
Laser cutting machine	16	2	6	4.63	0.885
CNC Milling machine	16	2	6	4.31	1.014
Vinyl cutter	16	2	6	3.37	1.310
Lathe	16	1	6	3.25	1.880
Quality control charts	16	2	6	2.69	1.250
Precision punch	16	1	6	3.56	2.308

Table 5. Frequency of services provided by Portuguese Fab Labs (α = 0.834).

	N	Minimum	Maximum	Average	Standard Deviation
Product printing	16	2	5	3.62	1.088
Prototypes creation support	16	2	5	3.13	1.147
New product design/support	16	1	5	3.25	1.238
Support to redefinition of production process	16	1	5	2.50	1.317
Materials consulting	16	1	4	2.25	0.931
Experimental courses	16	1	4	2.25	1.065

Table 6. Competences of Portuguese Fab Labs (α = 0.641).

	N	Minimum	Maximum	Average	Standard Deviation
Arduino programming	16	2	5	3.63	1.088
Software programming	16	1	5	3.00	1.211
Design software	16	3	5	4.31	0.793
Hardware	16	1	5	3.56	1.209
Materials	16	2	5	4.25	0.775
Business process	16	2	5	3.62	1.147
IoT	16	1	5	3.25	1.065
Digital manufacturing	15	3	5	4.47	0.640

Table 7. Factors considered in products design (α = 0.500).

	N	Yes	No
Design	16	14	2
Quality	16	11	5
Ergonomics	16	5	11
Security	16	9	7
Ecology	16	13	3

Table 2 presents data about the main industrial sectors that turn to Fab Labs. The wood industry sector, and electronic technology sector (2.56) are more present in Portuguese Fab Labs. The automotive sector (1.50) is the sector that least seeks the services offered by Portuguese Fab Labs.

Table 3 shows the number of types of products made. It appears that products made for a single customer are the majority, and these data are in line with those obtained in Table 1 where we analyze that in Portugal it is the individual customers that most attend Fab Labs. On the other hand, prototypes for companies are those that are made in smaller quantities by Fab Labs, again in line with data in Table 1, where it was possible to observe that manufacturing companies were the ones that least used the services of Portuguese Fab Labs.

Table 4 shows that, despite 3D printer being the 3rd most used equipment in Fab Labs, the most used equipment in Portuguese Fab Labs are, respectively, laser cutting machines and CNC milling machines, equipment that still belongs to subtractive manufacturing. This proves the reality of Portuguese companies that have not yet taken the leap towards a more ecological and sustainable manufacture as the additive manufacturing. The least used devices are the quality control charts, followed by the 3D scanner.

Data on Table 5 allow us to conclude that the service most frequently provided in Portuguese Fab Labs is the printing of products, proving the idea that 3D printers are one of the most used equipment in Portuguese Fab Labs. With the same value, the two services provided less frequently are materials consulting and experimental courses.

Regarding the Portuguese Fab Labs capabilities (Table 6), it is possible to observe that they are more oriented to the digital manufacturing area, followed by design materials and software. On the other hand, they have less capabilities in the field related with software programming.

Analyzing which factors are most considered in the design of products in the Fab Labs (Table 7), the most Fab Labs give great importance, in this order, on design, eco sustainability and product quality.

With regard to security, this field is balanced between those who give more importance to security and those who give less importance. However, there is a tendency towards those who place more importance on security.

Conversely, in terms of product ergonomics, there are still few who consider this field in the design of their products.

3.3. *Main Component Analysis*

For a better understanding of the structure of the different sectors in the Fab Labs, a Principal Analysis Components (PCA) was realized, for each study element, which is discussed below (Table 8).

Table 8. PCA—Main users of Portuguese Fab Labs (α = 0.862).

	Components	
	Concretization of Ideas	**Research**
Manufacturing companies	-	0.896
Individual customers	0.740	-
Professionals	0.906	-
Institutions/Schools	0.714	-
Universities	-	0.927
Artists	0.955	-
Designers	0.598	-
KMO	0.645	
% Cumulative Variance	55.087	74.672

Extraction Method: Principal Component Analysis. Rotation Method: Oblimin with Kaiser Normalization. Converged rotation in 5 iterations.

After conducting PCA related to the main sectors with which the Portuguese Fab Labs work (Table 9), the existence of 3 components can be observed. The KMO value is 0.481, which indicates that the proportion of the variance that the variables have in common is quite low.

Table 9. PCA—Sectors with which the Portuguese Fab Labs work (α = 0.523).

		Components	
	1	**2**	**3**
Fashion	-	0.784	-
Wood Industry	-	0.938	-
Mechanic	0.941	-	-
Automotive	-	-	0.732
Food	-	-	0.950
Electronic Technology	0.925	-	-
IoT	0.876	-	-
Software	-	0.655	-
KMO		0.481	
% Cumulative Variance	38.915	61.171	77.060

Extraction Method: Principal Component Analysis. Rotation Method: Oblimin with Kaiser Normalization. Converged rotation in 5 iterations.

After conducting a PCA applied to the type of products performed by the Portuguese Fab Labs (Table 10), we found that it was possible to group all types of products into just one component, called Product Development. The KMO value is 0.649, which indicates that the proportion of the variance that the variables have in common is reasonable.

Table 10. PCA—Type of products made (α = 0.770).

	Component Product Development
Products—commercialization	0.721
Product—single customer	0.703
Prototypes—enterprises	0.797
Prototypes—single customer	0.888
KMO	0.649
% Cumulative Variance	60.960

Extraction Method: Principal Component Analysis. 1 component extracted.

After executing a PCA on the frequency of use of the equipment in the Portuguese Fab Labs (Table 11), we found that it was possible to group all the equipment into 3 distinct components. The KMO value is 0.404, which indicates that the proportion of variance that the variables have in common is quite low.

Table 11. PCA—Frequency of use of equipment in Portuguese Fab Labs (α = 0.453).

	Component		
	1	2	3
3D Printer	-	-	0.834
3D Scanner	0.687	-	-
Laser cutting machine	-	-	0.604
CNC Milling machine	0.838	-	-
Vinyl cutter	0.654	-	-
Lathe	-	0.650	-
Quality control charts	-	-	0.680
Precision punch	-	0.900	-
KMO		0.404	
% Cumulative Variance	27.869	50.075	66.411

Extraction Method: Principal Component Analysis. Rotation Method: Oblimin with Kaiser Normalization. Converged rotation in 10 iterations.

After a PCA was carried out on the frequency of provision of services offered by the Portuguese Fab Labs (Table 12), it was verified that it was possible to group all types of products in just one component, called Support, training and product realization. Support and training is important, because people need to know how to work with machines to make their own product of their own design from a good idea. The KMO value is 0.727, which indicates that the proportion of the variance that the variables have in common is reasonable.

Table 12. PCA—Frequency of the provision of services offered by the Fab Labs (α = 0.834).

	Component Support. Training and Product Realization
Product printing	0.606
Prototypes creation support	0.807
New product design/support	0.700
Support to redefinition of production process	0.793
Materials consulting	0.780
Experimental courses	0.767
KMO	0.727
% Cumulative Variance	55.545

Extraction Method: Principal Component Analysis. 1 component extracted.

Related with skills of the Portuguese Fab Labs (Table 13) the PCA reveals that it was possible to group all equipment into 3 distinct components, the Skills in Hardware, Programming and Business (cumulative variance of 41.20%), Skills in the creation of ideas (cumulative variance of 63.08%), and Skills in materials (cumulative variance of 79.25%). The KMO value is 0.505, which indicates that the proportion of variance that the variables have in common is low.

Table 13. PCA—Competences of Portuguese Fab Labs (α = 0.641).

	Component—Skill		
	Hardware. Programming and Business	**Idea Creation**	**Materials**
Arduino programming	0.683	-	-
Software programming	0.724	-	-
Design software	-	0.841	-
Hardware	0.839	-	-
Materials	-	-	0.984
Business process	0.672	-	-
IoT	0.932	-	-
Digital manufacturing	-	0.879	-
KMO	0.505		
% Cumulative Variance	41.200	63.079	79.251

Extraction Method: Principal Component Analysis. Rotation Method: Oblimin with Kaiser Normalization. Converged rotation in 7 iterations.

By developing a PCA on the factors considered in the product design by the Portuguese Fab Labs (Table 14), we realize the existence of two components. The first group was called Quality in well-being (cumulative variance of 35.91%) and groups the type of quality tangible by the user in the use and handling of the product. The second group, called Indirect Quality (cumulative variance of 64.37%), is the group consisting of product quality other than ease of handling, that is, the quality of the design and the ecological quality of the products. The KMO value is 0.448, which indicates that the proportion of variance that the variables have in common is quite low.

Table 14. PCA—Factors considered when designing products (α = 0.500).

	Components—Quality	
	Well-Being Quality	**Indirect Quality**
Design	-	0.820
Quality	0.578	-
Ergonomics	0.786	-
Security	0.852	-
Ecology	-	0.700
KMO	0.448	
% Cumulative Variance	35.913	64.369

Extraction Method: Principal Component Analysis. Rotation Method: Oblimin with Kaiser Normalization. Converged rotation in 9 iterations.

3.4. Comparative Analysis of Results among Countries

In order to ascertain the existence of statistically significant differences between countries in relation to the factors analyzed, a Kruskal-Wallis test (for each variable) was performed, which is the non-parametric test corresponding to the ANOVA (Analyzes of Variance) parametric test that it was not used because the sample was small (N = 16 < 30). The test was carried out with 90% confidence, that is, the null hypothesis will be rejected in cases where the proof value ("sig." in SPSS) is less than 10%. At the same time, Bonferroni's post-hoc comparison tests were performed [79]. In the Post-hoc tests, we only present the lines where there were statistically significant differences from Portugal.

Table 15 shows the countries of all Fab Labs that entered the study. In a very succinct way, we can observe that Portugal is the country with the lowest number of Fab Labs spread throughout its territory, despite being the country with the highest response rate to the questionnaire.

Table 15. List of Fab Labs present in each country and those contacted.

	Portugal	Italy	France	Germany	Netherland	Spain	USA	Total
Fab Labs in country	25	134	151	46	32	46	158	592
Contacted Fab Labs	25	112	142	41	29	42	127	518
Responses	16	27	16	5	3	8	14	89
% of responses	64.00	24.11	11.27	12.20	10.34	19.05	11.02	17.18

Table 16 allows the analysis of a set of features of the generic profile of the Fab Labs for each country. Thus, in a first approach, it is possible to observe that in only 3 of the 7 countries under study there are more than 20 volunteers in at least one Fab Lab, Portugal being one of those that does not have a number of volunteers above that value. Regarding the Fab Labs' work area, we can see that Portugal has good work areas in its Fab Labs, compared to the other countries and is also well ranked in terms of the number of associated or registered users. However, considering the economic level, it is possible to detect that the Portuguese Fab Labs earn an annual revenue well below the European average, which is why it is the country with the lowest revenue both at European level and in comparison with the USA. It can be also observed that the average revenue in the USA is higher than the sum of the revenues of the European countries under study. With regard to investment in technology and machinery, Portugal is relatively well matched with the other European countries under study, with one of its Fab Labs having an investment between 300,000–500,000 €, whereas at European level only Germany has a Fab Labs with a higher investment. Comparing Europe with the USA, it turns out, once again, that the USA is quite distant from Europe, with two of its Fab Labs with investments exceeding €1,000,000. Finally, we find that Portugal is the country where Fab Labs receive the major number of state or European incentives, with 73.3% of Fab Labs receiving some type of these incentives.

Table 16. Fab Labs profile in Europe and USA.

	Portugal (16)		Italy (27)		France (16)		Germany (5)		Netherlands (3)		Spain (8)		USA (14)	
	N	%	N	%	N	%	N	%	N	%	N	%	N	%
							Volunteer Workers							
<20	16	100	17	100	11	68.7	3	60.0	3	100	8	100	12	85.7
>20	0	0.0	0	0.0	5	31.3	2	40.0	0	0.0	0	0.0	2	14.3
							Fab Labs dimension (square meters)							
5–24	0	0.0	1	3.7	1	6.3	0	0.0	0	0.0	1	12.5	3	21.4
25–74	4	25.0	9	33.3	8	50.0	2	40.0	0	0.0	1	12.5	2	14.3
75–200	8	50.0	12	44.4	5	31.3	2	40.0	3	100	5	62.5	4	28.6
>200	4	25.0	5	18.5	2	12.5	1	20.0	0	0.0	1	12.5	5	35.7
							Registry or associate users							
<50	9	59.3	16	59.2	7	53.8	2	40.0	1	33.3	6	75.0	5	35.7
50–100	2	12.5	5	18.5	5	31.3	0	0.0	0	0.0	0	0.0	1	7.1
>100	5	31.3	6	22.2	4	25.0	3	60.0	2	66.7	2	25.0	8	57.1
Fab Labs annual revenue	8.59 €		31.88 €		34.35 €		10.50 €		35.50 €		15.33 €		154.285 $	

Table 16. *Cont.*

	Portugal (16)		Italy (27)		France (16)		Germany (5)		Netherlands (3)		Spain (8)		USA (14)	
	N	%	N	%	N	%	N	%	N	%	N	%	N	%
Investment in technology and machinery (in thousands of euros)														
<10	7	43.8	14	51.9	9	56.3	1	20.0	2	66.7	4	50.0	0	0.0
10–50	4	25.0	7	25.9	5	31.3	2	40.0	1	33.3	3	37.5	3	21.4
50–100	3	18.8	4	14.8	2	12.5	1	20.0	0	0.0	0	0.0	5	35.7
100–300	1	6.3	1	3.7	0	0.0	0	0.0	0	0.0	1	12.5	2	14.3
300–500	1	6.3	1	3.7	0	0.0	0	0.0	0	0.0	0	0.0	1	7.1
500–1000	0	0.0	0	0.0	0	0.0	1	20.0	0	0.0	0	0.0	1	7.1
>1000	0	0.0	0	0.0	0	0.0	0	0.0	0	0.0	0	0.0	2	14.3
State or European funds														
Yes	11	73.3	11	40.7	6	37.5	2	40.0	1	33.3	1	12.5	6	42.9
No	4	26.7	16	59.3	10	62.5	3	60.0	2	66.7	7	87.5	8	57.1

A KrusKal-Wallis test was carried out to ascertain the possible existence of differences between different countries in relation to the main users of Fab Labs (Table 17). The test results show that there are some differences between countries. The differences are found in manufacturing companies, institutions/schools, artists, and designers, while for individual customers, professionals, and universities, no significant differences were found.

Table 17. Kruskal-Wallis test for Main users of Fab Labs analysis between countries.

Main Users	Sig.
Manufacturing companies	0.071
Individual customers	0.107
Professionals	0.521
Institutions/Schools	0.029
Universities	0.150
Artists	0.076
Designers	0.071

The same test was used to analyze differences between the sectors with which Fab Labs work (Table 18). It is possible to conclude that some evident differences between Fab Labs from different countries exist. These evidences were detected with respect to the fashion, mechanics, food, electronic technology, and IoT sectors. In contrast, in the wood industry sector, in the automotive sector and in software technology, no differences were found between Fab Labs of the various countries under study.

Table 18. Kruskal-Wallis—Sectors that Fab Labs work with.

Sectors	Sig.
Fashion	0.002
Wood Industry	0.421
Mechanic	0.026
Automotive	0.444
Food	0.016
Electronic Technology	0.009
IoT	0.019
Software	0.251

Subsequently, a Post-hoc test was performed, in order to detect where differences exist between different countries. However, only the differences in which Portuguese Fab Labs are involved were selected. Thus, it can be seen that, according to the responses to the questionnaires, there are differences between Portugal and Germany in the mechanical sector, as well as, in the electronic technology and IoT. In the IoT sector, Portugal is also different from France. It should be noted that in all these comparisons, Portugal is always at a lower level than other countries, as we can see in the column of the 90% confidence interval, in which both limits are negative. In addition to the aforementioned differences, there is a tendency for other differences to be seen, even though these are not so evident and therefore were not considered (the 90% confidence interval contains 0 between the lower and upper limit). To confirm the trend previously analyzed, Portugal in the electronic technology sector is also tending to be at a lower level compared to France and Italy. Moreover, it can be considered the case of the fashion sector, where Portugal in comparison with neighboring Spain is at a higher level, that is, the Portuguese Fab Labs work more for the fashion sector than the Spanish Fab Labs. These results can be observed in Table 19.

Table 19. Post-hoc Test—Sectors that Fab Labs work with.

Sectors	Country	Sig.	Lower Limit	Upper Limit
Fashion	Spain	0.133	−0.04	2.29
Mechanic	Germany	0.03	−3.29	−0.21
Electronic Technology	France	0.13	−2.16	0.03
	Germany	0.008	−3.63	−0.45
	Italy	0.165	−1.9	0.06
IoT	France	0.047	−2.4	−0.1
	Germany	0.089	−3.35	−0.02

A Kruskal-Wallis test was performed to determine differences between countries considering the types of products performed in Fab Labs (Table 20). The test indicates that there is a difference only in the prototypes for a single customer, and in the remaining hypotheses, no differences were found between the various countries.

Table 20. Kruskal-Wallis test—Type of products produced.

Products Type	Sig.
Products—commercialization	0.271
Products—single customer	0.112
Prototypes—enterprises	0.397
Prototypes—single customer	0.025

The difference found in Table 20 is in line with the difference found between Portugal and Italy, after the Post-hoc test was realized (Table 21). This difference confirms that Portuguese Fab Labs manufacture fewer prototypes for a single customer than Italian Fab Labs.

Table 21. Post-hoc Test—Type of products performed.

Prototypes	Country	Sig.	Lower Limit	Upper Limit
Prototypes for a single costumer	Italy	0.041	−1.92	−0.09

Using the Kruskal-Wallis test yielded the results that are shown in Table 22; differences were detected between countries considering the main equipment they use in Fab Labs, namely in CNC milling machines, vinyl cutters, lathe, and also in precision punching for printed circuits, while in the rest hypotheses no differences were found between different countries.

Table 22. Kruskal-Wallis test—Frequency of use of equipment in Fab Labs.

Frequency of Use of Equipment	Sig.
3D Printer	0.215
3D Scanner	0.816
Laser cutting machine	0.282
CNC Milling machine	0.019
Vinyl cutter	0.019
Lathe	0.004
Quality control charts	0.083
Precision punch	0.092

A Post-hoc test, presented in Table 23, was carried out, where it is possible to detect if there are differences between Portugal and the other countries. As it can be seen, Portugal is different from other countries in the use of CNC milling machines, lathe and precision punching for printed circuits. All of these differences show that Portugal uses this equipment more in its Fab Labs. This can be a negative indicator for Portugal, since both CNC milling machines and the lathe are subtractive manufacturing equipment, and the future depends on the use of machinery where additive technology is predominant.

Table 23. Post-hoc test—Frequency of use of equipment.

Frequency of Use of Equipment	Country	Sig.	Lower Limit	Upper Limit
Milling machines CNC	France	0.013	0.28	2.72
Lathe	Netherlands	0.059	0.14	4.49
	Germany	0.055	0.09	2.66
	Spain	0.076	0.05	3.2
	USA	0.009	0.35	3.01
	Italy	0.001	0.55	2.84
Precision punch	France	0.043	0.15	3.22
	USA	0.024	0.26	3.44

As for the frequency of provision of services offered by Fab Labs, the Kruskal-Wallis test performed (Table 24) allows us to conclude that there are differences between countries regarding materials consultancy and experimental courses, and in the remaining hypotheses, no differences were found between countries.

Table 24. Kruskal-Wallis—Frequency of provision of services offered by Fab Labs.

Frequency of Provision of Services	Sig.
Product printing	0.750
Prototypes creation support	0.144
New product design/support	0.331
Support to redefinition of production process	0.558
Materials consulting	0.035
Experimental courses	0.000

Once again, the differences found in the frequency of service provision are in line with the differences found in Portugal, after the Post-hoc test, is performed (see Table 25). These differences show us that Portugal, in relation to experimental courses, is unlike everyone else. This can also be a bad indicator for Portuguese Fab Labs, since Portugal provides less experimental course services. In addition to the mentioned differences, there are other differences, even though these are not so evident. Portuguese Fab Labs regarding materials consultancy are also tending to be at a lower level, that is, Portuguese Fab Labs tend to provide less services in this field, compared to Germany and USA. Possibly, this can be attributed to Portugal having a low conception of products and its industry being traditionally one of production instead of conception of goods.

Table 25. Post-hoc Test—Frequency of services provided by Fab Labs.

Frequency of Services Provided	Country	Sig.	Lower Limit	Upper Limit
Materials Consultancy	Germany	0.131	−3.56	0.06
	USA	0.131	−2.54	0.04
Experimental Courses	France	0.003	−2.28	−0.35
	Netherlands	0.085	−3.47	−0.03
	Germany	0	−3.75	−0.95
	Spain	0	−3.68	−1.32
	USA	0	−3.32	−1.32
	Italy	0	−2.46	−0.74

The KrusKal-Wallis test, carried out in relation to the skills of Fab Labs (Table 26), shows differences, between countries in skills in materials, business processes, IoT, and digital manufacturing, and in the remaining hypotheses, no differences were found between the various countries.

Table 26. Kruskal-Wallis test—Fab Labs skills.

Fab Labs Skills	Sig.
Arduino programming	0.673
Software programming	0.210
Design software	0.433
Hardware	0.189
Materials	0.097
Business process	0.072
IoT	0.059
Digital manufacturing	0.345

Table 27 shows the Post-hoc test, which allows to observe differences between Portugal and France in terms of skills in materials and business processes. However, the focus is on differences in IoT skills. It is visible that Portugal is different from all the others in a negative way, that is, the Portuguese Fab Labs have less skills in IoT than all the others. Since IoT, today, is one of the most innovative technologies and with a very high progression margin in the near future, this is certainly a bad indicator for Portuguese Fab Labs.

Table 27. Post-Hoc Test—Fab Labs skills.

Skills	Country	Sig.	Lower Limit	Upper Limit
Materials	France	0.097	0	1.87
Business Process	France	0.072	0.04	2.21
IoT	France	0	−3.69	−1.44
	Netherlands	0.059	−4.13	−0.12
	Germany	0	−4.15	−0.9
	Spain	0	−4	−1.25
	USA	0	−3.07	−0.75
	Italy	0	−3.39	−1.38

The Kruskal–Wallis test by ranks is a non-parametric method for testing whether samples originate from the same distribution. It is used for comparing two or more independent samples of equal or different sample sizes. A significant Kruskal–Wallis test indicates if at least one sample stochastically dominates another sample. The Kruskal-Wallis test performed in relation to the factors considered in the design of products (Table 28) shows the existence of differences, if they exist, in all the factors considered.

Table 28. Kruskal-Wallis test—Factors considered when designing products.

Factors to Designing Products	Sig.
Design	0.000
Product quality	0.000
Ergonomics	0.000
Security	0.000

The differences found in Table 28 confirm the differences found with Portugal. The Post-hoc test (Table 29) show that Portugal is different from all the others in all the factors considered. This is also a bad indicator for Portuguese Fab Labs, since Portuguese Fab Labs are the ones that least consider the different factors compared to all the others.

Table 29. Post-hoc Test—Factors considered in product design.

Factors	Country	Sig.	Lower Limit	Upper Limit
Design	France	0	−3.02	−1.23
	Netherlands	0	−4.38	−1.2
	Germany	0	−4.02	−1.43
	Spain	0	−3.97	−1.78
	USA	0	−4.26	−2.41
	Italy	0	−3.81	−2.22
Product Quality	France	0	−3.06	−1.06
	Netherlands	0.001	−4.43	−0.86
	Germany	0	−4.36	−1.46
	Spain	0	−3.79	−1.34
	USA	0	−3.99	−1.92
	Italy	0	−3.95	−2.16
Ergonomics	France	0	−3.6	−1.52
	Netherlands	0.009	−4.21	−0.5
	Germany	0	−4.6	−1.58
	Spain	0	−3.96	−1.41
	USA	0	−3.98	−1.82
	Italy	0	−4.36	−2.5
Safety	France	0	−3.17	−1.2
	Netherlands	0.09	−3.52	−0.02
	Germany	0	−4.26	−1.41
	Spain	0	−3.52	−1.11
	USA	0	−2.95	−0.92
	Italy	0	−3.5	−1.75

4. Conclusions

Fab Labs are known for being small workshops where anyone, institution, or company can develop or create something new. There are places where it is possible to do things, but there are other ones where it is difficult to do something, so Fab Labs offer a variety of very versatile equipment and a diverse range of services. When the sociodemographic and economic reality of Portuguese Fab Labs is compared with the Fab Labs of the main European countries (Italy, France, Germany, Netherlands, and Spain) and the USA, the results obtained show that, in the Portuguese reality, there are still some Fab Labs in an embryonic phase with few associated or registered users, but, on the other hand, others already have another maturity with more than 100 users. The number of volunteer workers also demonstrates that the Portuguese Fab Labs are not yet in the size of some of the Fab

Labs in other countries. Portuguese Fab Labs have areas of work and investment capacity in machinery and technology similar to those of other European countries. However, in terms of turnover, there is a big difference between Portugal and some of the other European countries, with American Fab Labs having completely different realities from the European ones, with a turnover of more than 6 times compared to the European average.

There are also many differences regarding experimental courses between Portugal and the other countries, and this indicator may be a barrier to innovation, information, and knowledge of new technologies. This indicator may be related to other results obtained, namely with the fact that Portuguese Fab Labs have less consideration for factors such as quality, ergonomics, safety in the design of their own products, which may be caused by a lack of knowledge. The Fab Labs should focus on ideas that can be transformed into new products. Hence, ideas capable of being turned into products are needed. Knowing what other countries are doing will help those who are further behind.

This work is relevant because it compares the FAB Labs of developed countries with those of Portugal. To overcome the difference in good practices existing in other countries, the Portuguese Fab labs need: (1) better publicity, more support for volunteer workers at FAB Labs, so that more ideas will appear and therefore more products; (2) the facilitation of the use of FAB Labs to have more volunteer workers, who must receive experimental courses, in order to make the best use of the available equipment; (3) evolution, from the current subtractive manufacturing to the additive manufacturing looking for innovation; (4) improved quality, ergonomics, and safety in the design of their own products; (5) and on the part of those responsible, the recognition, dissemination, and celebration of the best ideas that have turned into good products, in order to spread good practices.

However, it is pertinent to highlight the existence of some limitations in the research. In fact, the existence of Fab Labs in Portugal is still very small, being, compared to the other countries considered in the study, the country with the least number of Fab Labs. However, Portugal is also the country with the lowest population among the countries under study and the 2nd with the smallest territorial area. It is important to consider that the response rate of the Portuguese Fab Labs is the highest with 64% of respondents, Italy is the second highest with a response rate just above 24%, and therefore this can show the interest that Portuguese Fab Labs have in this study, maybe also as a yardstick, to compare their reality with the main European and American ones, and to be able to take advantage of this comparison to improve themselves. For the realization of future investigations, ideally the number of Fab Labs in Portugal should be greater, to close the discrepancy with the other countries under analysis. In short, considering the scarcity of studies on this topic, mainly in Portugal, this investigation becomes an important landmark for the literature and practice of Fab Labs.

Author Contributions: The authors confirm their contributions to this paper as follows: Conceptualization, S.C. (Sérgio Carqueijó), D.R., J.G. and G.S.; Data curation, S.C. (Sandro Carvalho) and L.B.; Formal analysis, M.D. and G.S.; Investigation, S.C. (Sandro Carvalho), D.R., J.G. and G.S.; Methodology, D.R., L.B. and F.M.; Writing—original draft, S.C. (Sérgio Carqueijó), D.R., J.G. and L.B.; Writing—review & editing, S.C. (Sandro Carvalho), D.R., K.Z. and G.S. All authors have read and agreed to the published version of the manuscript.

Funding: This research received no external funding.

Institutional Review Board Statement: Not applicable.

Informed Consent Statement: Not applicable.

Data Availability Statement: Data sharing not applicable.

Conflicts of Interest: The authors declare no conflict of interest.

References

1. Capdevila, I. Joining a collaborative space: Is it really a better place to work? *J. Bus. Strategy* **2019**, *40*, 14–21. [CrossRef]
2. Ferreira, F.; Guerra, H. Os desafios da metrologia por coordenadas nos processos de controlo das especificações dimensionais e geométricas de componentes técnicos, no novo paradigma da indústria 4.0. *Rev. Produção E Desenvolv.* **2018**, *4*, 125–132. [CrossRef]
3. Ferreira, F.; De Vicente, J.; Sanchez, A.M. Evaluation of the performance of coordinate measuring machines in the industry, using calibrated artefacts. *Procedia Eng.* **2013**, *63*, 659–668. [CrossRef]
4. Ni, H.; Zhuo, L.; Meng, Y.; Lv, S.; Wang, X.; Gu, H.; Li, B. Three-dimension Model and Rapid Prototyping of Toy Pen Holder Based on Reverse Engineering. In Proceedings of the 2018 3rd International Conference on Mechanical, Control and Computer Engineering, Huhhot, China, 14–16 September 2018; pp. 185–188.
5. Nishimura, P.; Rodrigues, O.; Júnior, G.; Silva, L. Prototipagem rápida: Um comparativo entre uma tecnologia aditiva e uma subtrativa. *Blucher Des. Proc.* **2016**, *2*, 4481–4491.
6. Rapid, I.L.S. *Rapid Prototyping & Manufacturing Technologies*; The Hong Kong Polytechnic University Industrial Centre: Hong Kong, China, 2012.
7. Schwenke, H.; Neuschaefer-Rube, U.; Pfeifer, T.; Kunzmann, H. Optical Methods for Dimensional Metrology in Production Engineering. *CIRP Ann.—Manuf. Technol.* **2002**, *51*, 685–699. [CrossRef]
8. Borlido, D. *Indústria 4.0—Aplicação a Sistemas de Manutenção*; Universidade do Porto: Porto, Portugal, 2017.
9. Coelho, P. *Rumo à Indústria 4.0*; Universidade de Coimbra: Coimbra, Portugal, 2016.
10. Dalenogare, L.S.; Benitez, G.B.; Ayala, N.F.; Frank, A.G. The expected contribution of Industry 4.0 technologies for industrial performance. *Int. J. Prod. Econ.* **2018**, *204*, 383–394. [CrossRef]
11. Del Val Román, J.L. Industria 4.0: La transformación digital de la industria. *Rev. Deusto Ing.* **2016**, 1–10. Available online: http://coddii.org/wp-content/uploads/2016/10/Informe-CODDII-Industria-4.0.pdf (accessed on 2 May 2019).
12. García Ferrari, T. Design and the Fourth Industrial Revolution. Dangers and opportunities for a mutating discipline. *Des. J.* **2017**, *20* (Suppl. 1), S2626–S2633. [CrossRef]
13. Fabela, A.M.R.; Flores, R.P. Desafios da formação profissional do designer industrial na Quarta Revolução Industrial (4RI). *Rev. Iberoam. Para La Investig. Y El Desarro. Educ.* **2018**, *8*, 1–22.
14. Sung, T.K. Industry 4.0: A Korea perspective. *Technol. Forecast. Soc. Change* **2018**, *132*, 40–45. [CrossRef]
15. Santos, G.; Bravi, L.; Murmura, F. Attitudes and behaviours of Italian 3D prosumer in the Era of Additive Manufacturing. *Procedia Manuf.* **2017**, *13*, 980–986.
16. Bravi, L.; Murmura, F.; Santos, G. Manufacturing labs: Where new digital technologies help improve life quality. *Int. J. Qual. Res.* **2018**, *12*, 957–974.
17. Lino, F.J.; Neto, R.J. A Prototipagem Rápida na Indústria Nacional. In Proceedings of the 2° Encontro Nacional do Colégio de Engenharia Mecânica da Ordem dos Engenheiros, Coimbra, Portugal, 15–16 December 2000.
18. Diez, T. Personal Fabrication: Fab Labs as Platforms for Citizen-Based Innovation, from Microcontrollers to Cities. *Nexus Netw. J.* **2012**, *14*, 457–468. [CrossRef]
19. Milara, I.S.; Georgiev, G.V.; Riekki, J.; Ylioja, J.; Mikko, J.Y. Human and Technological Dimensions of Making in FabLab. *Des. J.* **2017**, *20* (Suppl. 1), S1080–S1092. [CrossRef]
20. Zgodavova, K.; Bober, P.; Majstorovic, V.; Monkova, K.; Santos, G.; Juhaszova, D. Innovative methods for small mixed batches production system improvement: The case of a bakery machine manufacturer. *Sustainability* **2020**, *12*, 6266. [CrossRef]
21. Saura, C. *Aplicação da Prototipagem Rápida na Melhoria do Processo de Desenvolvimento de Produtos em Pequenas e Médias Empresas*; Universidade Estadual de Campinas: Campinas, Brazil, 2003.
22. Jacobs, P.F. *Stereolithography & Other RP&M Technologies: From Rapid Prototyping to Rapid Toolin*; American Society of Mechanical Engineers: New York, NY, USA, 1992.
23. Schneider, C.; Lösch, A. Visions in assemblages: Future-making and governance in Fab Labs. *Futures* **2019**, *109*, 203–212. [CrossRef]
24. Angrisani, L.; Arpaia, P.; Bonavolontá, F.; Moccaldi, N.; Moriello, R.S. A "learning small enterprise" networked with a FabLab: An academic course 4.0 in instrumentation and measurement. *Measurement* **2020**, *150*, 107063.
25. Ropin, H.; Pfleger-Landthaler, A.; Wolfram Irsa, W. A Fab Lab as integrative part of a Learning Factory. *Procedia Manuf.* **2020**, *45*, 355–360. [CrossRef]
26. Maravilhas, S.; Martins, J. Strategic knowledge management in a digital environment: Tacit and explicit knowledge in Fab Labs. *J. Bus. Res.* **2019**, *94*, 353–359. [CrossRef]
27. Di Roma, A.; Minenna, V.; Scarcelli, A. Fab Labs new hubs for socialization and Innovation. *Des. J.* **2017**, *20* (Suppl. 1), S3152–S3161. [CrossRef]
28. Love, T.S.; Roy, K.R.; Marino, M.T. Inclusive makerspaces, fab labs, and STEM labs. *Technol. Eng. Teach.* **2020**, *79*, 23–27.
29. Yülek, M.A.; Santos, G. Why Income Gaps Persist: Productivity Gaps, (No-) Catch-up and Industrial Policies in Developing Countries. *J. Econ. Issues* **2022**, *56*, 158–183. [CrossRef]
30. Alía, C.; Ocaña, R.; Caja, J.; Maresca, P.; Moreno-Díaz, C.; Narbón, J.J. Use of open manufacturing laboratories (Fab Labs) as a new trend in engineering education. *Procedia Manuf.* **2019**, *41*, 938–943. [CrossRef]
31. Santos, G.; Mandado, E.; Silva, R.; Doiro, M. Engineering learning objectives and computer assisted tools. *Eur. J. Eng. Educ.* **2019**, *44*, 616–628. [CrossRef]

32. Rejeb, H.B.; Roussel, B. Design and Innovation Learning: Case Study in North African Engineering new methodology to analyze the functional and physical architectures. *Procedia CIRP* **2018**, *70*, 331–337.
33. Santos, G.; Murmura, F.; Bravi, L. Fabrication laboratories: The development of new business models with new digital technologies. *J. Manuf. Technol. Manag.* **2018**, *29*, 1332–1357. [CrossRef]
34. Halassi, S.; Semeijn, J.; Kiratli, N. From consumer to prosumer: A supply chain revolution in 3D printing. *Int. J. Phys. Distrib. Logist. Manag.* **2019**, *49*, 200–216. [CrossRef]
35. Fontana, A.; Corti, D.; Alge, M.; Petrucciani, M.; Calvo, L.; Marangi, L. Integrating a LCA tool with a design platform towards a sustainable-aware PSS design: Application in a FabLAB environment. *IFAC Pap. OnLine* **2018**, *51*, 1125–1130. [CrossRef]
36. Félix, M.J.; Gonçalves, S.; Jimenez, G.; Santos, G. The contribution of design to the development of products and manufacturing processes in the portuguese industry. *Procedia Manuf.* **2019**, *41*, 1055–1062. [CrossRef]
37. Santos, G.; Gomes, S.; Braga, V.; Braga, A.; Lima, V.; Teixeira, P.; Sá, J.C. Value creation through quality and innovation—A case study on Portugal. *TQM J.* **2019**, *31*, 928–947. [CrossRef]
38. Félix, M.J.; Silva, S.; Santos, G.; Doiro, M.; Sá, J.C. Integrated product and processes development in design: A case study. *Procedia Manuf.* **2019**, *41*, 296–303. [CrossRef]
39. Doiro, M.; Fernández, F.J.; Félix, M.J.; Santos, G. Machining operations for components in kitchen furniture: A comparison between two management systems. *Procedia Manuf.* **2019**, *41*, 10–17. [CrossRef]
40. Santos, G.; Doiro, M.; Félix, M.J.; Mandado, E.; Sá, J.C.; Gonçalves, J.; Teixeira, P. On the concept of an integrated and lean model of product development proposed for intellectual property creation and competitive economies. *Int. J. Intellect. Prop. Manag.* **2020**, *10*, 409–435. [CrossRef]
41. Santos, G.; Rebelo, M.; Barros, S.; Silva, R.; Pereira, M.; Ramos, G.; Nuno Lopes, N. Developments regarding the integration of the Occupational Safety and Health with Quality and Environment Management Systems. In *Occupational Safety and Health—Public Health in the 21st Century*; Kavouras, I., Chalbot, M.-C.G., Eds.; The Nova Science Publishers: New York, NY, USA, 2014; Chapter 6; pp. 113–146. ISBN 978-1-63117-698-2. (eBook).
42. Bravi, L.; Santos, G.; Pagano, A.; Murmura, F. Environmental management system according to ISO 14001:2015 as a driver to sustainable development. *Corp. Soc. Responsib. Environ. Manag.* **2020**, *27*, 2599–2614. [CrossRef]
43. Rebelo, M.; Santos, G.; Silva, R. Integration of Standardized Management Systems: A Dilemma? *Systems* **2015**, *3*, 45–59. [CrossRef]
44. Teixeira, P.; Sá, J.C.; Silva, F.J.G.; Ferreira, L.P.; Santos, G.; Fontoura, P. Connecting lean and green with sustainability towards a conceptual model. *J. Clean. Prod.* **2021**, *322*, 129047. [CrossRef]
45. Ramos, D.; Fonseca, L.; Gonçalves, G.; Carvalho, R.; Carvalho, S.; Santos, G. Cost-Benefit Analysis of Implementing Circular Economy in a Portuguese Company: From a Case Study to a Model. *Qual. Innov. Prosper.* **2022**, *26*, 52–69. [CrossRef]
46. Teixeira, P.; Coelho, A.; Fontoura, P.; Sá, J.C.; Silva, F.J.G.; Santos, G.; Ferreira, L.P. Combining lean and green practices to achieve a superior performance: The contribution for a sustainable development and competitiveness—an empirical study on the Portuguese context. *Corp. Soc. Responsib. Environ. Manag.* **2022**, *in press*. [CrossRef]
47. Silva, V.; Sá, J.C.; Fonseca, L.; Santos, G. B Impact Assessment as a Sustainable Tool: Analysis of the Certification Model. *Sustainability* **2022**, *14*, 5590. [CrossRef]
48. Fonseca, L.; Silva, V.; Sá, J.C.; Lima, V.; Santos, G.; Silva, R. B Corp versus ISO 9001 and 14001 certifications: Aligned, or alternative paths, towards sustainable development? *Corp. Soc. Responsib. Environ. Manag.* **2021**, *29*, 496–508. [CrossRef]
49. Gershenfeld, N. Programming a New Reality. 2006. Available online: https://www.ted.com/talks/neil_gershenfeld_on_fab_labs (accessed on 9 June 2020).
50. Lô, A.; Diochon, P.F. Unsilencing power dynamics within third spaces. The case of Renault's Fab Lab. *Scand. J. Manag.* **2019**, *35*, 101039. [CrossRef]
51. COMPETE. FabLabs já Chegaram a Portugal. 2012. Available online: http://www.pofc.qren.pt/areas-do-compete/accoes-colectivas/os-projectos-em-que-apostamos/entity/fablabs-ja-chegaram-a-portugal?fromlist=1 (accessed on 8 February 2019).
52. DeNisco, A. Fab Lab Using Technology to Make (Almost) Anything! *Dist. Adm.* **2012**, *48*, 34–37.
53. Fonda, C.; Canessa, E. Making ideas at scientific fabrication laboratories. *Phys. Educ.* **2016**, *51*, 065016. [CrossRef]
54. Sá, J.; Vaz, S.; Carvalho, O.; Lima, V.; Fonseca, L.; Morgado, L.; Santos, G. A model of integration ISO 9001 with Lean Six Sigma and main benefits achieved. *Total Qual. Manag. Bus. Excell.* **2020**, *33*, 218–242. [CrossRef]
55. Santos, G.; Sá, J.C.; Félix, M.J.; Barreto, L.; Carvalho, F.; Doiro, M.; Zgodavová, K.; Stefanović, M. New Needed Quality Management Skills for Quality Managers 4.0. *Sustainability* **2021**, *13*, 6149. [CrossRef]
56. Barbosa, L.; Oliveira, O.; Machado, M.C.; Morais, A.C.; Bozola, P.M.; Santos, G. Lessons learned from quality management system ISO 9001:2015 certification: Practices and barrier identification from Brazilian industrial companies. *Benchmarking Int. J.* **2021**, *in press*. [CrossRef]
57. Fiaidhi, J.; Mohammed, S. Fab Labs: A Platform for Innovation and Extreme Automation. *IT Prof.* **2018**, *20*, 83–90. [CrossRef]
58. Santos, G.; Afonseca, J.; Lopes, N.; Félix, M.J.; Murmura, F. Critical success factors in the management of ideas as an essential component of innovation and business excellence. *Int. J. Qual. Serv. Sci.* **2018**, *10*, 214–232. [CrossRef]
59. Bravi, L. Additive Manufacturing: Analysis of the Economic Context and Evaluation of the Indoor Air Quality, with a Total Quality Management Approach. Ph.D. Thesis, Università Degli Studi Di Urbino Carlo Bo, Urbino, Italy, 2019.
60. Araújo, R.; Santos, G.; da Costa, J.B.; Sá, J.C. The quality management system as a driver of organizational culture: An empirical study in the Portuguese textile industry. *Qual. Innov. Prosper.* **2019**, *23*, 1–24. [CrossRef]

61. Santos, G.; Sá, J.C.; Oliveira, J.; Ramos, D.G.; Ferreira, C. Quality and safety continuous improvement through lean tools. In *Lean Manufacturing: Implementation, Opportunities and Challenges*; Nova Science Publishers: Hauppauge, NY, USA, 2019; pp. 165–188, ISBN 978-1-53615-725-3.
62. Bravi, L.; Murmura, F.; Santos, G. The ISO 9001:2015 quality management system standard: Companies' drivers, benefits and barriers to its implementation. *Qual. Innov. Prosper.* **2019**, *23*, 64–82. [CrossRef]
63. Costa, A.R.; Barbosa, C.; Santos, G.; Rui Alves, M. Six sigma: Main metrics and r based software for training purposes and practical industrial quality control. *Qual. Innov. Prosper.* **2019**, *23*, 83–100. [CrossRef]
64. Sá, J.C.; Amaral, A.; Barreto, L.; Carvalho, F.; Santos, G. Perception of the importance to implement ISO 9001 in organizations related to people linked to quality-an empirical study. *Int. J. Qual. Res.* **2019**, *13*, 1055–1070. [CrossRef]
65. Santos, G.; Murmura, F.; Bravi, L. Developing a model of vendor rating to manage quality in the supply chain. *Int. J. Qual. Serv. Sci.* **2019**, *11*, 34–52. [CrossRef]
66. Fleischmann, K.; Hielscher, S.; Merritt, T. Making things in Fab Labs: A case study on sustainability and co-creation. *Digit. Creat.* **2016**, *27*, 113–131. [CrossRef]
67. Barbosa, L.C.F.M.; de Oliveira, O.J.; Santos, G. Proposition for the alignment of the integrated management system (quality, environmental and safety) with the business strategy. *Int. J. Qual. Res.* **2018**, *12*, 925–940.
68. Kohtala, C. Making "Making" Critical: How Sustainability is Constituted in Fab Lab Ideology. *Des. J.* **2017**, *20*, 375–394. [CrossRef]
69. Byard, D.J.; Woern, A.L.; Oakley, R.B.; Matthew, J.; Fiedler, M.J. Green fab lab applications of large-area waste polymer-based additive manufacturing. *Addit. Manuf.* **2019**, *27*, 515–525. [CrossRef]
70. Carvalho, F.; Santos, G.; Gonçalves, J. Critical analysis of information about integrated management systems and environmental policy on the Portuguese firms 'website, towards sustainable development. *Corp. Soc. Responsib. Environ. Manag.* **2020**, *27*, 1069–1088. [CrossRef]
71. Talapatra, S.; Santos, G.; Uddin, K.; Carvalho, F. Main benefits of integrated management systems through literature review. *Int. J. Qual. Res.* **2019**, *13*, 1037–1054. [CrossRef]
72. Bravi, L.; Murmura, F.; Santos, G. Additive Manufacturing: Possible Problems with Indoor Air Quality. *Procedia Manuf.* **2019**, *41*, 952–959. [CrossRef]
73. FabFoundation. 2019. Available online: http://fabfoundation.org (accessed on 10 January 2019).
74. García-Ruiz, M.E.; Lena-Acebo, F.J. Fab Labs global survey: Characterization of Fab Lab phenomenon. In Proceedings of the Iberian Conference on Information Systems and Technologies, CISTI, Caceres, Spain, 1–6 June 2018.
75. Memon, A.B.; Meyer, K.; Thieme, M.; Meyer, L.P. Inter-InnoLab collaboration: An investigation of the diversity and interconnection among Innovation Laboratories. *J. Eng. Technol. Manag.* **2018**, *47*, 1–21. [CrossRef]
76. Jennrich, R.I.; Sampson, P.F. Rotation for Simple Loadings. *Psychometrika* **1966**, *31*, 313–323. [CrossRef]
77. Kaiser, H.F. An Index of Factorial Simplicity. *Psychometrika* **1974**, *39*, 31–36. [CrossRef]
78. Taber, K.S. The Use of Cronbach's Alpha When Developing and Reporting Research Instruments in Science Education. *Res. Sci. Educ.* **2018**, *48*, 1273–1296. [CrossRef]
79. Armstrong, R.A. When to use the Bonferroni correction. *Ophthalmic Physiol. Opt.* **2014**, *34*, 502–508. [CrossRef]

sustainability

MDPI

Article

B Impact Assessment as a Sustainable Tool: Analysis of the Certification Model

Vítor Silva [1], Vanda Lima [2,*], José Carlos Sá [1,3], Luís Fonseca [1,3] and Gilberto Santos [4]

1 School of Engineering, ISEP, Polytechnic of Porto, 4249-015 Porto, Portugal; 1081624@isep.ipp.pt (V.S.); cvs@isep.ipp.pt (J.C.S.); lmf@isep.ipp.pt (L.F.)
2 CIICESI, Escola Superior de Tecnologia e Gestão, Politécnico do Porto, 4610-156 Felgueiras, Portugal
3 Institute of Science and Innovation in Mechanical and Industrial Engineering (INEGI), 4200-465 Porto, Portugal
4 Design School, Polytechnic Institute Cavado Ave, 4750-810 Barcelos, Portugal; gsantos@ipca.pt
* Correspondence: vlima@estg.ipp.pt

Abstract: Currently, certification is an essential tool for a company's sustainability and a seal of trust for the stakeholders. The B Corporation (B Corp) certification system is in line with the leading indicators of sustainable development and social responsibility published by the general assembly of the United Nations, namely: environment, community, workers, customers, and governance. Nevertheless, it is essential that academic research should empirically assess the B Corp model's reliability for its validation and legitimization. In this study, we address the results of the B Impact Assessment of 2262 companies certified by B Corp from the beginning of 2017 to March 2021. The main objective is to analyze the B Impact Assessment, verifying the robustness and consistency of the model to measure and improve the economic, social, and environmental impact of companies. We analyzed the construct's validity through a confirmatory factorial analysis using AMOS statistical software. The results allowed us to identify some weaknesses and limitations of the B Impact Assessment. This certification system reflects an unadjusted model where the main assessment indicators have problems with regard to the measurement scale. The governance and customer indicators are the most vulnerable. The findings also allow us to state that there are apparently no minimum values established for each of the parameters evaluated, which may cause imbalances in the sustainable development process of B Corp companies. This research contributes to enhancing B Impact Assessment as a sustainability tool, highlighting areas for improvement concerning the indicators' measurement scales and the assessment process, including the monitoring of evaluators.

Keywords: B Impact Assessment; certified B Corp; sustainability; confirmatory factorial analysis (CFA)

Citation: Silva, V.; Lima, V.; Sá, J.C.; Fonseca, L.; Santos, G. B Impact Assessment as a Sustainable Tool: Analysis of the Certification Model. *Sustainability* **2022**, *14*, 5590. https://doi.org/10.3390/su14095590

Academic Editor: Carlos Rodríguez Monroy

Received: 25 March 2022
Accepted: 4 May 2022
Published: 6 May 2022

1. Introduction

It is widely recognized that certification is a relevant tool for the business sector to achieve sustainable development goals and a faithful commitment to stakeholders [1–4].

Created in 2006 in the United States as "B Lab," the B Corporation's certification model is intended to be an instrument for evaluating the social, economic, and environmental performance of companies. In addition to evaluating products and services, this certification model also evaluates operations, the sourcing and input of materials, charitable donations, business model, employees, community, environment, and customers [5].

This certification is perceived as a kind of seal of trust for stakeholders of a company's good performance [5]. The positive impact of this model is supported by transparency and accountability requirements. B Corporation certification is administered by B Lab which is a non-profit organization [5].

The certification process varies depending on the size and complexity of the company [5]. B Impact Assessment is used as a digital tool to assess a company's performance and encompasses five different indicators—governance, workers, community, environment, and customers—measured through a set of variables that express the practices and outputs achieved by companies [5].

This paper analyses the consistency and robustness of B Impact Assessment as a certification model that helps companies measure and improve their economic, social, and environmental impact in five different areas: governance, workers, community, environment, and customers. Considering the complexity of the model structure, which includes multiple indicators that are disaggregated into different variables, a confirmatory factor analysis was used.

Taking advantage of transparency as one of the most substantial aspects of this certification system, we study B Impact Assessment data from 2262 companies and organizations distributed worldwide. The database was taken directly from the B Corporation website.

By 2021, the B Corporation had more than 4000 certified companies [5]. The reason that leads us to focus attention on the B Corporation is the high growth of this certification throughout the world, especially in Europe.

To be eligible for certification, a company must demonstrate that it adheres to the "highest standards of verified social and environmental performance, public transparency and legal accountability to balance profit and purpose" [5]. Certification is obtained when a company reaches an 80-point threshold in the evaluation process [5].

Some of the various sustainable development objectives defined by the United Nations are in common with the B Impact Assessment.

Despite the success in the practical application of the B Impact Assessment and B Corp certification, there is a research gap in empirical academic research regarding the model's reliability, which is an essential issue for legitimizing any management model. Since this subject is essential to everyone, especially managers and entrepreneurs, we believe that this research is an asset to the community and in the right direction for ensuring sustainability.

According to [6], companies with a high level of initiatives related to corporate sustainability tend to obtain better market value and lower the cost of capital and cost of debt. Concerning adopting a B Corp policy, these effects on economic growth apparently need some time to be reflected [7]. However, due to the richness of our dataset and the techniques employed, our paper can address both deficiencies.

The remainder of this paper is structured as follows. In Section 2, a literature review is presented. Section 3 presents the research methodology. Section 4 provides the results of the study. In the final Sections 5 and 6, a summary of the results is presented and discussed.

2. Literature Review

Literature studies on the B Corporation reveal a subject with great potential emerging in the academic community [8], as it is in line with the leading sustainability demands.

The concept of sustainability has gained prominence in the scientific and business community in recent decades. For example, in the report "Our Common Future" (Brundtland Report), published in 1987, sustainable development was defined as "development that meets the needs of the present without compromising the ability of future generations to meet their own needs" [9] (p. 43). This principle underpins the 17 sustainable development goals (SDGs) defined by the United Nations (UN) in 2015, whose achievement depends on the commitment of companies, governments, and citizens [10].

The Brundtland Report (1987) definition of sustainability makes room for different sustainability models with diverse intensities of the relationship between the three sustainability dimensions: the social, the economic and the environmental.

Research by [11] is thought to be at the origin of the three-pillar conception of sustainability—social, economic, and environmental—usually represented by three intersecting circles with sustainability at the center. A review of the genesis and theoretical foundations of the concept of sustainability highlighted that "there is no single point of origin of this three-pillar conception, but rather a gradual emergence of various critiques in the early academic literature of the economic status quo from both social and ecological perspectives on the one hand, and the United Nations' quest to reconcile economic growth as a solution to social and ecological problems on the other" [12].

Theoretically, several relevant theories support sustainability and the adoption of B Corp certification. Stakeholder theory [13,14] highlights the relevance of a firm's relationships with its critical stakeholders with the integration of business and societal considerations fostering stakeholder value and leading to improved performance [15]. According to the resource-based view (RBV) theory [16], organizations should identify and use valuable, rare, difficult to copy, and non-substitutable resources to gain competitive advantages and abnormal profits. The adoption of sustainability and B Corp certification can generate these resources and support integration with stakeholders (an inimitable resource) in response to their demands. Additionally, institutional theory (the adoption of models from successful organizations) can also explain the destination of sustainability and B Corp certification as a search for organizational legitimacy [17].

The institutionalization of sustainable development by the UN, in 1987 in the Brundtland Report and the Earth Summit in 1992, drove a real international awareness and stance on the need to establish a global and effective sustainable development policy.

At the level of organizations, there has been a growing adoption of reporting and management practices for sustainability. According to [18], "given the complexity and multidimensionality of sustainable development, several standards have been proposed to address specific (environmental, social, economic) or practical (reporting mechanisms, management systems, etc.) issues" (p. 333).

Created in 2006 in the United States, B Lab is an already well known and rapidly growing organization in the certification field of corporate sustainability. B Corp certification is based on three essential pillars: social and environmental performance, public transparency, and legal accountability [5]. The five indicators included in the B Impact Assessment model can be equally associated with the three pillars of sustainability: economic (governance), social (workers, community, and customer) and environmental (environmental).

Some companies are still reticent about the financial impact resulting from implementing the B Corp certification. However, as shown by [19], the positive effects on the growth of companies' turnover in the short term, resulting from increased transparency and the positive socioenvironmental impact observed, outweigh the adverse effects due to the strict audit procedure. Furthermore, according to [20], the critical analysis of the potential of this certification compared to the basic principles of the circular economy allows an opportunity to portray multiple dynamic viewpoints concerning stakeholders.

B Corporation certification can activate circular economy pathways due to the actions of different stakeholders in an international context [20]. With the environment, for example, being one of the evaluation indicators during the B audit process, according to [21], companies' environmental impact processes have a direct and indirect association with their long-term development and financial performance. Therefore, B Corp certification can contribute fundamentally to economic growth of such companies. In a study [22] conducted on 851 certified B Corporation companies based in the United States, in the first 10 years of certification, it was possible to identify that institutional and economic resources support the diffusion of this certification model. According to [22], the author of this study, companies will be more likely to become certified the greater the number of organizations covered by this model, contributing to the social and environmental well-being of the planet. In addition, the economic resources of the external environment influence adherence to this certification.

The authors of [23] mention that the goal of the B Lab is to improve the alignment of the mission of companies and measure the impact of their business to meet the highest standards of society today. The focus on socioenvironmental performance, transparency, and legal responsibility is part of the strategy of this pro-social and entrepreneurial movement.

Some studies try to understand the motive and the most common reasons that lead companies and organizations to seek B Corp certification. For example, according to the analysis conducted by [24] the main reasons for seeking these certifications have to do with the fact that companies identify their mission with this model and the attempt to enhance their values and identity in the markets.

Another perspective that also deserves some attention is understanding the main reasons that lead stakeholders to seek out companies certified by B Corp. A study carried out on 20 consumers in Chile of products and services provided by B Corp companies reveals that consumers' main motivations were [25]:

✓ Socio-environmental responsibility
✓ Self-satisfaction
✓ Health and quality of life

Results from the same study [25] show that the image this certification conveys to the consumer is one of trust and effectiveness in the social, economic, and environmental fields. In addition, consumers also highlight the good quality and exclusive design when looking for B Corp companies or organizations. The world is increasingly dynamic and to achieve high levels of quality, everyone involved in this area must have new skills, such as creativity, teamwork, communication, and knowledge of new technologies [26].

Independently of this certification being a good tool for companies to identify more effective ways to integrate social values in markets and business, improvements in the future depend mainly on how founders and leaders relate to certification. The announcement of certifications in the sphere of sustainability is positively associated with the performance of companies [27]. In research conducted by [28], organizations and companies with a strong brand in markets often believe they have already achieved high ethical standards, yet they aspire to let stakeholders know this. Mitigating and reducing the ecological footprint, certainly one of the aspects valued by B Corp, cannot be considered by companies as a sacrifice but rather as an act of ensuring the quality of life and a more promising future. A growing trend of ecological consumerism has opened a vast market of opportunities for entrepreneurs to conduct more sustainable businesses. Companies should exploit this growing market by investing in innovative ecological processes and producing sustainable goods. Such policies will not be cost-free. However, the short-term cost would be outweighed by the long-term sustainability gain [29,30], and this message should be conveyed by agencies to top management.

In addition to this message, the B Corp must be seen as a faithful business partner, and its certification must provide a seal of legitimacy for external markets, investors, and customers [28].

The socio-economic and environmental values of B Corp act as driving and motivational forces on the overall assessment of the impact of benefits for the common good. Regarding environmental values, other tools, such as Lean, have also proven their contribution to improving sustainable business development due to the large reduction in waste produced [31] and this should be valued by the B Corp audit process. Within this framework, the B Corp movement competes alongside existing models as a new paradigm [32]. The search for competitive advantage causes those responsible for the marketing area, in particular, to start introducing corporate sustainability initiatives in search of differentiation in the markets and brand engagement with stakeholders [33]. Consequently, this may lead to high demand for certification systems such as B Corp. Tendentially, as explained by [34], female-gendered business owners are associated with a higher likelihood of obtaining B Corp certification. The authors of [35] have also stated this concerning adopting other certification systems such as ISO 9001. There seems to be a positive correlation between female managers and the adoption of tools for sustainable development.

To approach sustainability without mentioning the importance of social responsibility is an incomplete assessment of the sustainable development of companies. In addition to measuring sustainability, the B Impact Assessment indicators also measure social responsibility. In general, governmental funds and financial investment choices are allocated to companies that incorporate corporate social responsibility policies in their mission [36]. An organization with high status in terms of social responsibility and high ethical principles and practices, as is apparently the case with B Corp companies, is a reliable business partner and a reputable member of the business community [37]. Studies by [38,39] highlight that stakeholders in business markets increasingly value incorporating social responsibility measures. B Corp can act as a major partner in this, broadening commercial horizons.

According to the data available from B Lab, B Corp certification is mainly aimed at the service sector, with a small ecological footprint [5]. Quality improvement tools, such as Lean, have already demonstrated their capacity to develop sustainability indices [40,41]. Perhaps, by exploring different methodologies, B Corp can further expand its certification model to other sectors, such as manufacturing.

3. Method

3.1. Conceptual Model, Data, and Sample

The definition of the conceptual model consists of converting the elements that are intended to be analyzed in the research into a language that allows the systematic work of data collection and analysis [42]. This research aims to validate the certification model proposed by B Lab, which is used to measure companies' social, environmental, and economic impact. The B Impact Assessment comprises five indicators [5]—governance, workers, community, environment, and customers—measured through a set of variables that express the practices and outputs achieved by companies in terms of their economic, social, and environmental performance (see Figure 1).

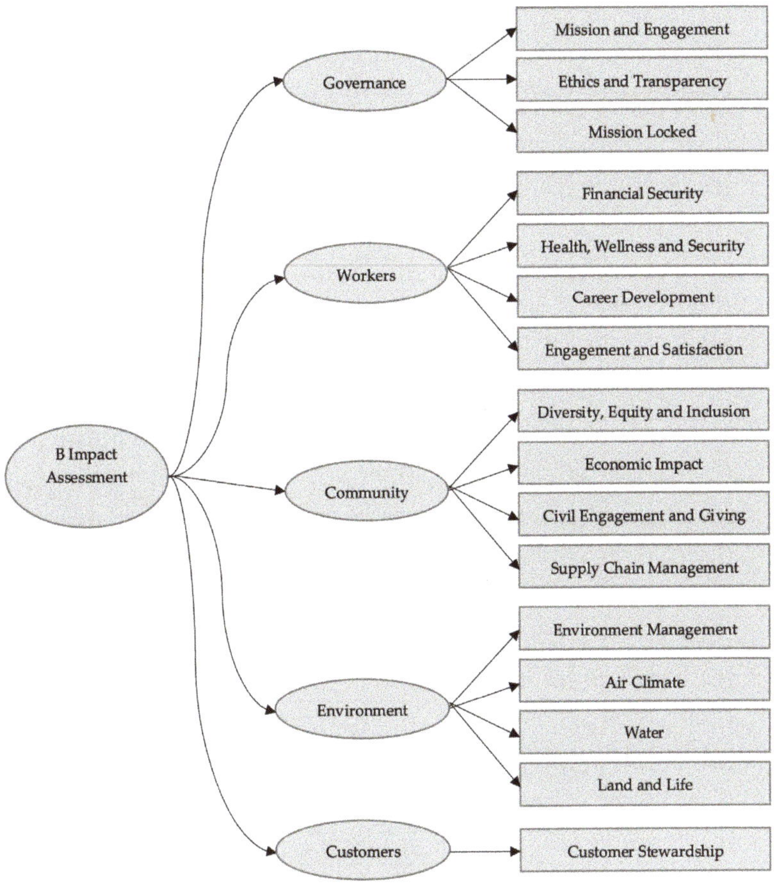

Figure 1. B Impact Assessment—measurement model.

The data from the assessment process of the certified B Corporation companies were accessed through the organization's official website (https://bcorporation.eu/directory, accessed on 28 April 2021), and this information was collected on the assessment of 2262 companies certified between January 2017 and March 2021. The data were extracted directly from the website during the period between March and May 2021, using Microsoft Excel software, which eliminates the probability of errors due to manual transcription of the data. Subsequently, a data processing method was developed through the Visual Basic programming language, where all data were neatly filled in an interactive sheet. The statistical analysis of the sample was performed using SPSS software and the confirmatory factorial analysis (CFA) using AMOS software. The variables extracted from the B Corp directory with significance for this research are mentioned in Table 1.

Table 1. B Impact Assessment (BIA) information extracted from B Corp directory.

BIA Audit Information (Variable or Indicator)		Designation
Name/ID		Commercial name of B certified company or organization.
Country		Country of origin of B certified company or organization.
City		City of origin of B certified company or organization.
Year of certification		Year in which the company or organization was certified.
Activity sector		Sector of activity in which the certified company or organization operates.
Governance scores:	"Mission and Engagement" "Ethics and Transparency" "Mission Locked"	Score attributed to the audit variables belonging to the governance indicator.
Workers scores:	"Financial Security" "Health, Wellness, and Safety" "Career Development" "Engagement and Satisfaction"	Score attributed to the audit variables belonging to the workers indicator.
Community scores:	"Diversity, Equity, and Inclusion" "Economic Impact" "Civic Engagement and Giving" "Supply Chain Management"	Score attributed to the audit variables belonging to the community indicator.
Environment scores:	"Environment Management" "Air and Climate" "Water" "Land and Life"	Score attributed to the audit variables belonging to the environment indicator.
Customers scores:	"Customer Stewardship"	Score attributed to the audit variables belonging to the customer indicator.
Final score		Summary of the final score for the 5 evaluation indicators.

The overall sample has a total of 2262 companies. However, we found that the observed variables changed over the period during which the certifications occurred, which led us to reduce their size to a constant observation period and preferably closer to the present. Therefore, this period portrays the data of B Corporation companies certified between the beginning of 2020 and March 2021.

Another reason we adjusted the initial sample size to a smaller one (556) was the missing values, since they exceeded 10%, which could create problems for us during the analysis [43]. In order not to distort the reality of the data collected on the B Impact Assessment, with the imputation of random data, the missing values were disregarded. These data are missing mainly because many of the B Corporation companies in the sample do not have employees, and therefore no point value was assigned to the observed variable.

3.2. Confirmatory Factor Analysis

B Impact Assessment is a model consisting of a set of indicators that represent complex concepts that cannot be measured directly and, as such, can be called latent variables. The measured scores are termed observed variables. This model comprises four latent variables (governance, workers, community, and environment) and one observed variable (customer stewardship), as can be seen in Table 2.

Table 2. B Impact Assessment—Latent and observed variables.

Latent Variables	Observed Variables
Governance	"Mission and Engagement" "Ethics and Transparency" "Mission Locked"
Workers	"Financial Security" "Health Wellness and Safety" "Career Development" "Engagement and Satisfaction"
Community	"Diversity Equity and Inclusion" "Economic Impact" "Civic Engagement and Giving" "Supply Chain Management"
Environment	"Environmental Management" "Air Climate" "Water" "Land and Life"
	"Customer Stewardship"

In order to validate the measurement model of B Impact Assessment, we performed confirmatory factor analysis using AMOS software, since this analysis allows us to assess the quality of the adjustment of a theoretical measurement model to the correlational structure examined between the observed variables [44]. Additionally, confirmatory factor analysis is the best-known statistical procedure for investigating relations between sets of observed and latent variables [45].

The confirmatory factor analysis influence not only the analytical aspects of the research but also the design and approach to data collection for decision making and problem solving [43]. According to the same author [43], it is essential to pay attention during the creation of the model to fundamental aspects that lead to a correct analysis, such as missing values, identification of outliers, and the construct's reliability and, most fundamentally, normality.

B Impact Assessment is assumed as a measurement model that allows the assessment of business in social and environmental dimensions. The specification of the measurement model is one of the most complex steps in multivariate analysis. The measurement model aims to identify the observed variables used to measure each of the latent variables (constructs/indicators). Thus, each latent variable is measured indirectly, reflecting consistency across multiple observed variables.

When building the measurement model, it is essential to follow some specification rules [44]: (i) the behavior of the observed variables results from the manifestation of the latent variables; (ii) the variance of the observed variables that is not explained by the latent variables is explained by specific latent factors (e.g., measurement errors); and (iii) measurement errors are usually independent.

4. Results

4.1. B Impact Assessment Global Scores

The BIA indicators have different dispersions that result from the indicators' use of different scales (see Figure 2). For example, the workers indicator presents the highest median value and the community indicator the highest score. In general, all the BIA indicators present a significant dispersion of the data, except for the governance indicator, which exhibits a higher concentration of the data.

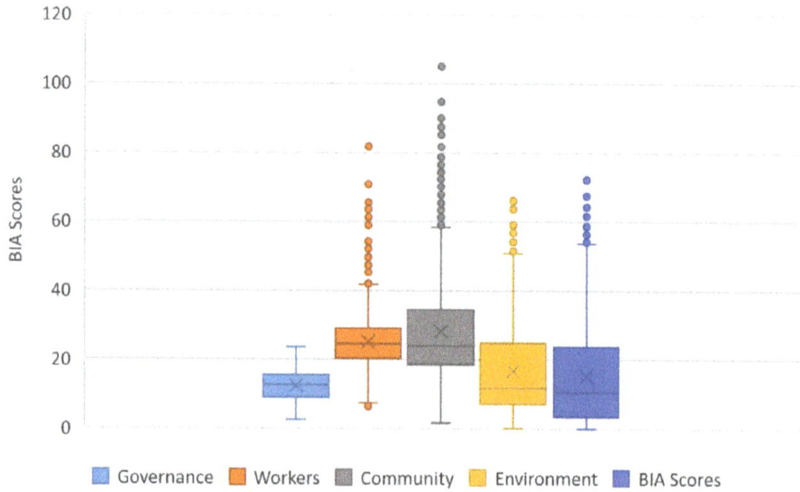

Figure 2. BIA scores (2262 companies).

Community is the evaluation B indicator with the highest number of discrepant values, followed by employees, customers, and the environment. Although there is a minimum final value for companies to obtain certification, it appears from the data analyzed that B Corp does not define an evaluation scale for its auditing process.

The data analyzed also allows us to generally verify the distribution of the final scores of the certified B companies (see Figure 3).

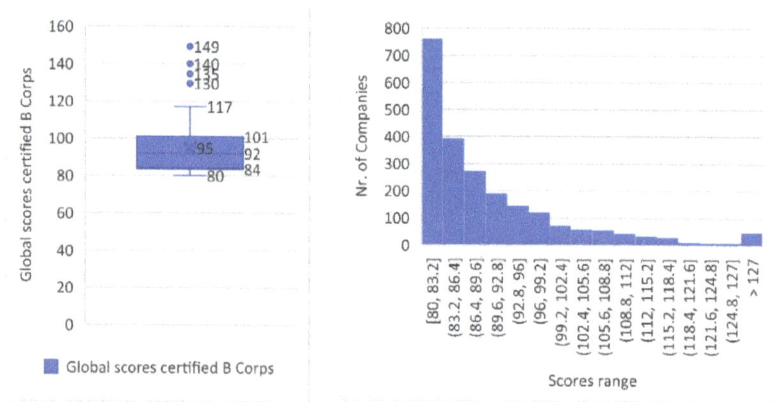

Figure 3. BIA final score (2262 companies).

As represented in the BIA final score chart above, all companies in our sample meet the minimum requirement of 80 points for B Corp certification, and most companies achieve a score between (80–83).

The countries with the highest number of certifications during our sample period are the United States with 681 companies, followed by the United Kingdom with 303 certified companies. Australia and Canada had 171 certifications during the same period. In Europe, besides the United Kingdom as mentioned above, France and Italy stand out with 87 certifications each. Portugal had 10 certifications during the period under observation.

In terms of the activity sector, organizations in the services sector with a minor ecological footprint predominate. The sectoral distribution of the B Corporation can be seen in Figure 4.

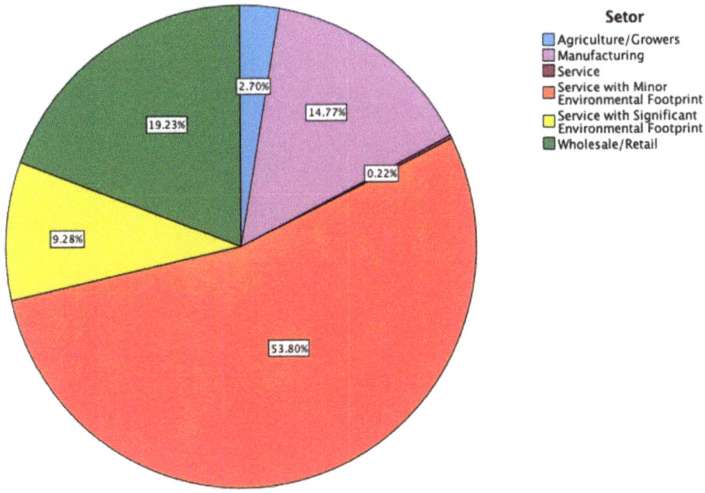

Figure 4. Sectorial distribution of B Corporation (2262 companies).

4.2. *B Impact Assessment—Measurement Model Validation*

As mentioned above, for reasons of data consistency and elimination of missing values, we reduced the initial sample to 556 cases, corresponding to companies certified in the period from January 2020 to March 2021. In Figure 5, we analyze the dispersion and median values of the observed variables present in this reduced sample.

The variable with the highest data dispersion is "Land and Life" from the environment indicator. The graph below shows that the audit B variables with a greater range of values are from community and environment indicators. The analysis of the "Mission locked" variable leads us to assume that a fixed rating scale with constant values is used to evaluate this item.

Since Figure 5 shows a high dispersion of the data, an analysis of extreme values was performed. Extreme values are "observations with a unique combination of characteristics identifiable as distinctly different from other observations" [43] (p. 64). Usually, an extreme value is an observation that presents an unusually high or low value.

To determine extreme values, we use univariate and multivariate detection methods:

(i) In terms of univariate analysis, we consider as an extreme value any observation with a value higher than $Q3 + 1.5 \times (Q3 - Q1)$ or lower than $Q1 - 1.5 \times (Q3 - Q1)$, where Q3 and Q1 represent quartiles 3 and 1, respectively [44]. When this proportion exceeds 5%, the impact on descriptive statistics is analyzed.

(ii) Concerning multivariate analysis, we used the Mahalanobis distance (D^2), which performs a multivariate assessment of each observation across a set of variables [43,44]. For large samples, an observation with a value greater than three when dividing the

Mahalanobis distance (D^2) by the degrees of freedom (df) is considered a possible multidimensional extreme value [43].

Table 3 presents univariate extreme values analysis.

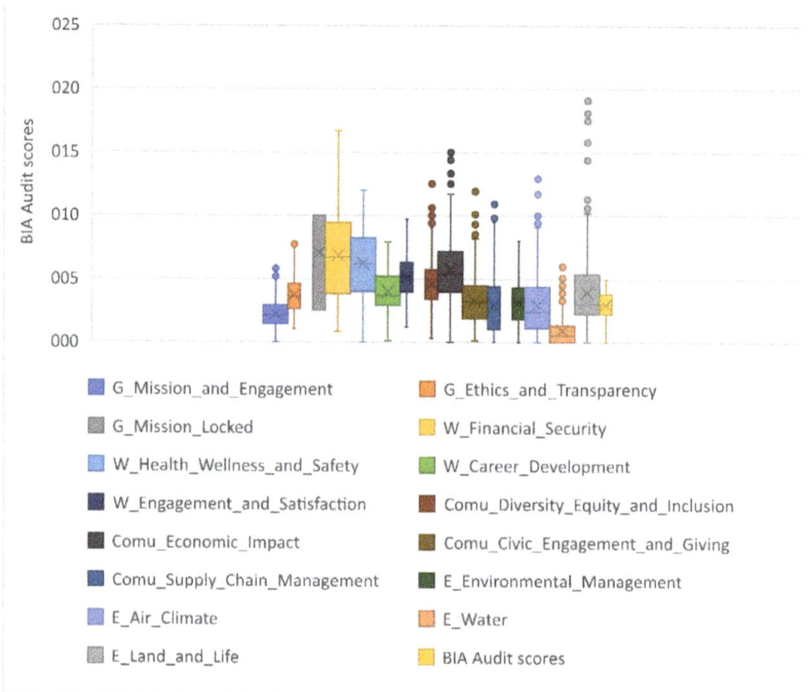

Figure 5. BIA observed variables (556 companies).

The variable water presents a percentage of extreme values higher than 5%. After analyzing the descriptive measures with and without extreme values (see Table 4), the 34 observations were removed, since the differences are significant.

Subsequently, the Mahalanobis distance (D^2) was calculated considering 15 degrees of freedom (16 observed variables minus 1). As can be seen in Table 5, there are two multivariate extreme values in the database.

For the validation of the measurement model, composed of four latent variables (governance, workers, community, and environment) and one observed variable (customer stewardship), a database of 520 companies was considered.

The five variables of the measurement model are intercorrelated. The four latent variables are measured through fifteen observed variables, and errors of measurement associated with each observed variable (e1–e15) are uncorrelated. Since latent variables are unobserved, their metric scale must be guaranteed by observed variables by setting at least one path coefficient of an observed variable or by setting the variance of the latent variable [43]. We have chosen to standardize the latent variables, setting their variance at 1.

There are several methods for adjusting measurement models. In this research, the maximum likelihood method was chosen. This method provides centered and consistent estimates and is assumed to be robust when the violation of the multivariate normality assumption of the manifest variables occurs [43].

Table 3. Univariate extreme values.

Observed Variables	N	Extreme Values [a]			
		Low	High	Total	Percentage
Mission and Engagement	556	0	10	10	1.8%
Ethics and Transparency	556	0	7	7	1.3%
Mission Locked	556	0	0	0	0.0%
Financial Security	556	0	0	0	0.0%
Health Wellness and Safety	556	0	0	0	0.0%
Career Development	556	0	0	0	0.0%
Engagement and Satisfaction	556	0	0	0	0.0%
Diversity Equity and Inclusion	556	0	11	11	2.0%
Economic Impact	556	0	12	12	2.2%
Civic Engagement and Giving	556	0	6	6	1.1%
Supply Chain Management	556	0	4	4	0.7%
Environmental Management	556	0	0	0	0.0%
Air Climate	556	0	8	8	1.4%
Water	556	0	34	34	6.1%
Land and Life	556	0	8	8	1.4%
Customer Stewardship	556	0	0	0	0.0%

[a] Number of cases outside the range (Q1 − 1.5 × (Q3 − Q1), Q3 + 1.5 × (Q3 − Q1)).

Table 4. Descriptive measures with and without extreme values.

Variable		N	Mean	Std Deviation
Water	With extreme values	556	0.9147	1.14065
	Without extreme values	522	0.7040	0.78993

Table 5. Mahalanobis distance.

Observations	D^2	D^2/df
404	81.33699	5.422466
237	47.41382	3.160921

Figure 6 presents the measurement model adjusted to a sample of 520 firms, including the values of the standardized factor weights and the individual reliability of each of the observed variables.

A summary table of regression weights with standardized coefficients and statistic tests for each of the observed variables of the B Impact Assessment is presented in Table 6.

Of all the variables observed, those that do not seem to contribute positively to the model due to the standardized factor weights are "mission locked" from the governance indicator and "supply chain management" from the community indicator. The negative standard coefficient shown in the model (see Figure 6 and Table 6) suggests that as the observed variable increases, the latent variable tends to decrease. All standardized estimates calculated are below 1.0. The governance indicator is reflected essentially in the observed variable "ethics and transparency," the workers indicator in the variable "career development," the community indicator in the variable "diversity equity and inclusion" and, finally, the environment indicator in the variable "land and life."

We then proceeded to evaluate the measurement model as a whole, using the adjustment indices (see Table 7). This analysis determines the goodness-of-fit between the hypothesized model and the sample data. The chi-square statistic with the respective degrees of freedom and the CFI and RMSEA indices are the most reported in the literature [46]. Additionally, the TLI index also stands out among other indices, with some incidence.

The CMIN/DF (chi-square/degree of freedom) is an absolute index that evaluates the quality of the model per se, without comparison with other models [44]. The CFI and TLI are relative indices of fit, since they assess the quality of the model relative to the model with the worst possible fit (independence model, in which there are no relationships between the observed variables) and/or the model with the best possible fit (saturated model, in which all the observed variables are correlated) [44]. Finally, RMSEA is a population discrepancy index that compares the model fit obtained with sample measures (sample means and variances) to the model fit that would be obtained with population measures (population means and variances) [44].

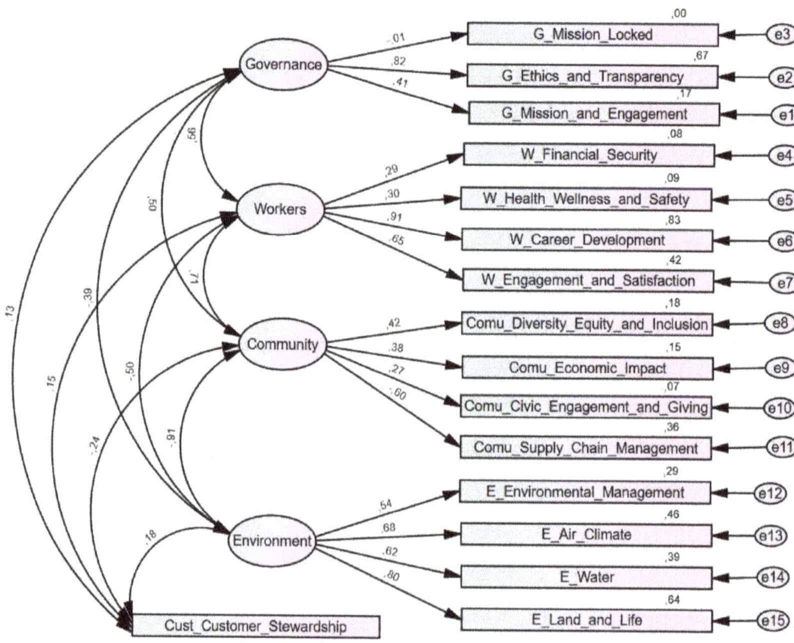

Figure 6. B Impact Assessment—Adjusted Measurement Model (original model).

Table 6. B Impact Assessment—Regression Weights (original model).

			Standard Estimates	*p*
Financial Security	<—	Workers	0.286	<0.001
Health, Wellness, and Safety	<—	Workers	0.299	<0.001
Career Development	<—	Workers	0.913	<0.001
Engagement and Satisfaction	<—	Workers	0.647	<0.001
Diversity, Equity, and Inclusion	<—	Community	0.424	<0.001
Economic Impact	<—	Community	0.384	<0.001
Civic Engagement and Giving	<—	Community	0.270	<0.001
Environmental Management	<—	Environment	0.542	<0.001
Air Climate	<—	Environment	0.679	<0.001
Water	<—	Environment	0.622	<0.001
Land and Life	<—	Environment	0.798	<0.001
Mission and Engagement	<—	Governance	0.411	<0.001
Ethics and Transparency	<—	Governance	0.819	<0.001
Supply Chain Management	<—	Community	−0.604	<0.001
Mission Locked	<—	Governance	−0.008	0.883

Table 7 presents the values obtained in the adjustment indices and the reference values referred to in the literature to consider a model with a good fit. In addition, the analyzed data allow us to verify that the BIA measurement model is outside the adjustment parameters, revealing an inferior quality of adjustment to the sample.

The variables "mission locked" from the governance indicator and "supply chain management" from the community indicator were eliminated to improve the model fit. Additionally, the modification indices were used, considering that values greater than 11 ($p < 0.001$) indicate local adjustment problems (see Table 8).

After assessing the theoretical plausibility of the modifications, the measurement errors were correlated, which led to a considerable improvement in the adjustment of the measurement model (see Figure 7 and Table 9).

Table 7. B Impact Assessment—Adjustment Indices (original model).

Adjustment Indices		Reference Values
CMIN/DF	4.992	Ratios on the order of 3:1 or lower are associated with models with good fit [43,47]
CFI	0.793	Values above 0.90 are usually associated with models with good fit [43]
TLI	0.738	Values above 0.90 indicate models with acceptable fit [47,48]
RMSEA	0.088	Values between 0.03 and 0.08 are associated with good fit, with 95% of confidence [43]

Table 8. B Impact Assessment—Modification indices.

			M.I.	Par Change
e7	<->	e13	14.155	0.378
e5	<->	e2	25.471	0.692
e5	<->	e14	17.292	0.306
e5	<->	e13	11.477	0.671
e5	<->	e10	12.260	−0.714
e5	<->	e7	31.367	0.818
e4	<->	e9	34.866	−2.270

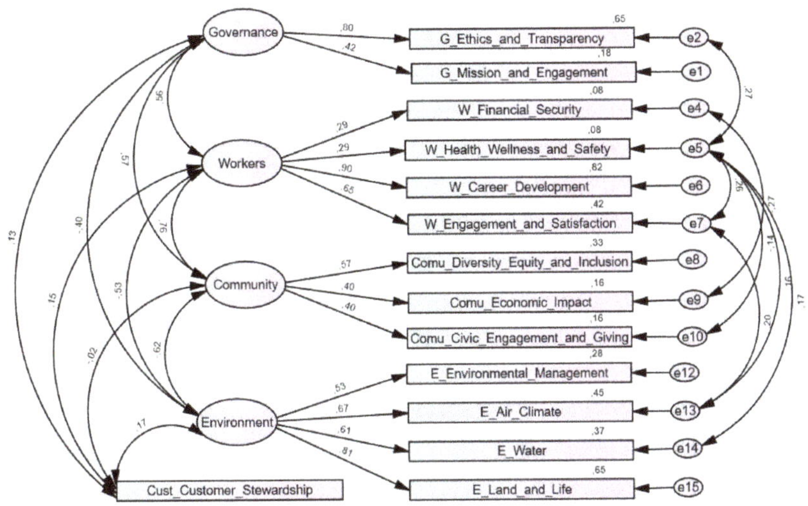

Figure 7. B Impact Assessment—Adjusted Measurement Model (modified model).

Table 9. B Impact Assessment—Adjustment Indices (modified model).

Adjustment Indices		Reference Values
CMIN/DF	3.388	Ratios on the order of 3:1 or lower are associated with models with good fit [43,47]
CFI	0.908	Values above 0.90 are usually associated with models with good fit [43]
TLI	0.863	Values above 0.90 indicate models with acceptable fit [47,48]
RMSEA	0.068	Values between 0.03 and 0.08 are associated with good fit, with 95% of confidence [43]

These modifications improved the adjustment of the BIA model, especially regarding the CFI and RMSEA indices. The remaining indices (CMIN/DF and TLI) reveal a sufferable fit of the model to data (see Table 9).

As explained in the method section, one of the fundamental assumptions of confirmatory factor analysis is data normality [43]. When the normality assumption is verified, the maximum likelihood method exhibits properties of consistency, asymptotic efficiency, and asymptotic null bias [44]. The univariate and multivariate normality analysis is shown in Table 10.

Table 10. Normality Assessment.

Variable	Min	Max	Skew	Kurtosis
Customers—Customer Stewardship	0.000	5.000	−0.293	−0.560
Governance—Mission Locked	2.500	10.000	−0.473	−1.734
Community—Supply Chain Management	0.000	10.900	0.855	0.292
Governance—Ethics and Transparency	1.000	7.800	0.563	−0.048
Governance—Mission and Engagement	0.000	6.000	0.620	0.178
Environment—Land and Life	0.000	10.200	0.989	0.383
Environment—Water	0.000	3.200	1.324	1.074
Environment—Air and Climate	0.000	12.900	1.018	0.911
Environment—Environmental Management	0.000	7.900	0.154	−0.416
Community—Civic Engagement and Giving	0.100	11.900	0.753	0.783
Community—Economic Impact	0.000	15.000	0.654	0.700
Community—Diversity, Equity, and Inclusion	0.300	12.500	0.689	0.815
Workers—Engagement and Satisfaction	1.200	9.700	−0.091	−0.444
Workers—Career Development	0.100	7.900	0.244	−0.625
Workers—Health, Wellness, and Safety	0.000	12.000	−0.148	−0.368
Workers—Financial Security	0.800	16.700	0.305	−0.817
Multivariate				3.088

According to [44], the normality assessment should be made through the analysis of the asymmetry (Sk), kurtosis (Ku), and multivariate kurtosis (KuMult) values. In confirmatory factorial analysis, we can assume that there is a severe violation of normality whenever $|Sk| > 2–3$, $|Ku| > 7–10$ and $|KuMult| > 10$ [44,47]. Only in an extreme scenario of violation of normality are the quality of the adjustment indices and parameter estimates questionable [44].

In this context, the variables fulfil the assumption of univariate and multivariate normality (skew values less than or equal to 1.3; kurtosis values less than or equal to 1 and multivariate kurtosis equal to 3.088).

Another aspect to consider in confirmatory factor analysis is the construct's reliability. Reliability is a measure of internal consistency, i.e., it measures the degree to which the different observed variables analyze the same aspect. Cronbach's alpha was used to determine the reliability of the latent variables, since it is one of the most commonly used measures to check the internal consistency of a set of items (see Table 11).

Table 11. Reliability analysis.

Latent Variables	Cronbach's Alpha
Governance	0.496 (2 items)
Workers	0.505 (4 items)
Community	0.403 (3 items)
Environment	0.723 (4 items)

Cronbach's alpha ranges between 0 and 1, and it is acceptable to aggregate items with a value greater than 0.6 [49]. However, as we can see, the latent variables present reliability problems since the values obtained for Cronbach's alpha are lower than 0.6, except for the "environment" variable, which presents a higher value (0.723). These reliability problems of the latent variables lead us to conclude that the B Impact Assessment model, as a measurement tool used by B Labs to assess companies' social and environmental impact, may not always give the same results when applied to structurally similar companies.

Several models were simulated in this research to understand which measurement model structure best fits the data. Of all the models studied, the one where it was possible to obtain the best adjustment indices was the following (see Figure 8):

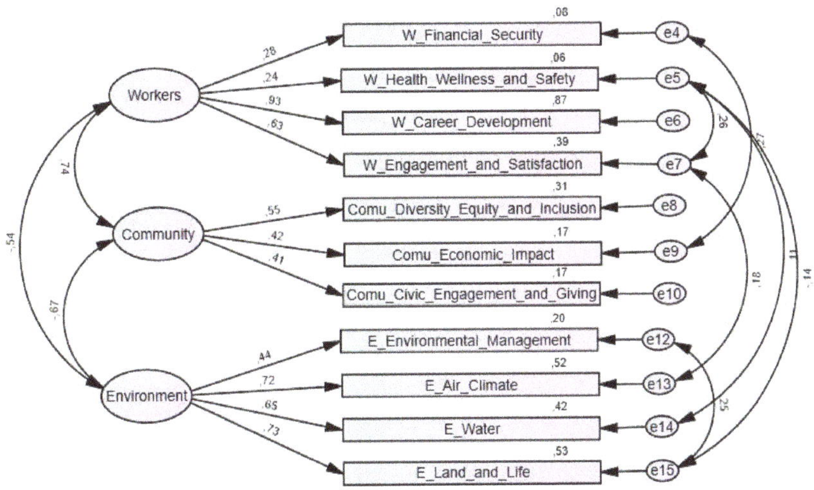

Figure 8. B Impact Assessment—Measurement Model (best fit model).

As can be seen, this model presents significant differences from the original model that was tested. First, there was a reduction in the number of latent variables from four to three by eliminating the governance indicator. Additionally, the observed variable (customer stewardship) was removed from the model.

These changes introduced in the B Impact Assessment model, according to Figure 8, allowed us to achieve the final adjustment values mentioned in Table 12. As a result, the simplified model exhibited a significantly higher quality of fit than the original model in the sample under study.

Table 12. B Impact Assessment—Adjustment Indices (best fit model).

Adjustment Indices		Reference Values
CMIN/DF	3.140	Ratios on the order of 3:1 or lower are associated with models with good fit [43,47]
CFI	0.941	Values above 0.90 are usually associated with models with good fit [43]
TLI	0.908	Values above 0.90 indicate models with acceptable fit [47,48]
RMSEA	0.064	Values between 0.03 and 0.08 are associated with good fit, with 95% of confidence [43]

5. Discussion

Public articles on B Corp certification have increased exponentially [8]. Some authors, such as Putnam Rankin and Matthews [22], challenge academics to focus their efforts on understanding this certification system by exploring all the perspectives surrounding it to answer fundamental questions about its evolution and importance in the pro-social field.

Although the body claims to be a pro-social movement that "meets the highest standards of social and environmental performance," in our research, we found companies certified by the B Corporation that achieved a high final score and were therefore classified as exceptional or extraordinary, but which then reveal weaknesses in key areas such as environmental performance.

According to the data collected, it appears that there are no minimum impact scores for the main sustainability pillars of the B Corp assessment system. This can lead to a considerable imbalance in companies that manage to achieve high values in certain assessment areas, thus obtaining the B Corporation brand seal, which, on the other hand, demonstrate profound weaknesses in important areas of sustainable development.

Our research is in line with Tabares [50]. This author states that the B Corp model is highly standardized and has similarities in different areas. However, of the five main pillars of B Impact Assessment, the only one that seems to follow a standardized assessment method is the one on governance (see Figure 2). Taking into account that the B Impact Assessment evaluates how a company's operations and business model influences their workers, community, environment, and customers, our research is also in line with Fonseca et al. [4]; Paelman, Van Cauwenberge and Vander Bauwhede [7]; Villela, Bulgacov and Morgan [28]; and Grimes, Gehman and Cao [34], among others.

Despite several attempts to adjust the B Impact Assessment model, we could not arrive at an adjusted model using all the audit variables that are part of the B Corporation certification process. Among all the models structured and studied, the only one that allowed us to reach a model with good adjustment and meeting the minimum number of three observed variables for each latent variable [43], was taking into account only the indicators of the environment, workers, and community, and therefore the governance and customers indicators are not part of this model.

Additionally, the reliability analysis allowed us to verify that the B Impact Assessment indicators present internal consistency problems, with the exception of environment. This finding should not be underestimated as it points to the possible lack of the capacity of the B Impact Assessment to provide consistent results when applied to structurally similar organizations.

Finally, it should be noted that the literature on sustainability tends to take for granted the reliability of certification practices [18]. According to [18] "we should be skeptical about the ability of certification auditing to ensure accountability of organizations with regard to sustainability, although auditing clearly plays a key symbolic role in producing order, promoting the emergence of rationality and legitimacy" (p. 345).

These findings suggest that restructuring actions should be implemented to improve the B Corp certification system, including both the measurement model and the evaluation process carried out by the evaluators of B Lab.

6. Conclusions

The announcement of certifications in sustainability is positively associated with the performance of companies [27]. In addition, B Corp certification conveys a perception of trust and effectiveness in the social, economic, and environmental fields [25], providing a seal of legitimacy for external markets, investors, and customers [28]. Nevertheless, the B Corp movement competes alongside existing sustainability models and certifications [32]. Therefore, academic research should empirically assess the B Corp model's reliability as essential for its validation and legitimization.

According to the main objective foreseen for our study, after analyzing the data related to the B assessment process of 2262 companies, we conclude that the B Corp certification model presents some weaknesses in the measurement constructs—mainly in the governance and customers indicators, which does not allow the creation of a solid and well-adjusted model. The only indicator that showed reliability, being somewhat suitable for the purpose of this research, was the assessment of the environment, with a Cronbach's alpha of 0.723.

The final values obtained on the adjustment of the B Impact Assessment model are outside the standards classified as good and suggested by the authors [43,47,48]. The only measurement model studied that could be classified as a good fit disregards the governance and customers indicators. These results further reinforce our argument about the weaknesses of the B Corp model with regard to the governance and customer evaluation indicators since it was possible to obtain a well-adjusted model without considering them. The variables observed in audit process B, "mission locked" and "supply chain management," present a negative covariance value which indicates the opposite direction of construct valuation. The variable "mission locked" has a *p*-value equal to 0.883, which statistically leads us to consider that this variable is insignificant for the model.

By interpreting the results, everything leads us to believe that B Impact Assessment does not follow a standard measurement scale with fixed maximum and minimum values. Except for the governance indicator, all other B indicators show a high dispersion of the data. The fact that there are apparently no minimum values established for each of the five B indicators can lead to a considerable imbalance in the companies that manage to attain high values in certain assessment areas. Thus, it is possible to obtain the B Corporation label, and, on the other hand, to show profound weaknesses in important areas of sustainable development, such as the environment. Factually, looking at the results, the most valued B indicator is community, and the least is governance, which reflects some personality and objectivity traits of this certification model.

The scores of B Corps are mostly close to the minimum value for obtaining the certificate (80 points), which demonstrates that a large proportion of companies have an overall performance close to the baseline requirements.

As final remarks, it can be noted that the B Corp certification model could be improved through a more accurate specification of the indicators' measurement scales. In addition, although the holistic perspective of the model can be pointed out as an advantage, it focuses on different social and environmental issues, which overall show measurement inconsistencies. Moreover, it is considered that the B Corp certification model can be strengthened by establishing procedures for monitoring the evaluators. It is known that the evaluators are a critical element of any certification process, so the implementation of systematic procedures for the selection, training, and monitoring of evaluators may also contribute to reducing measurement bias.

Furthermore, it should be acknowledged that models can have limitations, e.g., the context can be a significant source of influence, namely under highly unpredictable and unstable markets characterized by volatility, uncertainty, complexity, and ambiguity

(VUCA). Therefore, there are risks to oversimplifying the reality in which the organizations operate [51].

As future research recommendations, in-depth analyses of the certification processes of B Corp companies, collecting primary data, are suggested to find areas for improvement that may complement the results already achieved in this research.

Author Contributions: The authors confirm their contributions to this paper as follows: Conceptualization, V.S., V.L., J.C.S., L.F. and G.S.; Data curation, V.S. and V.L.; Formal analysis, V.S., V.L., J.C.S. and L.F.; Investigation, V.S., V.L., J.C.S., L.F. and G.S.; Methodology, V.S., V.L. and J.C.S.; Writing—original draft, V.S., V.L. and J.C.S.; Writing—review and editing, V.L., L.F. and G.S. All authors have read and agreed to the published version of the manuscript.

Funding: The work of the author Vanda Lima is supported by national funds, through the FCT—Portuguese Foundation for Science and Technology under the project UIDB/04728/2020.

Institutional Review Board Statement: Not applicable.

Informed Consent Statement: Not applicable.

Data Availability Statement: Data sharing not applicable.

Conflicts of Interest: The authors declare no conflict of interest.

References

1. Fonseca, L. ISO 14001:2015: An improved tool for sustainability. *J. Ind. Eng. Manag.* **2015**, *8*, 35–50.
2. Ferreira, C.d.S.; Poltronieri, C.F.; Gerolamo, M.C. ISO 14001:2015 and ISO 9001:2015: Analyse the relationship between these management systems standards and corporate sustainability. *Gest. Prod.* **2019**, *26*, e3906. [CrossRef]
3. Zimon, D.; Madzik, P.; Sroufe, R. The influence of ISO 9001 & ISO 14001 on sustainable supply chain management in the textile industry. *Sustainability* **2020**, *12*, 4282.
4. Fonseca, L.; Silva, V.; Sá, J.C.; Lima, V.; Santos, G.; Silva, R. B Corp versus ISO 9001 and 14001 certifications: Aligned, or alternative paths, towards sustainable development? *Corp. Soc. Responsib. Environ. Manag.* **2021**, *29*, 496–508. [CrossRef]
5. B Corporation, Certified B Corporation. 2021. Available online: https://bcorporation.net/ (accessed on 30 January 2021).
6. Lau, C.K. The economic consequences of business sustainability initiatives. *Asia Pacific J. Manag.* **2019**, *36*, 937–970. [CrossRef]
7. Paelman, V.; van Cauwenberge, P.; Bauwhede, H.V. The impact of B corp certification on growth. *Sustainability* **2021**, *13*, 7191. [CrossRef]
8. Diez-Busto, E.; Sanchez-Ruiz, L.; Fernandez-Laviada, A. The B corp movement: A systematic literature review. *Sustainability* **2021**, *13*, 2508. [CrossRef]
9. WCED (World Commission on Environment and Development). *Our Common Future*; Oxford University Press: New York, NY, USA, 1987.
10. United Nations. *Transforming Our World: The 2030 Agenda for Sustainable Development*; United Nations: New York, NY, USA, 2015.
11. Barbier, E.B. The Concept of Sustainable Economic Development. *Environ. Conserv.* **1987**, *14*, 101–110. [CrossRef]
12. Purvis, B.; Mao, Y.; Robinson, D. Three pillars of sustainability: In search of conceptual origins. *Sustain. Sci.* **2019**, *14*, 681–695. [CrossRef]
13. Freeman, R.E. *Strategic Management: A Stakeholder Approach*; Pitman: Boston, MA, USA, 1984.
14. Freeman, R.E.; Dmytriyev, S. Corporate Social Responsibility and Stakeholder Theory: Learning From Each Other. *Symph. Emerg. Issues Manag.* **2017**, *1*, 7–15. [CrossRef]
15. Fonseca, L.; Ramos, A.; Rosa, Á.; Braga, A.C.; Sampaio, P. Stakeholders satisfaction and sustainable success. *Int. J. Ind. Syst. Eng.* **2016**, *24*, 144–157. [CrossRef]
16. Barney, J. Firm Resources and Sustained Competitive Advantage. *J. Manag.* **1991**, *17*, 99–120. [CrossRef]
17. Di Maggio, P.J.; Powell, W.W. The iron cage revisited institutional isomorphism and collective rationality in organizational fields. In *Economics Meets Sociology in Strategic Management*; Emerald Group Publishing Limited: Bingley, UK, 2000.
18. Boiral, O.; Gendron, Y. Sustainable Development and Certification Practices: Lessons Learned and Prospects. *Bus. Strateg. Environ.* **2011**, *20*, 331–347. [CrossRef]
19. Paelman, V.; van Cauwenberge, P.; Bauwhede, H.V. Effect of B corp certification on short-term growth: European evidence. *Sustainability* **2020**, *12*, 8459. [CrossRef]
20. Poponi, S.; Colantoni, A.; Cividino, S.R.S.; Mosconi, E.M. The Stakeholders' perspective within the B corp certification for a circular approach. *Sustainability* **2019**, *11*, 1584. [CrossRef]
21. de Mendonca, T.; Zhou, Y. What does targeting ecological sustainability mean for company financial performance? *Bus. Strateg. Environ.* **2019**, *28*, 1583–1593. [CrossRef]
22. Rankin, C.P.; Matthews, T.L. Patterns of B Corps Certification: The Role of Institutional, Economic, and Political Resources. *Societies* **2020**, *10*, 72. [CrossRef]

23. Moroz, P.W.; Branzei, O.; Parker, S.C.; Gamble, E.N. Imprinting with purpose: Prosocial opportunities and B Corp certification. *J. Bus. Ventur.* **2018**, *33*, 117–129. [CrossRef]
24. Gehman, J.; Grimes, M. Hidden badge of honor: How contextual distinctiveness affects category promotion among certified b corporations. *Acad. Manag. J.* **2017**, *60*, 2294–2320. [CrossRef]
25. Bianchi, C.; Reyes, V.; Devenin, V. Consumer motivations to purchase from benefit corporations (B Corps). *Corp. Soc. Responsib. Environ. Manag.* **2020**, *27*, 1445–1453. [CrossRef]
26. Santos, G.; Sá, J.C.; Félix, M.; Barreto, L.; Carvalho, F.; Doiro, M.; Zgodavová, K.; Stefanović, M. New needed quality management skills for quality managers 4.0. *Sustainability* **2021**, *13*, 6149. [CrossRef]
27. Feng, Y.; Lai, K.h.; Zhu, Q. Legitimacy in operations: How sustainability certification announcements by Chinese listed enterprises influence their market value? *Int. J. Prod. Econ.* **2020**, *224*, 107563. [CrossRef]
28. Villela, M.; Bulgacov, S.; Morgan, G. B Corp Certification and Its Impact on Organizations Over Time. *J. Bus. Ethics* **2019**, *ahead of print*. [CrossRef]
29. Viswanathan, L.; Varghese, G. Greening of business: A step towards sustainability. *J. Public Aff.* **2018**, *18*, e1705. [CrossRef]
30. Alves, W.; Silva, Â.; Rodrigues, H.S. Green Practices as a Path towards the Sustainability: Evidence from Portuguese Companies. *Bus. Syst. Res. J.* **2020**, *11*, 7–20. [CrossRef]
31. Silva, S.; Sá, J.C.; Silva, F.J.G.; Ferreira, L.P.; Santos, G. Lean Green—The Importance of Integrating Environment into Lean Philosophy—A Case Study. *Lect. Notes Netw. Syst.* **2020**, *122*, 211–219.
32. Nigri, G.; del Baldo, M. Sustainability reporting and performance measurement systems: How do small- and medium- sized benefit corporations manage integration? *Sustainability* **2018**, *10*, 4499. [CrossRef]
33. Derqui, B. Towards sustainable development: Evolution of corporate sustainability in multinational firms. *Corp. Soc. Responsib. Environ. Manag.* **2020**, *27*, 2712–2723. [CrossRef]
34. Grimes, M.G.; Gehman, J.; Cao, K. Positively deviant: Identity work through B Corporation certification. *J. Bus. Ventur.* **2018**, *33*, 130–148. [CrossRef]
35. Sá, J.C.; Amaral, A.; Barreto, L.; Carvalho, F.; Santos, G. Perception of the importance to implement ISO 9001 in organizations related to people linked to quality—An empirical study. *Int. J. Qual. Res.* **2015**, *13*, 1055–1070. [CrossRef]
36. Hoepner, A.G.F.; Schopohl, L. State Pension Funds and Corporate Social Responsibility: Do Beneficiaries' Political Values Influence Funds' Investment Decisions? *J. Bus. Ethics* **2020**, *165*, 489–516. [CrossRef]
37. Memon, K.R.; Ghani, B.; Khalid, S. The relationship between corporate social responsibility and employee engagement-a social exchange perspective. *Int. J. Bus. Sci. Appl. Manag.* **2020**, *15*, 1–16.
38. Badía, G.; Cortez, M.C.; Ferruz, L. Socially responsible investing worldwide: Do markets value corporate social responsibility? *Corp. Soc. Responsib. Environ. Manag.* **2020**, *27*, 2751–2764. [CrossRef]
39. Fonseca, L. Strategic Drivers for Implementing Sustainability Programs in Portuguese Organizations–Let's Listen to Aristotle: From Triple to Quadruple Bottom Line. *Sustain. J. Rec.* **2015**, *8*, 136–142. [CrossRef]
40. Sá, J.C.; Vaz, S.; Carvalho, O.; Lima, V.; Morgado, L.; Fonseca, L.; Doiro, M.; Santos, G. A model of integration ISO 9001 with Lean six sigma and main benefits achieved. *Total Qual. Manag. Bus. Excell.* **2022**, *33*, 218–242. [CrossRef]
41. Teixeira, P.; Sá, J.C.; Silva, F.J.G.; Ferreira, L.P.; Santos, G.; Fontoura, P. Connecting lean and green with sustainability towards a conceptual model. *J. Clean. Prod.* **2021**, *322*, 129047. [CrossRef]
42. Quivy, R.; van Campenhoudt, L. *Manual de Investigação em Ciências Sociais*, 2nd ed.; Gradiva: Lisboa, Portugal, 1998.
43. Hair, J.F.; Black, W.C.; Babin, B.J.; Anderson, R.E. *Multivariate Data Analysis—A Global Perspective*, 7th ed.; Pearson Education: Upper Saddle River, NJ, USA, 2010.
44. Marôco, J. *Análise de Equações Estruturais—Fundamentos Teóricos, Software e Aplicações*; ReportNumber, Lda: Pêro Pinheiro, Portugal, 2010.
45. Byrne, B.M.; Taylor, R.; York, N. *Structural Equation Modeling with AMOS. Basic Concepts, Applications, and Programming*, 2nd ed.; Taylor & Francis: New York, NY, USA, 2010.
46. Jackson, D.L.; Gillaspy, J.A.; Purc-Stephenson, R. Reporting Practices in Confirmatory Factor Analysis: An Overview and Some Recommendations. *Psychol. Methods* **2009**, *14*, 6–23. [CrossRef]
47. Kline, R.B. *Principles and Practices of Structural Equation Modelling*, 4th ed.; Guilford Press: New York, NY, USA, 1998.
48. Xia, Y.; Yang, Y. RMSEA, CFI, and TLI in structural equation modeling with ordered categorical data: The story they tell depends on the estimation methods. *Behav. Res. Methods* **2019**, *51*, 409–428. [CrossRef]
49. Pestana, M.H.; Gageiro, J.N. *Análise de Dados para Ciências Sociais. A Complementaridade do SPSS*, 5th ed.; Edições Sílabo: Lisboa, Portugal, 2008.
50. Tabares, S. Certified B corporations: An approach to tensions of sustainable-driven hybrid business models in an emerging economy. *J. Clean. Prod.* **2021**, *317*, 128380. [CrossRef]
51. Fonseca, L. The EFQM 2020 model. A theoretical and critical review. *Total Qual. Manag. Bus. Excell.* **2021**, 1–28. [CrossRef]

sustainability

Article

Identifying Barriers in the Implementation of Agile Methodologies in Automotive Industry

Daniel Soares [1], Francisco José Gomes da Silva [1,2,*], Sandra Cristina Faria Ramos [1], Konstantinos Kirytopoulos [3], José Carlos Sá [1,2] and Luís Pinto Ferreira [1,2,*]

1 School of Engineering, Polytechnic of Porto, ISEP, 4249-015 Porto, Portugal; 1130482@isep.ipp.pt (D.S.); sfr@isep.ipp.pt (S.C.F.R.); cvs@isep.ipp.pt (J.C.S.)
2 INEGI—Instituto de Ciência e Inovação em Engenharia Mecânica e Engenharia Industrial, 4200-465 Porto, Portugal
3 School of Mechanical Engineering, National Technical University of Athens, 15780 Zografou, Greece; kkir@mail.ntua.gr
* Correspondence: fgs@isep.ipp.pt (F.J.G.d.S.); lpf@isep.ipp.pt (L.P.F.)

Abstract: Projects have grown into more dynamic and complex endeavors, and agile project management should be considered as a way to deal with them. This is a novel study in this field, because the implementation of agile project management in the automotive industry was not explored so far, thus, this work intended to fill this gap, by identifying barriers in the implementation of agile methodologies in project management regarding the automotive industry. This was conducted through a questionnaire survey, which was developed and distributed to 148 manufacturing companies of components for the automotive industry, out of a total number of companies of 240, and 56 complete answers were obtained (23.33%). Statistical analyses were performed using a Kruskal–Wallis test, a Mann–Whitney test, and Spearmen's correlation. A real picture of the implementation of agile project management in the Portuguese automotive industry is depicted through this work. 'Organizational', 'Knowledge and Technology', 'Institutional', and 'Financial' barriers are found to be the most important. However, in overcoming these barriers, companies can be more sustainable in economic, environmental, and social terms. Recommendations on how to overcome these barriers were presented, and a framework sequencing these recommendations was presented, leading to an effective implementation of agile methodologies. It starts with the willingness of the company, and all collaborators, to adopt the agile methodologies, looking for the agile values as an input in order to achieve a competitive advantage. It is followed by an initial investment, which intends to attain the deployment of an agile team, composed of highly skilled collaborators with a clear understanding of the agile objectives, who disseminate knowledge about agile methodologies to the other collaborators, increasing their ability to implement agile methodologies in project management. This team should work and develop frameworks and workflows, according to each company's characteristics and environment. The studied aspects can be replicated in other countries, and allow a comparison of the situation between countries, trying to correlate the culture of each country with the ability to implement agile methodologies, among other aspects, such as economical level of the companies, type of production, and the commitment of the collaborators to improve processes and create competitive advantages with which to face competitors.

Keywords: agile project management; agile methodologies; barriers; enablers; automotive industry; sustainability in industry

Citation: Soares, D.; da Silva, F.J.G.; Ramos, S.C.F.; Kirytopoulos, K.; Sá, J.C.; Ferreira, L.P. Identifying Barriers in the Implementation of Agile Methodologies in Automotive Industry. *Sustainability* **2022**, *14*, 5453. https://doi.org/10.3390/su14095453

Academic Editors: Felix T. S. Chan and Ali Bahadori-Jahromi

Received: 23 February 2022
Accepted: 29 April 2022
Published: 1 May 2022

1. Introduction

The automotive sector is a vital sector in the worldwide economy. It contributes to the growth of other related activities, such as metallurgy, plastic/rubber processing, electronics, textiles, etc., and consecutively empowers global industrialization [1]. The outstanding contribution of the automotive industry (AI) to technological advancement was the establishment of full-scale mass production. The increasing competitive pressure of the market emphasized the ability of industrial firms to improve the indicators of quality, cost, and time-to-market of new products. The management of the increasingly complex products became more and more important. This brought attention to investment in human resources to manage tasks following previously defined concepts and organizational frameworks, through project management (PM) activities, even under rigid, stable, and inflexible capabilities [2,3]. The globalization and expanded markets increased the dynamism and complexity of projects, leading to the shift from traditional PM to agile project management (APM) [4,5]. The implementation of agile methodologies (AM) faced various barriers, and there is extensive literature regarding this matter in the information technology (IT) field, where APM emerged [6]. However, the literature background is much less extensive regarding the manufacturing sectors, and almost nothing exists regarding the automotive industry (AI) [7].

Despite an intense search, no works were found that focused directly on identifying barriers in the implementation of APM in the automotive industry, or exploring eventual gains in sustainability for the companies. There are several works focused on agile software development [8,9]. Thus, this was the main motivation and novelty of this work. This research focused on this literature gap, by seeking to define and characterize the AI regarding the implementation of AM through a questionnaire survey. In particular, the aim was to answer the following research questions:

- Do the AI companies need to implement AM?
- Are the AI companies predisposed to AM?
- What are the main barriers for implementing AM in the AI?
- What are the main enablers for implementing AM in the AI?

Thus, it was intended to identify the main barriers impeding component-producing companies from implementing AI, and establish new strategies to overcome the difficulties of these companies in adopting agile project management methodologies. Thus, it is possible to ensure a more sustainable future for their organization, and corresponding relationship with stakeholders.

2. Literature Review

2.1. *Agile Project Management*

The AI dramatically increased the pace of launching new products. As a direct consequence, automotive companies face an emerging challenge: to increase the frequency, reliability, and profitability of the innovations resulting from research. At the same time, they need to keep their ability to develop more vehicles than ever, and maintain tight constraints on quality, cost, and lead time. Fritzsche [10] argues that such a strategic challenge called for a deep transition in car manufacturers' product design processes. However, Beaume et al. [2] state that these organizations became reluctant to apply disruptive concepts regarding organizational structure. Taking this into attention, San Cristóbal et al. [4] claim that as projects have grown in complexity, the ways of managing them should be reconsidered. This evolution is studied by Sanchez et al. [5], and these authors refer to some projects that streamlined the traditional PM, and transitioned to a new agile approach, namely agile project management (APM). This is studied by Špundak et al. [11] and Highsmith et al. [12], who point out that agility is the word that differentiates APM from traditional PM. In fact, agility deals with constant innovation, product adaption, shortening delivery times, adjustment of processes and people, quality, and reliable results. Hence, Beck et al. [13] show that APM evolved since the creation of an agile manifesto for software development, in 2001. Furthermore, APM proposes a series of agile methods, practices, and tools used today, as pointed out by Conforto et al. [6].

Agile project management is also studied in terms of sustainability in several other industrial and service sectors, but not in the automotive industry. Silva et al. [14] develop frameworks creating the corresponding linkages between the tools and artifacts of the

APM, and the three pillars of sustainability. They conclude that implementing APM practices leads to increased sustainability of the companies in the economic, environmental, and social aspects. Obradović et al. [15] also study the relationship between APM practices and sustainability, concluding that the first has direct repercussions on the second. Žužek et al. [16], taking into account the SMEs Slovenian panorama, also study the implications of APM practices on sustainability, but outside of the software sector. They conclude that, even in small companies and out of the software area of work, the effect is clearly felt. Thus, it is assumed that implementation of APM practices also induces clear gains in the sustainability of companies. In the work by Sharma et al. [17], who study the effect of agility in the industry, several definitions of agility linked to production systems are found, as well as some barriers to the implementation of these agile systems in manufacturing.

2.2. *Agile Methodologies*

As traditional PM has its own characteristics, methods, and tools (program evaluation and review technique (PERT), critical path method (CPM), etc.), APM also presents some specific methods. There are various techniques under the AM umbrella, with some characteristics in common: they are all iterative, incremental, and evolutionary; and customer involvement is mandatory [18]. Scrum is one of the most common agile techniques, being a single team iterative process framework used to manage product development. Scrum is based on a variety of concepts such as customer feedback, daily scrum meetings, product backlog, sprint backlog, sprints, and being delivery-ready after each sprint [19]. As defined by PMI and Agile Alliance [20], these sprints are timeboxes of one month or less, with regular durations, where the potentially releasable increment of a product is developed. The team responsible for this method consists of the product owner, who tries to maximize the value of the product; the scrum team, which is a self-organized and cross-functional team, who have everything they need inside the team to deliver a working product; and the scrum master, who ensures that the scrum process is sustained and guarantees the scrum team complies with the rules. The scrum method allows for a reduction in planning overhead, due to its flexibility and easy adaptability to any changes in stakeholders' needs, at any stage in product development. It focuses on developing customer relationships to increase product quality, and improvement in performance. Each short cycle/sprint enables the release of short prototypes, so that the customer can monitor the development and provide continuous feedback [14,21].

There is a relation between lean, agile, and kanban methods, since these last two are descendants of lean thinking. Freitas et al. [22] give a good description of this methodology, stating that the word kanban is translated to 'card' or 'visual sign', because kanban boards promote the visualization of the system workflow for everyone. This information is structured in columns that represent the states of the work, for example, 'to do', 'doing', and 'done'. However, it can adapted to any other state needed by the team. The cards used in the kanban method provide a clear understanding of the workflow, as well as bottlenecks, in overall conditions. Both scrum and kanban have the customer as a focus when responding quickly to requests, are highly adaptive, and collaborative, fitting self-managing teams very well. Originally, the transition from the scrum to kanban was designed as scrumban. As new agile methods emerged, it evolved into a hybrid framework where teams used scrum as a framework, and kanban for process improvement. In scrumban, the work was structured into small sprints, and took advantage of the kanban boards to visualize and monitor the work. The tasks were placed on the kanban board, and the team managed its work within the work in progress limits. Daily meetings happened to maintain cooperation between the team, removing impediments. There were no predefined roles in this methodology—the team maintained their current roles. It eliminated the planning activities and velocity measurement, and focused on smooth flow, minimizing work in progress.

2.3. Barriers and Enablers in the Implementation of Agile Methodologies

A barrier means any factor that hinders, affects, or resists the implementation of a certain action, resulting in its delay or obstruction. Conversely, enablers are factors that facilitate, help, accelerate, or encourage the completion of that action. An agile methodology is any practice or method correlated to an APM action that contributes to the execution of a process, and that may employ one or more methods and tools [22]. Despite the extensive literature regarding the application of APM in the software industry, there is a lack of empirical studies on other types of industries, sectors, and projects [23]. Bohem and Turner [24] explore management challenges in implementing agile processes in traditional development organizations and frame the question: 'How do you merge agile, lightweight processes with standard industrial processes without either killing agility or undermining the years you have spent defining and redefining your systems?'. In order to expand the research from the IT field to the AI, it is necessary to investigate similarities between these sectors. Moreover, it is also necessary to explore eventual barriers that are observed in the AI, and previously identified in the IT sector. Two fields are considered as common regarding IT and the AI: cleaner production and lean manufacturing. As defined by Silva and Gouveia [25], cleaner production is an approach that intends to reduce environmental impacts, while lean manufacturing focuses on the reduction of costs and time. Both aspects have a direct influence on the product and market. These tools tend to help manufacturing processes become more sustainable.

From the available literature, it is possible to conclude that agile methods are theoretically applicable in most industries, and prove their success in practice, specifically in large organizations [7,26]. The concept of agile manufacturing evolved from lean management, and enabled the organizations to react and proact to the unpredictable and diversified market changes [27]. The barriers found in the literature regarding the above-mentioned sectors are presented in Table 1. The enablers, possible solutions, or recommendations found in the literature are presented in Table 2.

Table 1. Barriers found in the literature and employed in the questionnaire.

Literature Reference	Barrier(s)
Vermunt et al. [28]	Internal; external; financial; organizational; knowledge and technology; supply chain and market; institutional.
Potdar et al. [29]	Improper competency management.
Silva and Gouveia [25]; Potdar et al. [29]; Sindhwani et al. [30]	Governmental policies and support.
Ghani and Bello [31]; Salinas et al. [32]; Potdar et al. [29]; Sindhwani et al. [30]	Ineffective customer relationship.
Vermunt et al. [28]; Caldera et al. [33]; Ghani and Bello [31]	Lack of knowledge and skills.
Silva and Gouveia [25]; Vermunt et al. [28]; Potdar et al. [29]	Ineffective supply chain.
Ghani and Bello [31]; Caldera et al. [33]	Existent organizational culture.
Caldera et al. [33]; Tan et al. [34]	Time constraints.
Caldera et al. [33]; Kostić [35]	Risk.
Ghani and Bello [31]; Potdar et al. [29]	Stakeholders' attitude.
Added through the authors' personal experience and industrial knowledge	Absence of immediate quantifiable benefits; lack of project team flexibility; project lead time not critical; not applicable to our product; organization not able to apply AM; flexibility is not a priority; there is no time to think about that; staff not prepared to AM; lack of government benefits; change predisposition.

Table 2. Enablers, possible solutions or recommendations, employed in the questionnaire.

Literature Reference	Enablers, Possible Solutions or Recommendations
Ghani and Bello [31]; Vermunt et al. [28]	Organizational support
Ghani and Bello [31]; Silva and Gouveia [25]	Stakeholder collaboration
Ghani and Bello [31]	Investment in training; lean relationship

2.4. Hypotheses Formulation

In the production of components for the automotive industry, a relatively high degree of automation is normally used [36], which must be adapted to the quantity produced per batch, and the variety of products carried out on each production line [37]. Thus, given the series produced and the diversity of products manufactured, each company must adopt the most appropriate manufacturing system [38]. This has direct implications for project management, as a mass-manufactured product may have a line dedicated only to that product, with fixed automation. On the other hand, a family of products requires a flexible production system [39]. Agile project management is much more useful in the latter case, and this gives rise to the first hypothesis considered in this work: "The company production type influences AM implementation". The lifetime of each product in production is also an extremely important factor in defining the manufacturing process and, consequently, in the project [40]. Avery short life cycles implies the need for greater agility, while longer cycles are managed through a conventional management system. This depends on the relationship with stakeholders, and the time-to-market required by customers [41]. Moreover, this defines the second hypothesis considered in this work: "The AM implementation has an association with the company's degree of change?". Decision making for agile project management implies that top management have a deep understanding of the tools normally used, and know how to manage them with dexterity [29]. Thus, a third hypothesis is formulated: "Improper competence management affects AM implementation". However, there has to be an entire team (or teams) that conveniently operationalizes the implementation of agile management. To this end, the teams should be initially sensitized, and presented with the advantages and methodologies to be adopted, so that there is a predisposition to change from a traditional philosophy to an agile methodology [16]. Thus, a fourth hypothesis is formulated: "The change predisposition influences AM implementation". However, the results of adopting an agile methodology do not always appear immediately, and are not always perfectly quantifiable. The doubt regarding the level of benefit that results for the company and its employees could be a factor that negatively affects the teams. In fact, it is theoretically impossible to predict what level of benefits everyone will see, and when, through the implementation of agile methodologies [22]. Thus, a new hypothesis emerges: "The absence of immediate quantifiable benefits influences AM implementation". In sum, the following hypotheses were considered in this study:

H1. The company production type influences AM implementation.

H2. The AM implementation has an association with the company's degree of change.

H3. The improper competency management affects AM implementation.

H4. The change predisposition influences AM implementation.

H5. The absence of immediate quantifiable benefits influences AM implementation.

3. Methodology

This study focused on the automotive industry, because this sector was responsible for the sharp growth in Portuguese exports. In addition, the automotive industry is highly dynamic, in terms of launching new products, which implies continuous management of new products and projects. The sector still presents high standards of demand, thus benefitting more intensely from the advantages brought by agile project management. The automotive industry is subject to strict quality standards and high supplier accountability, which is also in line with some of the benefits brought by agile project management. Furthermore, it is a sector extremely concerned with sustainability, which is why it fits perfectly into the principles that guide agile project management. The study focused on this sector, as there was no study in this area for this sector. Furthermore, as it is an industry that forms the basis of several highly industrialized countries, it allows similar studies to be developed for the same sector in other countries, thus, sustaining its scalability.

The adopted research methodology utilized a well-established pattern used in similar studies [7,32,42]. Figure 1 shows the methods used, the research flow, and the expected outcomes. In sum, after careful investigation work, the gap that gave rise to the work was determined. Barriers to AM implementation in other sectors were identified, namely in software development. Based on these barriers, a questionnaire was developed. A compilation of the results, and corresponding statistical treatment, was carried out. Based on these results, a flowchart was designed for an easier implementation of AM and the recommendations and enablers are pointed out.

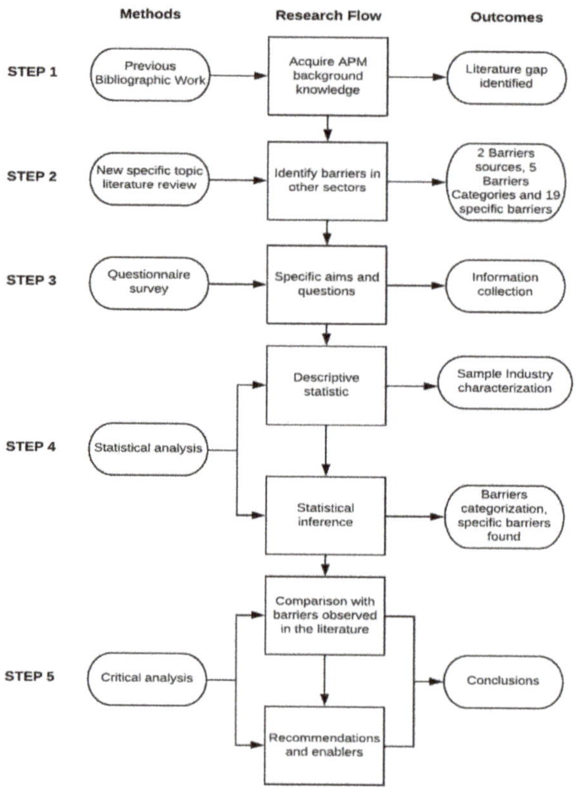

Figure 1. Flowchart with the methodology used, research flow, and expected outcomes.

3.1. Research Instrument

The questionnaire included 38 questions of multiple-choice, checkboxes, scale, and open type questions. The scale questions were drawn using a Likert-type scale. The questionnaire also presented some cross-check questions in order to verify the coherence of the answers. Table 3 shows details about the questions and variables in the questionnaire.

Table 3. Questionnaire framework.

Groups	Questions	Variables
Respondent characterization	1–5	Familiar with APM; age; department; job function.
Company characterization	6–11	Region; employees (company size), organizational structure; exportation volume; product requirements.
Manufacturing process characterization	12–19	Production type; production changes and purposes influence; improve flexibility; management involvement; manufacturing system; influence factors; quality criteria.
Product development process and PM	20–23	Product development process; PM approach; project and product development criteria; project changes.
APM environment	24–32	APM transition; APM culture; AM application; departments applying AM; APM certification; PM team; minimum qualification; agile techniques and tools; agile techniques; tools contribution.
Barriers in the implementation of AM	33–37	Barriers category; barriers source; specific barriers; other impediments; AM implementation difficulty.
Enablers for the implementation of AM	38	AM implementation enablers.

To ensure its compliance, the questionnaire was evaluated by a panel of five experts in the fields of mechanical engineering, PM, and statistical analysis.

3.2. Survey Sample

The sample size used in the study was determined using the equation [43]:

$$n \geq \frac{\frac{z^2 \cdot p \cdot (1-p)}{e^2}}{\frac{N-1}{N} + \left(\frac{z^2 \cdot p \cdot (1-p)}{N \cdot e^2}\right)} \quad (1)$$

where n is the sample size, N is the population size, p represents the population proportion of companies that implement agile methodologies, e is the sampling error, and z is the $(1 - \propto/2) \times 100\%$ percentile of a standard normal distribution, with $(1 - \propto) \times 100\%$, representing the confidence level. The sample size obtained with the previous equation ensures that the estimate of the proportion of the population of companies implementing agile methodologies are within e of the true population proportion.

According to the Portuguese Manufacturers Association for the AI sector, in 2019, there were 240 companies operating in Portugal related to the manufacture of components for the AI [44], who represent the theoretical study population (sample). As no pilot study was conducted, nor was there prior information on the percentage of companies implementing AM, in the calculation of the sample size it was considered $p = 0.5$. This value leads to the largest sample size and, consequently, the highest probability of hitting the target.

The formula presented above, with $N = 240$, $p = 0.5$, $z = 1.96$ (for 95% confidence level), and a maximum sampling error of 5%, leads to $n \geq 148$. Of the 148 questionnaires sent out, 56 were completed and returned to the authors. All responses were valid for further analysis. Unfortunately, the high non-response rate reduced sample size for 56 affected the sampling error, which became 11.5%.

3.3. Data Analysis

Data collected was analyzed using two different approaches: descriptive analysis and inferential analysis. In descriptive analysis, bar graphs with percentages were used. The inferential analysis was used to generalize the sample. This analysis included comparisons among groups, and a correlation analysis. Group comparisons according to the findings related to the barriers in the implementation of AM in the AI were performed using the Mann–Whitney U test, when comparing two groups based on ordinal variables. Otherwise, the Kruskal–Wallis test (with multiple comparisons if the test results in a significant result) was used when the number of groups to be compared was at least three. Spearman's correlation analysis was performed to assess the association between two ordinal variables.

Statistical analyses were performed using R/RStudio software version 1.0.153. Hypothesis tests with a p value of 0.05 or lower were considered as statistically significant results.

4. Results

4.1. Respondents' Characterization

In this survey, almost 85% of the respondents are familiar with APM, which is a very substantial percentage. Almost all respondents (98.2%) have a higher education degree, and about 80% of respondents belong to project management and product development (PM & PD), management, and production departments. These departments are directly linked to this study. The summary of the results regarding the respondents' characterization is presented in Figure 2.

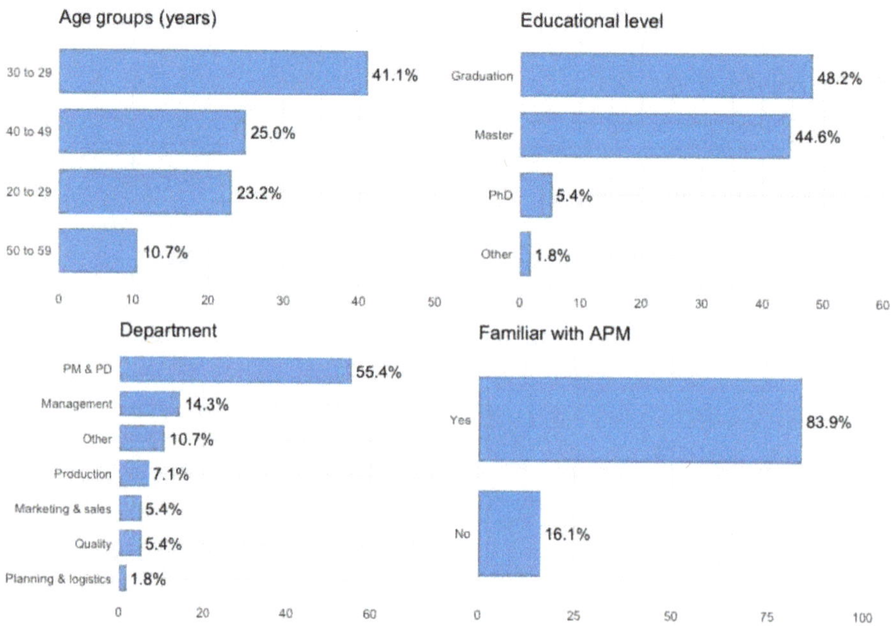

Figure 2. Respondents' characterization.

4.2. Companies' Characterization

Figure 3 depicts a summary of the results for variables directly linked to the company' characterization. About 70% of the inquired companies are large and relevant companies in this sector, with almost 90% claiming high levels of exportation (>75%). Moreover, about 84% are subjected to imposed requirements from original equipment manufacturers (OEMs). This reveals that these companies present both international relevance and dependency. Regarding the organizational model, the functional structure is the most commonly adopted approach by Portuguese companies within the automotive sector. In this organizational model, specialties do not interfere each other, usually being used in stable environments that do not experience quick and constant changes, either internally and externally. This structure causes difficulties of integration, communication, and coordination between sectors, since each team is independent in terms of decision making and activities performed. In addition, the company's employees do not retain a global vision of the business, believing that they are uniquely focused and specialized in their work area. This circumstance avoids better alignment of them with the company strategic objectives, and is not the best departure point for AM implementation.

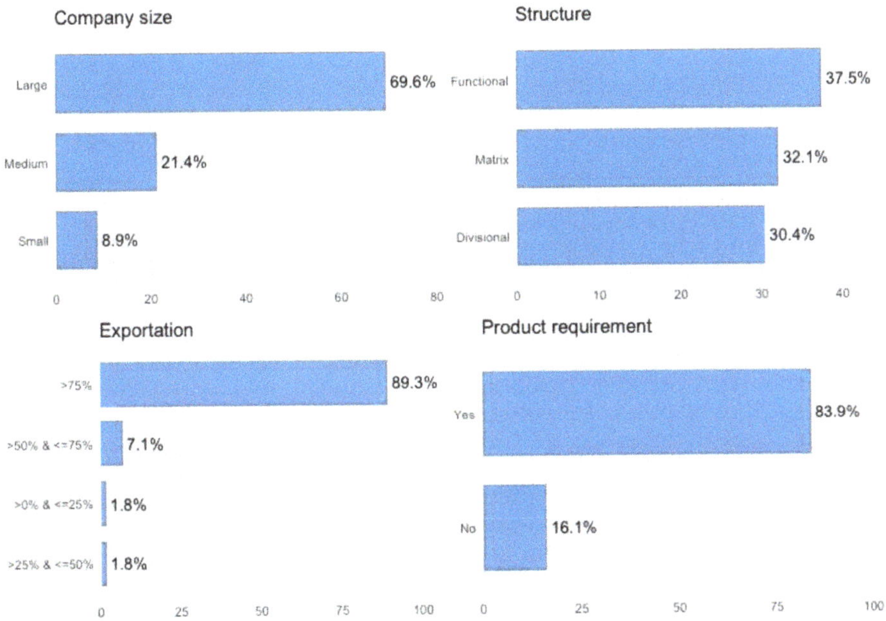

Figure 3. Companies' characterization.

4.3. Manufacturing Processes' Characterization

Regarding the production type, mass production is the main manufacturing system mentioned (51.8%), as it is the most common process used in the AI. Batch production also represents a significant percentage (23.2%), corresponding partially to the activity of metalworking sector in the AI. More than 62% of the companies experience manufacturing changes in the last three years. For more than 80% of the respondents, 'Increase production' and 'Cost reduction' are the main factors that dictate these changes. 'Imposed changes by OEMs', 'Design' and 'Attempt to implement AM' are highlighted as the main reasons of the need for these changes by more than 60% of the respondents. However, "Environmental issues" and "Governmental policies" do not significantly account for the above-referred changes. Furthermore, the results suggest that almost 93% of the companies intend to improve manufacturing processes, in order to increase flexibility. This flexibility might improve companies' responsiveness, because of the changes occurring in the production process. Considering the degree of involvement between top management and production departments, around 85% of the respondents claim that the top management is highly or extremely involved. This shows that there is no lack of support from the top management team regarding AM implementation. About 81% of the companies declare that they possess an agile manufacturing system, meaning they can produce a planned range of product models within a certain family of products. This type of system is more capable of dealing with unplanned changes. Indeed, this system possesses enough ability to support different processes in a different sequence, instead of just being prepared for one certain cycle of processes in one certain sequence.

The companies' quality concerns were evaluated, on a scale of one (none) to six (very high), aiming to identify factors companies give more attention to. The responses illustrate that all factors are important, although some differences are observed. Primarily, the main quality concern is 'Customer satisfaction', which is a crucial feature in AM implementation. On the other hand, 'Improvement of internal communication' did not attract the respondents' attention. This matches with the functional organization type adopted by the companies, demonstrating that they are focused exclusively on their work, and that communication and interaction with other sectors does not matter. Thus, employees end up losing interest in the overall goals of the company. This idea is completely opposite to the principles of agile management. Indeed, employees are just focused on 'Productivity', 'Financial performance', and 'Consistency of product', leaving aspects such as 'Implementation of best practices', and 'Improvement of internal communication' in the background.

4.4. Product Development Processes and Project Management

In order to understand and characterize the product development process, four questions were defined. Firstly, it was sought to understand if the companies develop this process by themselves, or through subcontracting external services. It was possible to verify that more than 91% of the companies develop their projects mostly in house. It was then necessary to find out what kind of approach is usually applied in the PM field. It was noticeable that a lot of PM teams experience the agile approach or, at least, iterative, or incremental methods. These methods allow for customer feedback, enabling changes during the product development phase to adjust and improve it. Thus, they guarantee customer satisfaction using this process. The main criterium taken into account during the project and product development process is 'Follows the customer requirements. This confirms the high degree of dependence that these companies are subjected to. This is also corroborated by another fact: more than 80% of the companies follow the product requirements imposed by the OEMs. It is also important to emphasize the low percentage of the 'Origin of materials' (17.9%), and 'Sustainability' (23.2%) criteria, indicating that the environmental aspect is one of the least important during the project and product development phases. This depicts how the AM implementation is useful regarding the desired and needed increase in sustainability, and also shows a lack of knowledge in how AM empowers all sustainability pillars.

4.5. Agile Project Management Environment

Figure 4 shows how familiar the companies are with AM practices. Regarding the APM culture of the companies, more than 73% of them are studying the implementation of, or are working under, AM, which is a very reasonable indicator. Approximately the same percentage have departments where agile techniques (kanban, scrum, or scrumban) are applied. The APM certification process is still difficult and less requested by companies. Nevertheless, a significant percentage of responses (30%) claim they have certified collaborators in APM inside the company. Also, almost 60% of the companies have more than ten workers devoted to the PM team. Lastly, most of the companies have a minimum qualification of a bachelor's degree for the project and product development teams. People with secondary education or master's degree represent 19.6% and 14.3% of respondents, respectively. The situation is not the best, and represents a lack of skills and knowledge regarding the implementation of AM, which constitutes a serious barrier. Besides that, in a general way, the companies demonstrate they are aware of what APM is. However, it is likely they do not devote the necessary efforts to adopt a fully agile approach.

Figure 5 depicts a stacked bar chart where some specific agile techniques are illustrated. The purpose of this question was to understand the level of knowledge of the respondents regarding these techniques and, at the same time, if some of the techniques are applied inside the companies. Once again, one of the most referred topics is 'Customer integration', confirming the importance of customer satisfaction for these companies. Another concept often referred to as being usually applied is the 'Daily meeting', which is part of the scrum technique. Nonetheless, other tools are part of the 'Daily meeting' technique, such as 'Product backlog', 'Sprint iteration', and 'Incremental development'. However, those tools are not used in the same way. This could mean that companies could apply one or another tool, without fully applying the concept, such as the scrum technique.

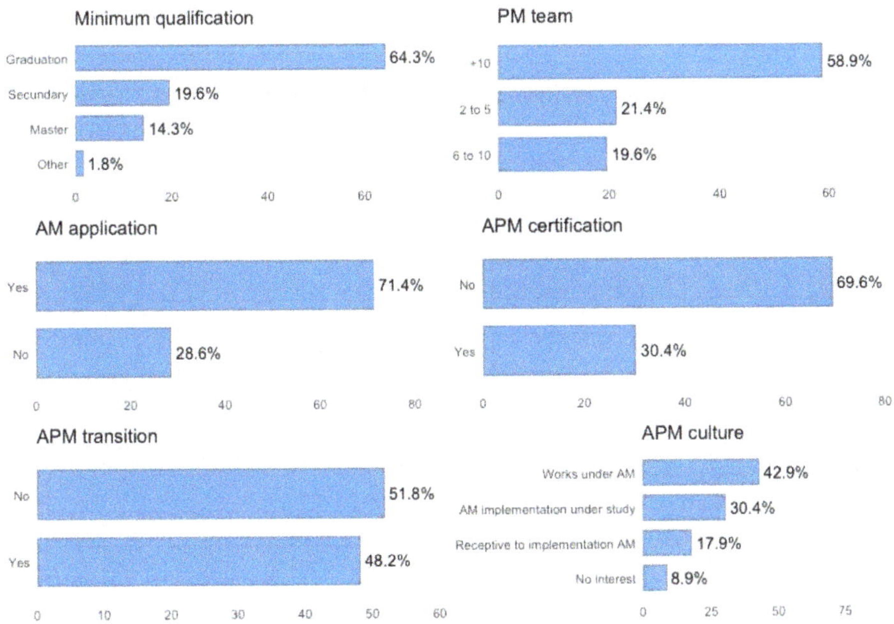

Figure 4. Characterization of the APM environment into the companies.

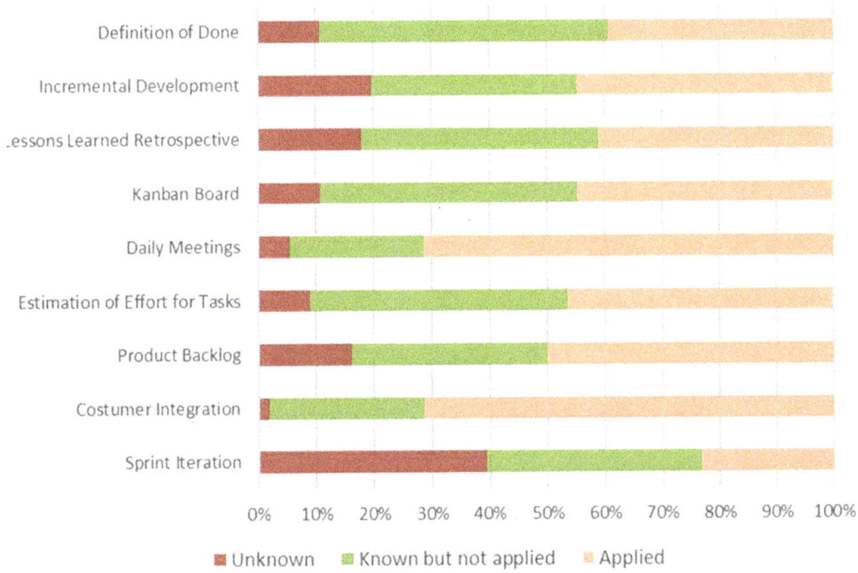

Figure 5. Percentages of agile methods and application of tools.

Moreover, 82.1% of the respondents claim that the contribution of agile techniques and tools to the companies vary between highly and extremely important. Respondents clearly point out the importance of these methods in their workplaces, even without applying them in full. It should be noted that the methods described in this study do not represent most of the practices, tools, and techniques comprising APM theory, but only a small and selected sample.

4.6. Barriers in the Implementation of Agile Methodologies

Figure 6 presents factors that affect AM implementation, as well as their influence from the respondents' point of view. These factors are considered as categories for the barriers in the implementation of AM, and barriers further described fit these categories. The most notable categories of barriers in the stacked bar chart are 'Technical knowledge and capacity' and 'Organizational culture', both with an influence of between moderate and very high in more than 60% of the responses. These factors could have an extreme impact on the success of AM implementation. The first is imperative, since everyone dedicated to the job needs to be aware of what they are supposed to do. Moreover, all workers need to possess the knowledge to develop their own tasks. The second is the factor with the greatest influence, which could be the biggest hurdle for AM implementation in the companies. This means that a company established for several years is less flexible in changing the way it is organized, being less able to change these roots rapidly and effectively. Also, it is worth mentioning that the 'Lack of financial support' does not seem to be an impediment to the implementation of AM.

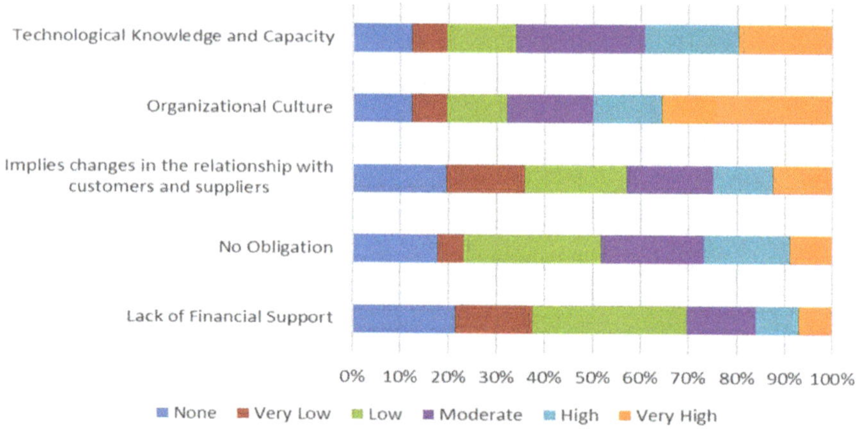

Figure 6. Barriers' categories and their influence for non-implementation of AM.

The main sources of barriers from the respondents' point of view are internal (61%), external (5%), or both (34%). Thus, external environment presents a low influence on the AM implementation, while internal impediments are the most significant. This correlates with the answers given in the previous question and depicted in Figure 6, since the two most observed factors belong to barriers arising from an internal source.

Figure 7 represents specific barriers and their influence on the implementation of AM. Each specific barrier fits into a barrier category and barrier source, previously referenced. 'Lack of knowledge and skills', 'Stakeholders' attitude', 'Time constraints', 'Existent organizational culture', 'Change predisposition', 'Staff not prepared to AM', and 'There is no time to think about that', are barriers with an influence between a moderate and very high level for more than 50% of the respondents. It is worth highlighting the low influence of factors such as 'It is not applicable to our product' and 'Organization not able to apply AM' indicating that, indeed, the implementation of AM in the AI is conceivable.

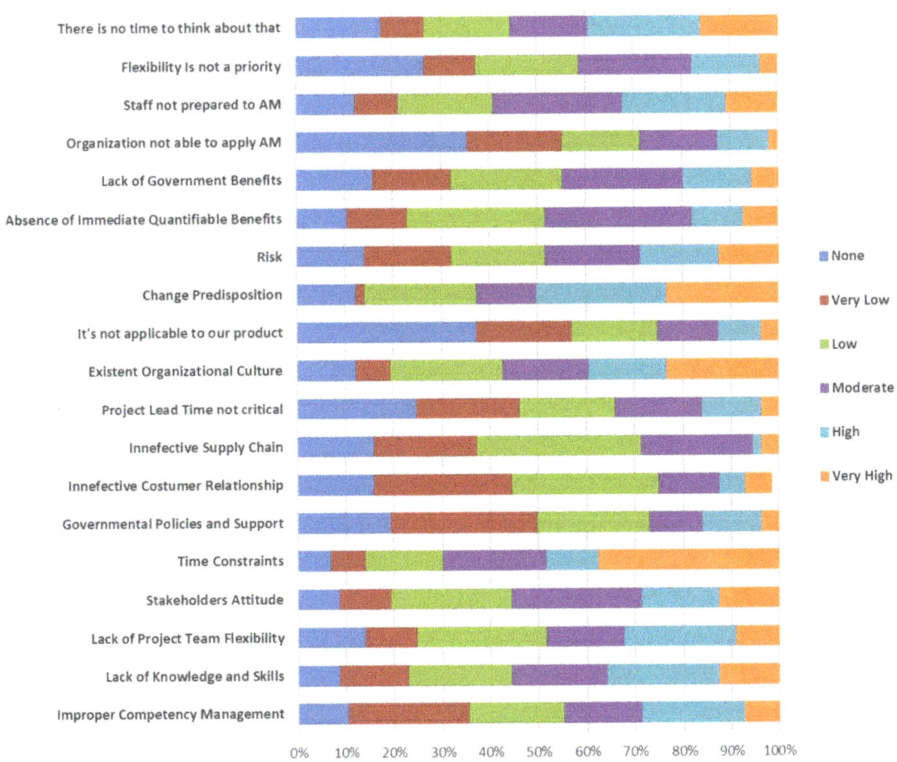

Figure 7. Specific Barriers and their influence on AM implementation.

Barriers with a lower impact are those such as 'Lack of knowledge and skills', 'Stakeholders' attitude', 'Time constraints', 'Existent organizational culture', 'Change predisposition', 'Staff not prepared to AM', and 'There is no time to think about that'. Otherwise, barriers such as 'It is not applicable to our product' and 'Organization not able to apply AM' are almost disregarded by respondents. Regarding the open-ended question, two factors are highlighted: 'Certification procedures' and 'Lack of collective motivation'.

4.7. Enablers for the Implementation of Agile Methodologies

The last question of the questionnaire, representing enablers found in the literature for the implementation of AM, intended to understand which factors have more influence on the automotive sector. The obtained results are as follows: 'Organizational support' (78.6%), 'Stakeholders collaboration' (60.7%), 'Investment in training' (76.8%), and 'Lean relationship' (28.6%). 'Organizational support' and 'Investment in training' are the most referenced enablers. The first aims to acquire the adequate support from the top management, while the second claims that formation and training in the field are essential to implement AM. No extra enablers are pointed out by respondents in the open-ended question.

4.8. Statistical Analysis—Hypotheses' Investigation

Data analysis was performed through charts and the evaluation of results so far. However, the extrapolation of results to the population studied requires the application of statistical inference. Thus, the following five investigation hypotheses were defined for this purpose:

H1. The company production type influences AM implementation.

H2. The AM implementation has an association with the company's degree of change.

H3. The improper competency management affects AM implementation.

H4. The change predisposition influences AM implementation.

H5. The absence of immediate quantifiable benefits influences AM implementation.

It is worth mentioning that, for all tests of hypotheses, only data corresponding to responses in which the respondent claimed to be familiar with AM are considered. Also, only the results related to tests statistically significant are presented.

Hypothesis 1 (H1) intends to evaluate if the 'Production type' adopted in each company is an 'Organizational' barrier that inhibits the AM implementation. Thus, the null hypothesis (H0) is: the production type does not influence the AM implementation in the company. Based on the Kruskal–Wallis test, there is evidence that there are differences between the groups (χ^2 (2) = 10.052; p value = 0.040), which supports the statement that the 'Production type' in each company influences the AM implementation through 'Organizational' barriers. According to multiple pairwise comparison, it is observed that the batch production type represents statistically significant differences between the mass (median = 3.5; min. = 1; max. = 6), continuous process (median = 2.5; min. = 2; max. = 3), and other production types (median = 2.5; min. = 1; max. = 4). It is possible to conclude that the companies producing through a continuous process, or another production type, seem to encounter fewer impediments in AM implementation from an organizational point of view. This means that if product manufacturing is divided into processes, the organizational interactions, and the complexity of the production process, is simplified. In contrast, companies under batch production type reveal statistical differences among the other groups. These companies, despite allowing for changes or modifications between batches, require more intense and accurate planning, scheduling, and control over the process, i.e., a higher complexity. Also, just one step is performed at a time on multiple items, which is a disadvantage if the product requirements are constantly changing. As a result, the organizations applying this production type could face 'Organizational barriers' in implementing AM, due to a higher management complexity.

Hypothesis 2 (H2) aims to investigate if the 'AM implementation' has a relationship with the 'Company's degree of change'. Thus, H0 is as follows: AM implementation does not have a relationship with the company's degree of change. According to the Mann–Whitney test, there is a relationship between the 'AM implementation' and the 'Company's degree of change' (W = 110.500; p value = 0.048; median = 2.5; min. = 2; max. = 6 for no AM implementation; median = 4.5; min. = 2; max. = 6 for AM implementation). This confirms that the organizations with high levels of change need to adopt AM.

In order to evaluate if the 'Degree of management involvement' affects the implementation of AM, hypothesis 3 (H3) is investigated. Thus, H0 is as follows: the improper competency management does not affect AM implementation. The Spearmen's correlation is studied for each dependent variable ('Organizational culture' and 'Knowledge and technology'). It is possible to observe that 'Improper competency management' has a positive correlation with the 'Organizational culture' barrier (rs = 0.465; p value = 0.001), as well as the 'Knowledge and technology' barrier (rs = 0.450; p value = 0.002). For that reason, it is possible to state that 'Improper competency management' affects the AM implementation through both the 'Organizational culture' and 'Knowledge and technology' barriers. This influence could arise either from a lack of management involvement, or due to a lack of expertise and skills of the managers. This depends on if the problem is being analyzed from the organizational point of view or from the knowledge and technological point of view, respectively.

Hypothesis 4 (H4) emerged intending to understand if the 'Change predisposition' influences the AM implementation. Accordingly, H0 is drawn as: the change predisposition does not influence the AM implementation. The Spearmen's correlation evaluates the correlation between the 'Change predisposition' and the 'No obligation' barriers, both fitting in the 'Institutional' barrier category. There is statistical evidence (rs = 0.373; p = 0.010) to state that the 'Change predisposition' influences the AM implementation, because it is correlated to 'No obligation' barrier. Moreover, these variables are framed in the 'Institutional' barrier. Thus, it is possible to state that the 'Institutional' barrier has influence on the AM implementation. It is confirmed that one of the biggest challenges in the application of AM is the resistance from society, and its aversion to change.

Lastly, the study intended to verify if the 'Absence of immediate quantifiable benefits' influences the AM implementation. Thus, hypothesis 5 (H5) is investigated, with H0 being defined as follows: the absence of immediate quantifiable benefits does not influence the AM implementation. Spearmen's correlation is performed, trying to understand if there is an association between the 'Absence of immediate quantifiable benefits', and the 'Lack of financial support' barriers. Both barriers fit into the 'Financial' barrier category. There is statistical evidence (rs = 0.380; p value = 0.009) to state that the 'Absence of immediate quantifiable benefits' influences AM implementation, since it is correlated to the 'Lack of financial support' barrier. These variables are associated, as they also fit into the 'Financial' barrier category; thus, it is possible to state that the 'Financial' barrier has direct influence on AM implementation.

5. Discussion

The 56 questionnaires with valid responses allow this survey to provide a group of interesting results concerning the implementation of AM in the AI in Portugal. Through statistical inference, it was possible to draw conclusions about the identified barriers in the AI sector, as well as the variables that have a direct influence on them.

5.1. *Analysis about the Barriers Found*

Across this study, 'Financial', 'Organizational', 'Knowledge and technology', and 'Institutional' were identified as the main categories of barriers in AM implementation. These results are in line with what is mentioned in [31] and [33]. However, the 'Market and supply chain' barrier is identified through the performed statistical tests, despite the high level of exportation, and the imposed requirements from OEMs regarding more than 80% of the companies. This corresponds to a high external dependence of the companies on the market and supply chain. The specific variables found to have influence on the implementation of AM in the AI are 'Production type', particularly 'Batch production', 'Organizational culture', 'Companies' degree of change', 'Improper competency management', 'Change predisposition', 'No obligation', 'Absence of immediate quantifiable benefits' and 'Lack of financial support'. These barriers are also identified in other sectors, namely, in the software development industry [32]. Indeed, Vermunt et al. [28], Ghani and Bello [31], and Caldera et al. [33] reference 'Lack of knowledge and skills' as one of the main barriers on the integrated development of the companies in different fronts: lean practices implementation, circular economy, and IT use. On the other hand, 'Organizational' barriers are mainly related to lack of time to implement new methodologies and improper management, which is corroborated by Potdar et al. [29], Caldera et al. [33], and Tan et al. [35], through observations made on the software development industry. Problems related to supply chain are referenced previously by Silva and Gouveia [25], Vermunt et al. [28], and Potdar et al. [29]. Thus, it is observed that some barriers cross different sectors of activity, from software development to the production of components for the automotive industry. However, when detailing the difficulties that each sector refers to as significant, there is a clear difference: while the software development sector needs a strong connection with its stakeholders, understanding their needs and reacting in an agile way, the producers of components for the automotive industry are essentially dependent on their main customer, an original equipment manufacturer (OEM). These OEMs set quality standards, timeframes, and maximum costs, imposing a much more demanding level of restrictions on the producers of components for the automotive industry. This fact limits the creativity of automotive components manufacturers, and imposes development paces difficult to manage using traditional management technics. Thus, the adoption of agile methodologies would be of great interest for this sector. The difference regarding the relationship between the automotive sector and the software development sector is observed in the work by Ciriello et al. [45], which essentially addresses the relationship between the software industry and its stakeholders. However, beyond the traditional barriers to the agile methodologies already identified in other sectors, the AI needs to overcome other important identified barriers, such as: (a) the difficulty in quantifying the real benefits of the agile implementation; (b) the non-obligation of adoption of these methodologies by the current standards and rules; (c) the rooted culture of conducting the projects in a traditional way; and (d) the type of production. However, because automotive models are changing quickly, agile adoption would help this industry present increased responsiveness in answering the OEM requests. This can be accomplished as shown by what is already conducted by the aircraft repair sector, as described by Freitas et al. [22]. Cultural barriers are something equally common to several sectors, but it is particularly observed in the automotive industry, due to the specificity of standards and technologies involved. In a recent work, Smite et al. [46] mention that the cultural barrier is also felt in offshore projects. In fact, the cultural aspects are very different between these sectors, and there is another crucial difference: software can be distributed digitally, and manufacturing projects usually need complex hardware. This implies an adequate supply chain, as referred by Shashi et al. [47], and the present study also confirms it as a barrier to adopting agile methodologies. However, as argued by Brandl et al. [48], and confirmed in this work, manufacturing includes other barriers not considered in the software industry. In order to standardize the taxonomy used in the designation of barriers identified around agile methodologies, Shameem et al. [49] created five categories. These categories integrate 22 different barriers identified around the software industry, but that same taxonomy left out some barriers from other sectors, namely those discussed in this work.

5.2. *Analyzing Possible Enablers*

After the analysis of the obtained results, along with the extensive literature review previously performed, some recommendations and enablers were pointed out to overcome the identified barriers and improve the AM effectiveness, in line with the enablers pointed out in [33]. 'Organizational barrier' is influenced by the 'Production type' followed in the manufacturing and, usually, it is not an easy factor to change once it is rooted in the company [50]. However, the differences between the production types bring different advantages, as well as challenges. In the case of batch production, it is a process of higher complexity that requires intense planning, scheduling, and control [51,52]. This complexity, and the constantly changing environment, can create great organizational obstacles. This is particularly true if there is no knowledge, skills, technological capacity, strong level of intercommunication, or tune among the team involved. Hariyani and Mishra [53] study the main organizational barriers and enablers to some manufacturing companies becoming more sustainable. Indeed, some of them fit the results obtained through the survey very well, namely, competence, flexibility, and quickness. Moreover, Ahmed et al. [54] argue that managers must focus on the human capital to allow their organization more flexible and ready to change.

Regarding the 'Companies' degree of change', the best approach to take is adopting APM, since it focuses on agility, adaptation, response to unpredictable changes, continuous improvement, and innovation [11,12]. A recent study by Bottani [55] concludes that the introduction of agile management in companies has a very small number of enablers, and that they vary markedly between different types of industries. Given the results obtained in this work, some enablers are identified. However, the number of barriers encountered is substantially higher, and more difficult to overcome. It is necessary for industries usually characterized by intrinsic rigid, predictable, and stable architectures to rapidly adapt to the new requirements of society and constantly changing environments.

Another identified barrier is 'Improper competency management', having influence on both the 'Organizational' and 'Knowledge and technological' aspects. This is in line with the conclusions obtained by Potdar et al. [29]. Building closer relationships, such as strong management support, is crucial to prevent a lack of management involvement, and encourage education and training. This guarantees a skilled and competent staff, with a clear understanding of the agile objectives from a knowledge and technological perspective, as pointed out in [33]. Additionally, Sarangee et al. [55] reference organizational issues and technological knowledge as barriers in the development of new products in universities. This means that both in services and industry, these barriers are identified as some of the main factors affecting the implementation of agile methodologies. In the study performed by Loiro et al. [56], addressing the APM use in manufacturing, the deployment of a team to deal with APM (AGILE team) and a communicational workflow is proposed, in order to improve the agile manufacturing performance. The deployment of a team specifically dedicated to APM could be a massive enabler from an 'Organizational' and 'Knowledge and technology' points of view. Firstly, because this team oversees complex organizational communications, such as a customer making a request until it is successfully approved, this drastically improving the organizational dynamism. Then, the required knowledge for APM implementation resides in a skilled team with different persons assuming different roles, being uniquely dedicated to this process.

Perhaps one of the most difficult barriers to overcome is the 'Institutional' aspect. In fact, it entails a predisposition to change, no obligation thinking, resistance from society, and aversion to change. This factor is also pointed out in the work developed by Vermunt et al. [28], and by Gonçalves et al. [57]. It is vital that companies, and all people involved, show willingness to adopt AM, embracing a friendly, agile organization, and a team environment philosophy with ambition and motivation. This cultural change is mandatory to increase the liveliness of companies.

Regarding the 'Financial' aspect, there is no formula to realize the economic advantages of adopting AM. Beyond this work, this barrier is pointed out by Vermunt et al. [28]. The authors believe that this barrier is more prone to upsurges in the automotive industry and countries where the productivity is lower, because the business yield is usually reduced. These facts constraint the use of a company's own financial resources to support agile transition, even when agile methodologies bring back the investment needed in a reduced period. Indeed, the APM intends to reduce the impact of unpredictable changes, improve time-to-market, respond and adapt to complexity, and focus on customer satisfaction. These factors help to bring back the initial investment necessary for staff education and training, as well as all the psychological and organizational changes required.

5.3. *Flowchart*

Thus, considering all these recommendations and enablers, it was possible to sequence and represent them in a flowchart, having an effective AM implementation as the outcome. Initially, and to overcome the 'Institutional' barrier, it is vital that the company and all people involved have the necessary willingness and predisposition to adopt AM. Then, agile values should be deemed as the input to achieve competitive advantage and overcome the 'Financial' barrier. Subsequently, an initial investment is necessary to achieve the next step, the deployment of an agile team with highly skilled collaborators. The agile team should develop frameworks and workflows according to each company's characteristics and environment, in order to drastically improve the 'Organizational' and 'Knowledge and technology' aspects. As a result, an effective AM implementation is achieved, as represented in Figure 8.

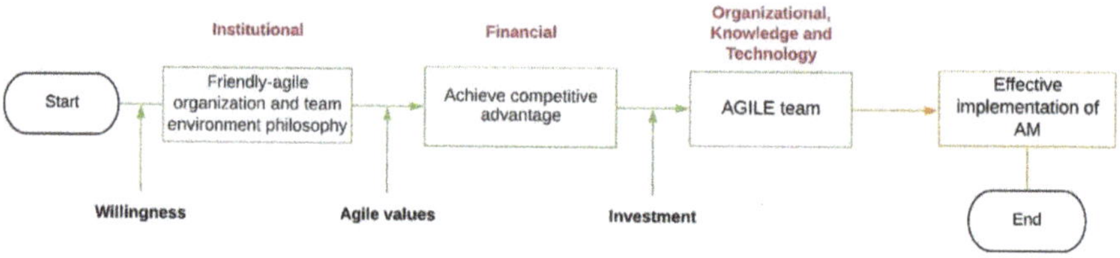

Figure 8. Flowchart to improve AM implementation effectiveness.

6. Conclusions

This study intended to identify barriers in AM implementation in the AI, after this gap was identified in the literature. The approach follows the proposal described by Sandberg and Alvesson [58].

First, intense bibliographic research was carried out in order to identify possible barriers previously identified in the implementation of agile methodologies in other sectors, such as software development. Based on these initial data, a survey was prepared, which was validated by experts in this field. The survey was distributed to 148 companies, with 56 valid responses. The results were compiled and treated statistically, incorporating descriptive statistics and statistical inference. The generic data obtained from the responses collected allowed us to understand that:

- 'Companies' degree of change' has a direct relation to AM implementation;
- 'High levels of change in the products' induce the need to adopt AM;
- Just 9% of the companies have no interest in implementing APM, and 70% are already applying agile methods;
- Most of respondents do not have skills in APM; their qualifications may be enough to apply agile tools, but insufficient to apply a fully agile approach;
- The main identified barriers are 'Organizational culture' and 'Knowledge and technology', and 61% of respondents state that the main barriers' source is solely internal;
- 'Organizational', and 'Knowledge and technology' barriers in the AI were identified due to the 'Improper competency management' factor;
- 'Lack of management involvement' seems to be a problem. The same factor is related to the 'Knowledge and technology' barrier, showing a lack in managers' skills;
- 'Change predisposition', and its correlation with the 'No obligation' barrier act as 'Institutional' barriers for the AM implementation;
- 'Lack of financial support' does not seem to be a barrier, however, a relation between the 'Absence of immediate quantifiable benefits' and 'Lack of financial support' barriers is established;
- A total of 90% of the respondents pointed out two barriers not included in the questionnaire: 'Certification procedures' and 'Lack of collective motivation';
- They also consider that the main enablers for AM implementation are 'Organizational support' and 'Investment in training'.

Based on these data and regarding the research questions pointed out, it was possible to draw the following conclusions:

- Do the AI companies need to implement AM? It became clear that the type of production followed by the AI requires agile methodologies to respond in due time to market requests. There is a constant modification and updating of products, requiring greater flexibility and reaction speed on the part of automotive component manufacturers. Thus, AM could significantly contribute to reducing the time needed to introduce products into the market and improve the dialogue between stakeholders. With these improvements, the efficiency of the processes improves, reducing costs and avoiding waste in terms of time and consumption of natural resources. This increases the sustainability of the processes.

- Are the AI companies predisposed to AM? Yes, companies are willing to adopt AM, but there is still some lack of preparation in both top and middle management, in order to facilitate an effective implementation of AM. However, the preparation in terms of skills still seems to be insufficient, and it is still possible to perceive that companies wait for governmental financial support (targeted support programs). This minimizes the investment necessary for the adoption of AM practices.
- What are the main barriers for implementing AM in the AI? 'Organizational culture' and 'Knowledge and technology' are identified as the main barriers. To understand this result, it is necessary to know that most Portuguese companies in the sector are of family origin, and that there is a considerable deficit of training in engineering and management in many companies. Due to international competitiveness and the need to impose themselves on the market, most companies significantly improved this aspect, but it appears that there are still some gaps.
- What are the main enablers for implementing AM in the AI? Based on the previous topic, it is not surprising that the main enablers focus on 'Organizational support' and 'Investment in training'. The organizational structure of the companies still shows a lack of training in the area of management and, despite the mandatory training of employees in a certain percentage of their annual working time, this training is still not strategically programmed or directed to some aspects, such as the adoption of AM. Indeed, the implementation of AM is not yet a priority for companies.

The framework developed will certainly facilitate the adoption of AM practices in companies in this sector, and the same can be adopted in other sectors presenting similar characteristics.

The realization of this work, and the conclusions drawn, show a real picture of the current situation in a European country about the implementation of AM in the field of the automotive components manufacturing. Project management in the automotive industry is a constant, but the introduction of agile methodologies is still scarce. Although there are numerous studies on sustainability in the automotive area, none, so far, have addressed how sustainability can be improved in this sector through the adoption of agile methodologies. This is the main novelty of this work, leaving open future topics for research, and allowing the extension of the scope of this work to other countries, taking into account the specificity of the automotive industry. As lines for further investigation, the following topics can be considered:

- What are the barriers and enablers related to other important Portuguese industries, such as the cork industry and the textile industry?
- Are there similarities between the barriers and enablers found within the Portuguese AI compared to other European countries? What about other industrialized countries on other continents?
- Can the developed framework be applied in other types of Portuguese, European, or world industries?
- How will this situation evolve over time?

It is hoped that this study is read and understood by political and industrial decision-makers, and they realize that the sustainability they are looking for, in economic, social, and environmental terms, can be achieved more quickly and effectively if the key barriers are quickly overcome, and priority is given to identified enablers. Since there is no study in this area and in this country, it is a pioneering study, which can now be replicated in other European countries, or in other regions of the world, allowing a comparison between different realities and approaches.

Author Contributions: Conceptualization, D.S., S.C.F.R., K.K., and F.J.G.d.S.; methodology, F.J.G.d.S., S.C.F.R. and J.C.S.; investigation, D.S., F.J.G.d.S., L.P.F. and K.K.; supervision: L.P.F., F.J.G.d.S., K.K. and S.C.F.R.; writing original draft—re-viewing and editing, D.S., F.J.G.d.S., L.P.F., S.C.F.R., J.C.S. and K.K. All authors have read and agreed to the published version of the manuscript.

Funding: This research received no external funding.

Institutional Review Board Statement: Not applicable.

Informed Consent Statement: Not applicable.

Data Availability Statement: No data is available.

Acknowledgments: The authors would like to thank LAETA/INEGI/CETRIB for all support provided.

Conflicts of Interest: The authors declare no conflict of interest.

References

1. OICA. Economy Contributions. Available online: http://www.oica.net/category/economic-contributions/rd/# (accessed on 7 January 2022).
2. Beaume, R.; Maniak, R.; Midler, C. Crossing innovation and product projects management: A comparative analysis in the automotive industry. *Int. J. Proj. Manag.* **2009**, *27*, 166–174. [CrossRef]
3. Jeffs, J.; Papillon, B.M. Globalization, the new economy and project management: A graph theory perspective. *J. Mod. Proj. Manag.* **2019**, *7*, 30–42.
4. San Cristóbal, J.R.; Carral, L.; Díaz, E.; Fraguela, J.A.; Iglesias, G. Complexity and project management: A general overview. *Complexity* **2018**, *2018*, 4891286. [CrossRef]
5. Sanchez, F.; Micaelli, J.P.; Bonjour, E.; Monticolo, D. A step for improving the transition between traditional project management to agile project management using a project management maturity model. *J. Mod. Proj. Manag.* **2019**, *7*, 88–105. [CrossRef]
6. Conforto, E.C.; Amaral, D.C.; da Silva, S.L.; di Felippo, A.; Kamikawachi, D.S.L. The agility construct on project management theory. *Int. J. Proj. Manag.* **2016**, *34*, 660–674. [CrossRef]
7. Moradlou, H.; Asadi, M. Implementation of agile manufacturing principles in small and medium enterprises (SMES). *J. Mod. Process. Manuf. Prod.* **2015**, *4*, 31–44.
8. Souza Bermejo, P.H.; Zambalde, A.L.; Tonelli, A.O.; Souza, S.A.; Zuppo, L.A.; Rosa, P.L. Agile principles and achievement of success in software development: A quantitative study in Brazilian organizations. *Procedia Technol.* **2014**, *16*, 718–727. [CrossRef]
9. Dingsøyr, T.; Nerur, S.; Balijepally, V.; Moe, N.B. A decade of agile methodologies: Towards explaining agile software development. *J. Syst. Softw.* **2012**, *85*, 1213–1221. [CrossRef]
10. Fritzsche, A. Implications of agile manufacturing in the automotive industry for order management in the factories-evidence from the practitioner's perspective. *Procedia CIRP* **2018**, *72*, 369–374. [CrossRef]
11. Špundak, M. Mixed agile/traditional project management methodology—Reality or illusion? *Procedia Soc. Behav. Sci.* **2014**, *119*, 939–948. [CrossRef]
12. Highsmith, J. *Agile Project Management: Creating Innovative Products*; Addison-Wesley Professional: Boston, MA, USA, 2009; ISBN-13 978-0321658395.
13. Beck, K.; Beedle, M.; van Bennekum, A.; Cockburn, A.; Cunningham, W.; Fowler, M.; Martin, R.C.; Mellor, S.; Thomas, D.; Grenning, J.; et al. Agile Manifesto 2001, Agile Alliance. Available online: https://www.agilealliance.org/agile101/the-agile-manifesto/ (accessed on 19 December 2021).
14. Silva, F.J.G.; Kirytopoulos, K.; Ferreira, L.P.; Sá, J.C.; Santos, G.; Nogueira, M.C. The three pillars of sustainable development and agile project management: How do they influence each other. *Corp. Soc. Responsib. Environ. Manag.* **2022**, *29*. *Accepted for publication*. [CrossRef]
15. Obradović, V.; Todorović, M.; Bushuyev, S. Sustainability and Agility in Project Management: Contradictory or Complementary? In *Advances in Intelligent Systems and Computing III (CSIT 2018)*; Shakhovska, N., Medykovsky, M., Eds.; Springer: Cham, Switzerland, 2019; Volume 871. [CrossRef]
16. Žužek, T.; Gosar, Ž.; Kušar, J.; Berlec, T. Adopting agile project management practices in non-software SMEs: A case study of a Slovenian medium-sized manufacturing company. *Sustainability* **2020**, *12*, 9245. [CrossRef]
17. Sharma, S.; Oberoi, J.S.; Gupta, R.D.; Saini, S.; Gupta, A.K.; Sharma, N. Effect of agility in different dimensions of manufacturing systems: A review. *Mater. Today Proc.* 2022, *in press*. [CrossRef]
18. Malhotra, R.; Chug, A. Comparative Analysis of Agile Methods and Iterative Enhancement Model in Assessment of Software Maintenance. In Proceedings of the 3rd International Conference on Computing for Sustainable Global Development (INDIACom), New Delhi, India, 16–18 March 2016; pp. 1271–1276.
19. Rising, L.; Janoff, N.S. The Scrum Software Development Process for Small Teams. *IEEE Softw.* **2000**, *17*, 26–32. [CrossRef]
20. Project Management Institute; Agile Alliance. *Agile Practice Guide*; Project Management Institute: Campus Boulevard, PA, USA, 2017; ISBN-13 978-1628251999.
21. Schwaber, K.; Sutherland, J. The Scrum Guide: The Definitive Guide to Scrum: The Rule of the Game. 2020, p. 19. Available online: http://www.scrumguides.org/docs/scrumguide/v20 (accessed on 14 December 2021).
22. Freitas, F.; Silva, F.J.G.; Campilho, R.D.S.G.; Pimentel, C.; Godina, R. Development of a suitable project management approach for projects with parallel planning and execution. *Procedia Manuf.* **2020**, *51*, 1544–1550. [CrossRef]
23. Livermore, J.A. Factors that Impact Implementing an Agile Software Development Methodology. In Proceedings of the 2007 IEEE SoutheastCon, Richmond, VA, USA, 22–25 March 2007; pp. 82–86. [CrossRef]

24. Boehm, B.; Turner, R. Management challenges to implementing agile processes in traditional development organizations. *IEEE Softw.* **2005**, *22*, 30–39. [CrossRef]
25. Silva, F.J.G.; Gouveia, R.M. *Cleaner Production: Toward a Better Future*; Springer Nature Publishing: Cham, Switzerland, 2020; ISBN-13 978-3030231675.
26. Rose, A.N.M.; Deros, B.M.; Ab Rahman, M.N.; Norani, N. Lean Manufacturing Best Practices in SMEs. In Proceedings of the 2011 International Conference on Industrial Engineering and Operations Management, Kuala Lumpur, Malaysia, 22–24 January 2011; pp. 872–877.
27. Bottani, E. Profile and enablers of agile companies: An empirical investigation. *Int. J. Prod. Econ.* **2010**, *125*, 251–261. [CrossRef]
28. Vermunt, D.A.; Negro, S.O.; Verweij, P.A.; Kuppens, D.V.; Hekkert, M.P. Exploring barriers to implementing different circular business model. *J. Clean. Prod.* **2019**, *222*, 891–902. [CrossRef]
29. Potdar, P.K.; Routroy, S.; Behera, A. Addressing the agile manufacturing impediments using interpretive structural modeling. *Mater. Today Proc.* **2017**, *4*, 1744–1751. [CrossRef]
30. Sindhwani, R.; Mittal, V.K.; Singh, P.L.; Aggarwal, A. Modelling and analysis of barriers affecting the implementation of lean green agile manufacturing system (LGAMS). *Benchmark. Int. J.* **2019**, *26*, 498–529. [CrossRef]
31. Ghani, I.; Bello, M. Agile adoption in IT organizations. *KSII Trans. Internet Inf. Syst.* **2015**, *9*, 3231–3248. [CrossRef]
32. Salinas, M.R.N.; Neto, A.G.S.S.; Emer, M.C.F.P. Concerns and limitations in agile software development: A survey with paraguayan companies. *Commun. Comput. Inf. Sci.* **2018**, *802*, 77–87. [CrossRef]
33. Caldera, H.T.S.; Desha, C.; Dawes, L. Evaluating the enablers and barriers for successful implementation of sustainable business practice in 'lean' SMEs. *J. Clean. Prod.* **2019**, *218*, 575–590. [CrossRef]
34. Tan, T.; Chen, K.; Xue, F.; Lu, W. Barriers to building information modeling (BIM) implementation in China's prefabricated construction: An interpretive structural modeling (ISM) approach. *J. Clean. Prod.* **2019**, *219*, 949–959. [CrossRef]
35. Kostić, D.M. Challenges of agile practices implementation in the medical device software development methodologies. *Eur. Proj. Manag. J.* **2017**, *7*, 36–44.
36. Costa, M.J.R.; Gouveia, R.M.; Silva, F.J.G.; Campilho, R.D.S.G. How to solve quality problems by advanced fully-automated manufacturing systems. *Int. J. Adv. Manuf. Technol.* **2018**, *94*, 3041–3063. [CrossRef]
37. Silva, F.J.G.; Soares, M.R.; Ferreira, L.P.; Alves, A.C.; Brito, M.; Campilho, R.D.S.G.; Sousa, V.F.C. A Novel automated system for the handling of car seat wires on plastic over-injection molding machines. *Machines* **2021**, *9*, 141. [CrossRef]
38. Magalhães, A.J.A.; Silva, F.J.G.; Campilho, R.D.S.G. A novel concept of bent wires sorting operation between workstations in the production of automotive parts. *J. Braz. Soc. Mech. Sci. Eng.* **2019**, *41*, 25. [CrossRef]
39. Costa, R.J.S.; Silva, F.J.G.; Campilho, R.D.S.G. A novel concept of agile assembly machine for sets applied in the automotive. industry. *Int. J. Adv. Manuf. Technol.* **2017**, *91*, 4043–4054. [CrossRef]
40. Chen, K.-S.; Chiou, K.-C.; Yu, C.-M. Lifetime performance index of electronic products. *Microelectron. Reliab.* **2020**, *113*, 113941. [CrossRef]
41. Wang, G.; Shao, M.; Lv, S.; Kong, X.; He, Z.; Vining, G. Process parameter optimization for lifetime improvement experiments considering warranty and customer satisfaction. *Reliab. Eng. Syst. Saf.* **2022**, *221*, 108369. [CrossRef]
42. Monteiro, J.; Silva, F.J.G.; Ramos, S.F.; Campilho, R.D.S.G.; Fonseca, A.M. Eco-design and sustainability in packaging: A survey. *Procedia Manuf.* **2019**, *38*, 1741–1749. [CrossRef]
43. Singh, A.S.; Masuku, M.B. Sampling techniques & determination of sample size in applied statistics research: An overview. International. *J. Econ. Commer. Manag.* **2014**, *2*, 1–22.
44. AFIA. PORTUGAL—Auto Components Industry. 2020. Available online: https://afia.pt/en/statistics/ (accessed on 9 January 2022).
45. Ciriello, R.F.; Glud, J.A.; Hansen-Schwartz, K.H. Becoming agile together: Customer influence on agile adoption within commissioned software teams. *Inf. Manag.* **2022**, *59*, 103645. [CrossRef]
46. Šmite, D.; Moe, N.B.; Gonzalez-Huerta, J. Overcoming cultural barriers to being agile in distributed teams. *Inf. Softw. Technol.* **2021**, *38*, 106612. [CrossRef]
47. Shashi; Centobelli, P.; Cerchione, R.; Ertz, E. Agile supply chain management: Where did it come from and where will it go in the era of digital transformation? *Ind. Mark. Manag.* **2020**, *90*, 324–345. [CrossRef]
48. Brandl, F.J.; Roider, N.; Hehl, M.; Reinhart, G. Selecting practices in complex technical planning projects: A pathway for tailoring agile project management into the manufacturing industry. *CIRP J. Manuf. Sci.* **2021**, *33*, 293–305. [CrossRef]
49. Shameem, M.; Kumar, R.R.; Nadeem, M.; Khan, A.A. Taxonomical classification of barriers for scaling agile methods in global software development environment using fuzzy analytic hierarchy process. *Appl. Soft Comput.* **2020**, *90*, 106122. [CrossRef]
50. Stettina, C.J.; Hörz, J. Agile portfolio management: An empirical perspective on the practice in use. *Int. J. Proj. Manag.* **2015**, *33*, 140–152. [CrossRef]
51. Araújo, W.; Silva, F.J.G.; Campilho, R. Manufacturing cushions and suspension mats for vehicle seats: A novel cell concept. *Int. J. Adv. Manuf. Technol.* **2017**, *90*, 1539–1545. [CrossRef]
52. Moreira, B.M.D.N.; Gouveia, R.M.; Silva, F.J.G.; Campilho, R.D.S.G. A novel concept of production and assembly processes integration. *Procedia Manuf.* **2017**, *11*, 1385–1395. [CrossRef]
53. Hariyani, D.; Mishra, S. Organizational enablers for sustainable manufacturing and industrial ecology. *Clean. Eng. Technol.* **2022**, *9*, 100375. [CrossRef]

54. Ahmed, A.; Bhatti, S.H.; Gölgecic, I.; Arslan, A. Digital platform capability and organizational agility of emerging market manufacturing SMEs: The mediating role of intellectual capital and the moderating role of environmental dynamism. *Technol. Forecast. Soc. Chang.* **2022**, *177*, 121513. [CrossRef]
55. Sarangee, K.; Schmidt, J.B.; Srinath, P.B.; Wallace, A. Agile transformation in dynamic, high-technology markets: Drivers, inhibitors, and execution. *Ind. Mark. Manag.* **2022**, *102*, 24–34. [CrossRef]
56. Loiro, C.; Castro, H.; Ávila, P.; Cruz-Cunha, M.M.; Putnik, G.D.; Ferreira, L. Agile project management: A communicational workflow proposal. *Procedia Comput. Sci.* **2019**, *164*, 485–490. [CrossRef]
57. Gonçalves, D.; Bergquist, M.; Alänge, S.; Bunk, R. How digital tools align with organizational agility and strengthen digital innovation in automotive startups. *Procedia Comput. Sci.* **2022**, *196*, 107–116. [CrossRef]
58. Sandberg, J.; Alvesson, M. Ways of constructing research questions: Gap-spotting or problematization? *Organization* **2011**, *18*, 23–44. [CrossRef]

sustainability

Article

Analyzing the Implementation of Lean Methodologies and Practices in the Portuguese Industry: A Survey

Carla M. A. Pinto [1,2], Jorge Mendonça [1,*], Lurdes Babo [3], Francisco J. G. Silva [1,4] and José L. R. Fernandes [1]

1 ISEP—School of Engineering, Polytechnic of Porto, 4200-072 Porto, Portugal; cap@isep.ipp.pt (C.M.A.P.); fgs@isep.ipp.pt (F.J.G.S.); 1040698@isep.ipp.pt (J.L.R.F.)
2 CMUP—Center for Mathematics, University of Porto, 4169-007 Porto, Portugal
3 CEOS.PP and Porto Accounting and Business School, Polytechnic of Porto, 4265-004 São Mamede de Infesta, Portugal; lbabo@iscap.ipp.pt
4 INEGI—Instituto de Ciência e Inovação em Engenharia Mecânica e Engenharia Industrial, 4200-465 Porto, Portugal
* Correspondence: jpm@isep.ipp.pt

Abstract: The mass production paradigm on which much of the industry was based has changed. The market is increasingly demanding, requesting diversity and products that are more and more adapted to personal wishes and requirements. This implies producing a greater diversity of products in smaller quantities. Competitiveness is enormous, which forces most companies to be truly effective and efficient, taking care of product quality, delivery time, and final cost. Lean methodologies have been a valuable aid in this field. The diversity of Lean tools has been shown to have answers to the most diverse challenges, and companies are aware of this, increasingly adopting methodologies and processes that aim to progressively reduce waste and adapting their production paradigm to what the market requires. This work intends to provide a vision, as global as possible, of the pathway of Lean implementation in the Portuguese industry. For this purpose, a survey was carried out with a significant sample of Portuguese industrial companies from a wide range of activity sectors. The data collected through the survey were treated statistically, and then a SWOT analysis of the results was performed, which provided a collection of precious information on the evolution of industrial companies in Portugal.

Keywords: lean manufacturing; lean practices; SWOT analysis; survey; questionnaire; Portuguese industry; lean implementation analysis

Citation: Pinto, C.M.A.; Mendonça, J.; Babo, L.; Silva, F.J.G.; Fernandes, J.L.R. Analyzing the Implementation of Lean Methodologies and Practices in the Portuguese Industry: A Survey. *Sustainability* **2022**, *14*, 1929. https://doi.org/10.3390/su14031929

Academic Editor: Paulo Peças

Received: 17 January 2022
Accepted: 6 February 2022
Published: 8 February 2022

1. Introduction

Market demands have changed in recent decades, forcing companies to quickly adapt to new production paradigms [1]. The growing competitiveness between companies in the same sector increased the need for companies to look for differentiating factors to act in the market. This induced the improvement of the quality and diversity of products and a reduction in the waste that usually exists throughout the production processes. Companies tried to increase as much as possible their effectiveness and efficiency [2]. The later are pivotal factors for the permanence of any industrial company in the current market. In this context, Lean tools have shown to be a valuable aid. Initially developed in Japan to adapt automobile production to market needs after World War II, Lean methodologies were adopted in the 1970s and 1980s by the American industry, having gained particular prominence worldwide with the launch of the book, The Machine That Changed the World [3]. Currently, Lean thinking has literally invaded both industry and service sectors, eager to eliminate waste and make companies economically healthier. One of the factors that has strongly contributed to this vast dissemination is the diversified set of tools that fall under the Lean thinking, which can be applied individually or together, depending on the specific needs of each company [4]. From the most basic tools such as the 5S

(Sort, Set in order, Shine, Standardize, Sustain), to more complex tools like VSM (Value Stream Mapping), among others, they allow for the identification and analysis of problems, enabling them to be later overcome. However, the success of its application implies that top management makes the entire structure of the company aware, and that everyone has the objective of properly applying these tools and contributing to the company's improvement [5]. There are many successful stories, though this success is not global. Winning stories are found through case studies in several sectors, from the health care services to the industry.

Moreover, Lean methodologies have been associated in recent decades with other equally powerful and necessary tools, such as Six Sigma and Sustainability [6–8]. In fact, the purposes of these concepts can be perfectly coherent. The Six Sigma methodology provides a set of five steps that essentially aim to analyze a quality problem in a process and find solutions for it, implementing and creating the necessary cycle to solidify the problem's resolution through constant monitoring [9]. This methodology contributes to the reduction of quality problems, i.e., reduction in the generation of non-conforming products, which is in line with Lean thinking, and reducing waste. On the other hand, when Lean is associated with Sustainability, it has commonly assumed the name Lean-Green, and it also aims to reduce waste, especially that which impacts the environment, such as reducing energy consumption [10]. Lean thinking has also been increasingly associated with ergonomics at work. In fact, there have been studies that prove the connection between a better positioning in the workstation and the performance that a worker can perform on his workstation [11,12]. Thus, there has been a growing interest in improving workstations in order to make the workers' work easier, which translates into greater productivity, that is, the reduction of wasted time and greater satisfaction in performing functions. More recently, Lean thinking has also been strongly linked to the concept of Smart Manufacturing, or Industry 4.0, as this concept aims to make production as flexible as possible, which strongly reduces waste both in terms of the process and throughout the logistics involving the different stages of production [13]. In [14], the authors perform a systematic review of the integration of four continuous improvement tools (CI), namely Lean, Len Six Sigma (LSS), Kaizen, and Sustainability, and of frameworks of Industry 4.0, from 1989 until 2015. A general framework is then proposed to combine CI strategies and technologies supporting Industry 4.0, to be applied to a wide range of industries. Are incorporated the essential differentiators of the four CI above, namely waste elimination and process optimization from Lean, the excellence of product quality from LSS, production upgrade through CI from Kaizen, and ecofriendly products and minimization of environmental waste from Sustainability. The realization of the Internet of Things and Cyber-Physical Systems aids in connecting the real world with virtual setups, which boosts monitoring and control of manufacturing systems and product development. It is thus possible to see that extremely positive developments have been carried out on several fronts and using different methodologies, so that the effectiveness and efficiency of processes could be vastly improved, both in the industry and in the services sector. Automotive assembly lines and companies that produce components for the automotive industry are in a sector that has consistently applied Lean methodologies as well as its most diverse developments, as it is a highly competitive industry where it is necessary to be constantly thinking of reducing waste and applying the most diverse Lean tools for this purpose. Next, some examples of theoretical and practical developments related to the application of Lean methodologies in various sectors of industry and services are reviewed. Rosa et al. [15] investigated Bowden cables production lines usually applied in automotive vehicles, with a view to increasing their efficiency, allowing these lines to meet the new levels of demand required by the market. In a first phase, they applied the VSM tool to identify the places where it was most urgent to act, then applied a PDCA (Plan-Do-Check-Act) cycle to implement the solutions identified as likely to be applied to improve processes. Through this study, it was possible to increase productivity by 41%, requiring a low investment, which had a return in 4 months of activity. The lines thus came to comply with the new demand requirements, with no need to replicate the line to match that demand. Also based on similar production lines for Bowden cables, Rosa et al. [16] applied the SMED tool in order to minimize setup time in the production of different references for those cables, as there is a tendency to reduce the quantity of cables associated with each reference and order and there is a growing number of references to be produced. The study allowed saving around 58.3% in the weekly time consumed in setups, which was time that reverted to productive time, thus increasing production. Also, in a company that produces pipes for air conditioning systems for automobiles, Antoniolli et al. [17] applied the standard work tool and obtained significant productivity gains, which translated into an increase of 16% in the OEE, going from 70% to 86%, thereby reaching the world reference standard established by Nakajima (85%) [18]. Sivaraman et al. [19] also used Lean tools, this time in conjunction with the DMAIC (Six Sigma) methodology to improve the efficiency of an assembly line of engines for motor vehicles. After identifying the problems, measuring waste, seeking and implementing solutions, and correspondingly monitoring the improvements introduced, it was possible to register an increase in the efficiency of the assembly line of 7%. In turn, Yuik and Puvanasvaran [20] sought to develop a framework for using Lean tools, which could be applied generically in the metalworking industry. The framework was also based on a Lean tool, PDCA. The framework essentially aimed to help SME practitioners to implement Lean tools quickly and effectively, trying to take full advantage of them in terms of efficiency of the production process. Also in the metalworking sector, Pinto et al. [21] applied some pillars of the TPM (Total Productive Maintenance) methodology in a machining sector consisting of CNC lathes and CNC milling machines, having achieved a 23% decrease in breakdowns for CNC lathes and 38% for CNC milling machines. Thus, it was possible to induce a 5% increase in the OEE of this sector of the company due to the increase in equipment availability. Using the Level Scheduling and Takt-time tools, Klašnja et al. [22] promoted a detailed analysis of a harness production line, using those tools and developing a redefinition of the line. It was possible to obtain a better balance of the different tasks performed along the line, showing that it is perfectly possible to use Lean tools to increase the productivity of manufacturing lines. Correia et al. [23] promoted an in-depth study of an assembly line of complex electronic products (surveillance cameras) subject to low order volumes and high model variability, with a view to increasing ergonomics and productivity. Initially applying VSM to identify all tasks and corresponding times and sequence, they then applied Lean Line Balancing (LLB), also known as Lean Line Design (LLD), which promoted a significant improvement in line efficiency through a better balance of activities on the line. It was also registered that after this phase, other Lean tools should be applied, namely Standard Work, SMED, 5S, and TPM, with a view to eliminating some of the wasted time identified in the production line. Based on the textile sector, which is normally highly pressured in terms of competitiveness and produces many components with low added value, Neves et al. [24] carried out a multifaceted study that aimed to increase the efficiency of trimming production lines. Applying tools such as the PDCA cycle, 5S, and 5W2H, together with other tools normally linked to quality improvement such as Pareto's Diagrams and Ishikawa's Diagram, it was possible to reduce the time allocated to certain tasks and increase the availability of equipment, achieving a 10% increase in worker availability, which translates into very high productivity gains for the sector. On the other hand, Martins et al. [25] used only the SMED tool to increase the efficiency of a set of three electron beam welding equipment. After a very exhaustive study of the internal and external tasks, it was possible to verify that the tasks that could be external were already external, and thus, to improve the efficiency of the process it would be necessary to make some slight modifications to the equipment, which induced a decrease of 50% in the time needed to perform the setup. In addition, the application of the SMED methodology also allowed for the identification of the causes responsible for the generation of some quality problems, eliminating them almost entirely through actions taken to reduce setup time. It was, thus, a successful work on two different fronts.

SWOT analysis has been a widely used tool in the assessment of situations intended to be improved, or in which there has already been an intervention and the aim is to verify the implemented improvements. However, having already been used in completely different situations, such as those reported below, the tool has shown its multifaceted character. In fact, Zhou et al. [26] successfully used the SWOT analysis to develop a selection model for the suitable strategy of prefabrication implementation in rural areas, proving the versatility of the tool in decision models. Indeed, facing four different strategies with different grades of aggressivity and describing the positive and negative factors regarding the four fields of the SWOT analysis, it was possible to define and select the best strategy regarding the faced objectives. However, Rocha and Caldeira-Pires [27] also used this tool, but to analyze how the Environmental Product Declaration promotion could be carried out. The objective is different, but the principle used to carry out the analysis is exactly the same: a brainstorming is carried out where all the strengths, weaknesses, opportunities, and threats are listed, allowing them to consider the best solution or situation among the different sets listed, or just become aware of the situation or state of a particular process or system. In this regard, it is important that all ideas are extremely objective and perfectly accepted as consensual by the group that is preparing this SWOT analysis, and the constitution of the group and the mentality that should preside over the brainstorming are very important to the final result. Cui et al. [28] used SWOT analysis to assess which key factors are taken into account in terms of need, location, and requirements when planning to build new underground pedestrian systems. The consistent analysis of the different ideas that resulted from the respective brainstorming allowed a greater discussion to be established around what is important when such a system starts to be planned and gave rise to guidelines that will allow for assessment of when such a system is really necessary and what its requirements are. In order to eliminate ambiguities, Phadermrod et al. [29] applied an Importance-Performance Analysis (IPA) to identify SWOT based on customer satisfaction surveys. This analysis was later applied in a case study in a higher education institutions in Thailand, revealing a high reliability of results in the assessment of the institution's situation, thus validating the application of the IPA to SWOT analyses and drastically reducing the subjectivity that is often related to ideas resulting from brainstorming.

Some studies resorted to SWOT analysis to identify the best Lean implementation practices. In particular, Sodhi and colleagues [30] conducted a detailed SWOT analysis and identified LSS (Lean Six Sigma) as the best waste management technique in Indian manufacturing organizations. Abu et al. [31] performed a systematic literature review on Lean manufacturing in wood and furniture SME and used a SWOT analysis to establish an extensive list of major factors influencing the strengths, weaknesses, opportunities, and threats, helping better decision making in organizations. Abu et al. [31] found that although industrials are familiar with Lean concepts and methodologies, their implementation in the area of wood and furniture is still low. In line with this, Tomioka et al. [32] revised publications over a period of 10 years and applied SWOT analysis to launch main conclusions about the implementation of Lean philosophy in Brazilian manufacturing. Several internal and external factors conditioning success in this emerging industry were identified.

Even though several gaps and problems concerning Lean deployments were found in Brazilian industry, these authors concluded that Lean tools contribute to the evolution, performance improvement, and competitiveness of the sector. In a different setting, De et al. [33] investigated the combined effect of Lean and sustainability-oriented innovation (SOI) in SMEs. The SWOT tool was used to compare inefficient SMEs with the efficient ones and notice required improvements in less successful companies. Mishra and Chakraborty [34] applied SWOT analysis to examine several possible frameworks for companies struggling to implement a Lean philosophy. This strategic decision-making tool can be used to help companies to choose the framework that provides competitive advantage in the sector.

This work intends to provide a global view of the implementation status of Lean thinking in Portuguese companies. A SWOT analysis is performed, supported by statistically worked data. A corresponding discussion is carried out over the results, trying to contribute to a better perception of the evolution of the Lean practices implementation in Portugal. The results can be further compared with other similar studies to be carried out in other countries, allowing for a comparison between the evolution of the implementation in different countries, dissecting the main reasons for different grades of evolution. A similar but very small study, involving fewer companies, was done in 2011 [35].

The present study was guided by two research questions:

RQ1: What is the pathway of Lean implementation in the Portuguese industry, since 2011?

RQ2: What are the main ongoing challenges and disadvantages for Portuguese companies in the dawn of Industry 4.0 implementation?

This article is divided into five sections. In the first, the theoretical contextualization that supports the development of the work is performed. In the second, the adopted methodology is explained. This is followed in Section 3 by the results and relevant discussion. In Section 4, the SWOT analysis and main outcomes are described. This work ends with the conclusions in which the main ideas drawn from this study are highlighted, and future work is sketched.

2. Materials and Methods

This work intended to investigate the state of the adoption and implementation of Lean tools in the Portuguese industry. The instrument used for this purpose was a survey carried out among companies belonging to diverse sectors of activity.

A quantitative data analysis using descriptive statistics, inferential statistical procedures, and SWOT analysis was performed.

2.1. Population and Sample

The population of the present study consisted of companies operating in Portugal. Firstly, a search on the Internet allowed us to collect a list of firms of interest for the study, and an email was sent to more than 1000 companies. This method proved to be fruitless, and consequently it was decided to select the target audience using the social network of professionals and companies, LinkedIn. In a total of 1045 messages and emails sent, 119 responses were obtained (a percentage of 11.4%). The sample considered in this study was based on a set of 98 companies from different sectors of activity, corresponding to the validated responses to the questionnaire.

2.2. Survey

To collect data, a survey was designed based on several studies related to Lean implementation [36–39]. This survey was developed taking into account two sets: companies that implemented the Lean methodology and companies that did not. In relation to companies that have implemented the Lean methodology, the intention was to understand the level of evolution and the differences in the implementation models between the various types of companies, the main characteristics and difficulties experienced, and to enumerate the factors most influencing the success of its implementation. In the second set of companies, the main goal was to know the reasons for not implementing the Lean philosophy. The online questionnaire performed in the Google Forms® (Montain View, CA, U.S.A.) application ran online between January and April 2020. This format was chosen because online questionnaires present numerous advantages in terms of cost, time, easiness of administration, data organization, and analysis [40,41]. The data file with the answers obtained was downloaded in MS Excel® (Redmont, WA, U.S.A.) format.

The first two sections of the questionnaire consisted of generic questions aiming at characterizing the respondents and the companies, namely department, years of work and position held in the company. The third section addressed respondents whose companies had already implemented Lean tools. The last question was intended for respondents whose companies did not yet apply the Lean philosophy.

2.3. Methods

The questions asked in the questionnaire were translated into qualitative variables (nominal and ordinal) and quantitative variables containing Likert-type scale questions from 1 to 6, and the Lean impact on companies was measured. The study of companies was grouped in terms of role performed by the respondent, size of the company (turnover and number of employees), sector of activity, years of Lean implementation, phase of Lean progress, company labor market (international or national), and types of support applied in Lean implementation. Based on these groups, statistical tests were carried out to verify any standardization or differentiation.

A quantitative data analysis was performed using descriptive statistics to summarize the information collected, followed by inferential statistical procedures based on non-parametric tests for independent samples, namely Mann–Whitney and Kruskal–Wallis [42,43]. In order to provide a holistic picture of the problem under study, the quantitative data analysis was used as the basis for the SWOT analysis. The SWOT analysis technique is a widely used tool in strategic planning, allowing for the identification of the Strengths, Weaknesses, Opportunities, and Threats for a company [44]. The statistical analysis was completed using Microsoft Excel and IBM SPSS Statistics® (Chicago, IL, U.S.A.) for Macintosh, Version 25.0.

3. Results

This chapter presents the characterization of the sample and the discussion of the results obtained by the application of non-parametric statistical tests for independent samples. These samples result from the response groups identified by company size, function performed, time of activity in the company, time of implementation of Lean methodology, internationalization of the company, training in Lean methodology, certified company, and models of Lean implementation. Mann–Whitney tests (for two groups of independent variables) and Kruskal–Wallis tests (for three or more groups of independent variables) were applied to each variable (item) of each of the questions: "What are the results obtained with the implementing Lean in the company?", "What are the main difficulties experienced in implementing Lean?" and "What factors do you consider important for the successful implementation of Lean Methodology?"

The tests were performed assuming the non-normality of the variables and considering the null hypothesis of the non-existence of differences in the responses between the groups. By performing a hypothesis test, it was possible to ascertain whether this difference was significant from the test p values or test significance [43].

3.1. Data from the Respondent

The first part of the questionnaire characterizes the sample distributed by the following components: department, years of work, and position held in the company.

3.1.1. Department

Most respondents (22.45%) reported that they belong to the Production department. A proportion of 22.45% belong to "Other" department, followed by 18.37% that belong to the Quality department, 11.22% to the Project Management department, 9.19% to Research and Development, 6.12% to Commercial, and 4.08% to the Finance Department. A very small percentage of respondents belong to the Purchasing (1.02%), Logistics (1.02%), and Sales (1.02%) departments.

3.1.2. Years of Experience and Position in the Company

Most of the respondents (60.20%) have been working in the company for less than 5 years, and 51.02% hold an engineering position. A proportion of 24.49% of respondents occupy Board positions, and the remainder are in Supervision (12.24%) and Management (10.20%) positions.

There is no significant association between "Years of experience" and "Position performed" in the company, as a result of the chi-square test for the association between two variables, ($\chi^2 = 16.918$ and $p = 0.153$). Therefore, the position is independent of the years of experience in the company.

3.2. *Characterization of the Company*

The second part of the questionnaire focuses on the characterization of the company, according to the points below.

3.2.1. Number of Employees and Turnover

Approximately forty-nine percent (48.97%) of respondents belong to companies with 250 or more employees, 32.65% work in companies with a turnover between 10 million and 50 million euros per year, and 31.63% work in those earning more than 50 million euros annually. The number of employees is significantly associated with the company's turnover ($\chi^2 = 76.599$ and $p = 0.000$) indicating a possible dependence of turnover on the number of employees (Table 1).

Table 1. Number of employees and turnover (M€).

Employees	<2	2–10	10–50	>50	Don't Know/Reply	TOTAL
1–9	2.04%	0.00%	3.06%	0.00%	0.00%	5.10%
10–49	2.04%	5.10%	7.14%	0.00%	4.08%	18.36%
50–249	0.00%	8.16%	15.31%	2.04%	1.02%	26.53%
≥250	0.00%	1.02%	7.14%	29.59%	11.22%	48.97%
TOTAL	4.08%	14.28%	32.65%	31.63%	16.32%	98.96% *

* Validated responses.

It was observed that a slight majority (51.01%) of the companies participating in this study are LE (Large Enterprises), as they have an annual turnover of more than 50 million euros or a number of employees equal to or greater than 250. The other 48.99% of the companies are outside this group and are Small and Medium Enterprises (SMEs), which include micro, small, and medium enterprises. They employ less than 250 people and simultaneously have an annual turnover that does not exceed 50 million euros or a total annual balance that does not exceed 43 million euros [45].

3.2.2. Company's Sector of Activity

The automobile industry (16.33%) and the metallurgical industry (10.20%) are the most representative sectors of activity in this study. The most diverse sectors have considerably lower proportions, from "Manufacture of machinery and equipment" (8.16%) and Environment (4.08%) to Banking, Electrical equipment, and Health (2.04%). Indicated by "Other" are the 22 sectors from the most diverse areas, including Aeronautics, Consulting, Market Studies, Information Technology, Services, and Telecommunications, among others.

3.2.3. Activity Sector, Internationalization, Lean Training, and Certification

In this regard, it is observed that the secondary sector, with a proportion of 53.06%, and the tertiary sector, with 38.78%, are the most expressive sectors among the companies present in the sample. Fifty-seven percent of participating companies only operate in the domestic market, and 86% are certified. It should also be noted that 71% of respondents have had training in Lean methodology or Lean tools.

3.2.4. Proportion of Companies That Apply Lean Methodology

This is the key issue in the questionnaire, which separates companies applying the Lean philosophy from those that do not. It was observed that, from the 98 companies participating in the study, 68% were practitioners of the Lean philosophy.

3.3. *Factors That Prevent the Implementation of the Lean Philosophy*

A key issue in this study was to identify factors preventing Lean implementation in the 32% companies that do not apply it. Most of them (41%) consider that the company's management is the most impeding factor (Table 2). The answer that stands out right after is "The current system works, and we don't need Lean" (28%). However, some respondents (24%) mentioned that the Lean implementation is being planned. With a slightly lower percentage of responses (17%) is the employees' culture. Indeed, the respondents think that with the current employees' mindset, they would never succeed in implementing Lean.

Table 2. Factors that prevent Lean implementation.

Factors that Prevent Lean Implementation	%
Company management	41
The current system works and we don't need Lean	28
Lean implementation is currently planned	24
Human resources have no mindset or culture	17
Financial resources	3
Service company, this need not having been identified	3
Management's ignorance of its existence	3
I have no idea	3

In the dissertation by Moreira [35], the main factor preventing the introduction of Lean was the Financial Resources of the company, with 45.1%. In the current study, this factor has registered a significant decrease to 3%. In the same study, the factor with the highest percentage of responses was Top Management, with 39.2%. This result is in line with the current study, which supports the idea that the greatest effort for change will have to come from the Company Management.

3.4. *Lean Implementation*

A specific section of the questionnaire was addressed only to companies that practice the Lean philosophy. The questions were intended to assess the development they have had with its implementation.

3.4.1. Years of Lean Implementation

The vast majority of companies (59.71%) implemented Lean "less than 5 years ago", 19.41% applied Lean "between 5 and 10 years ago", and only 17.91% implemented it "more than 10 years ago". This is a sign that the application of Lean methodology is still relatively recent in Portugal.

3.4.2. Lean Methodology and Lean Tools Application by Department

The departments where the implementation of the Lean methodology is more expressive are Quality (63%) and Logistics (47%), while the Project Management department (10%) is the less experienced. In the middle, one can find Production (32%), Research and Development (27%), Maintenance (19%), Commercial (18%), and Financial (17%), among others (Figure 1).

The 5S tool continues to lead the preferences of the companies, with 87%, similar to what occurred in 2011, when it was the preference of 88.6% of Lean companies [35]. Kaizen was the second option with the highest number of responses; it had an increase from 62.9% to 77% since 2011 [35]. This shows that Kaizen has been progressively used in companies operating in Portugal. The PDCA cycle is the following choice, with 73% of respondents claiming to use it in their companies. In another study carried out by students on Industrial Management and Engineering Master in ISEP – School of Engineering, Polytechnic of Porto, Portugal, between 2001 and 2010, in a small sample of 18 companies, 20 Lean tools were identified as being used in industrial projects, with Kanban as the main choice [46]. Finally, the application of Quality Function Deployment (QFD) was chosen by 18% of respondents. It is also verified that 8% of the companies indicate "Other", which may suggest the existence of other tools created by the companies themselves (Figure 2).

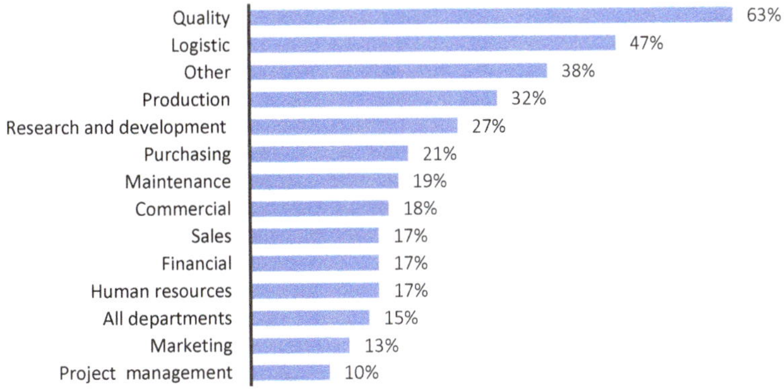

Figure 1. Lean application by department.

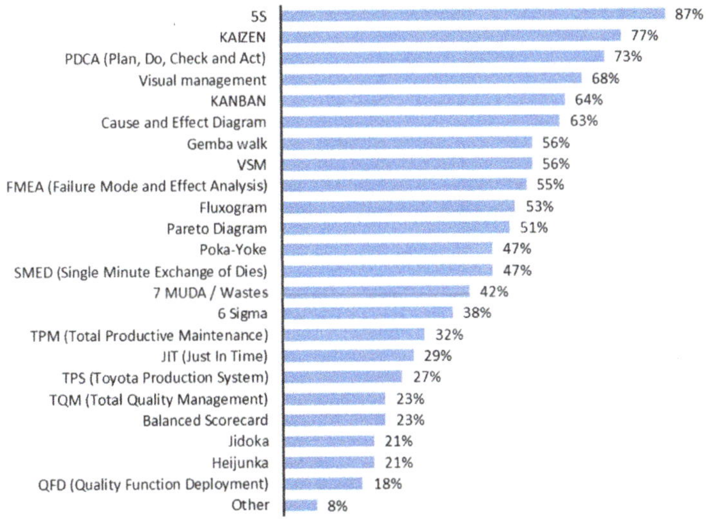

Figure 2. Lean tools and continuous improvement methodologies applied in companies.

3.4.3. Stage of Lean Implementation

Most of the companies are at an intermediate phase (44.78%), followed by the advanced and the initial phases, with 26.87% and the 25.38%, respectively. Approximately 3% of the respondents did not indicate the progression stage.

In the aforementioned Lean study in Portugal [35], respondents chose the intermediate stage, with 51%, the advanced stage with 31%, the introduction stage with 9%, and the planning stage with 9%. We remark that, in this study, the number of companies is 98 and in the previous one [35] this number is considerably lower (18), and thus, the comparison is somewhat negligible. Nevertheless, we observe an increase in the number of companies applying Lean, especially those at the initial stages. This and other responses to the questionnaire, namely the one with respect to the need for improvement of Lean practices in their company, with 95% of the respondents saying "yes", suggest that companies progressively recognize Lean implementation's benefits for the company as a whole.

Table 3 shows the distribution of the progress of the implementation of Lean methodology by years. It is observed that most of the companies (59.71%) have implemented Lean "less than 5 years ago" and approximately half of these companies are in an intermediate stage. From the perspective of the progress phase, most of the companies (44.78%) are in the "Intermediate" phase. For companies with Lean implemented "For more than 10 years" (17.91%), approximately 58% located in this category are in the "Advanced" stage.

Table 3. Stage of Lean implementation per number of years of implementation.

Years of Lean Implementation	Initial	Intermediate	Advanced	TOTAL
Less than 5 years	20.90%	29.85%	8.96%	59.71%
Between 5 and 10 years	2.99%	8.96%	7.46%	19.41%
More than 10 years	1.49%	5.97%	10.45%	17.91%
TOTAL	25.38%	44.78%	26.87%	97.03% *

* Validated responses.

It is verified that there is no significant association between "Years of Lean implementation" and "Stage of Lean implementation", according to the result of the chi-square test for association between two variables with a significance $p > 0.05$, ($\chi^2 = 13.770$ and $p = 0.077$), which indicates that the number of years of implementation is independent of the Lean progress stage.

Most of the companies involved in this study (56.72%) are international, with 25.37% in the intermediate phase and 20.90% in the advanced stage of Lean implementation. It is also verified that most of the companies operating within the national market are in an intermediate phase. There is no significant association between "National/International" and "Stage of Lean implementation", according to the results of the chi-square test for association between two variables, with a significance of $p > 0.05$, ($\chi^2 = 5.211$ e $p = 0.136$). This indicates that companies' internationalization is an independent factor of the "Stage of Lean implementation".

As for the classification of a company as SME or LE, the analysis shows that most of the companies are LE (61.20%), with 29.85% in the Intermediate phase and 25.37% in the Advanced stage of application of Lean methodology (Table 4). SMEs are mostly in the Initial phase of implementation (19.40%). There is a significant association between "SME/LE classification" and "Stage of Lean implementation", as a result of the chi-square test for association between two variables, with a significance of $p < 0.05$, ($\chi^2 = 18.510$ e $p = 0.000$). The latter suggests a likely relationship between the classification of SME/LE and the development phase of Lean application.

Table 4. Stage of Lean implementation and classification of the company.

Employees	Initial	Intermediate	Advanced	TOTAL
SME	19.40%	14.93%	2.99%	37.32%
LE	5.98%	29.85%	25.37%	61.20%
TOTAL	25.38%	44.78%	28.36%	98.52% *

* Validated responses.

The previous results are in line with the fact that 61.20% of the companies present in this study that implement Lean methodology are LE, with 46.27% operating in the international market (Table 5). Again, we verified a significant association between these two characteristics ($\chi2 = 15.363$ e $p = 0.000$). Thus, we perceive the fact that companies with an applied Lean methodology act mainly in the international market, combined with the accentuated "internationalization" in large companies.

Table 5. PME/LE vs National/International.

	National	International	TOTAL
SME	28.35%	10.45%	38.80%
LE	14.93%	46.27%	61.20%
TOTAL	43.28%	56.72%	100%

The distribution of the three phases of Lean implementation by the three sectors of activity (Table 6) indicates that 61.20% of the companies belong to the Secondary sector, with a percentage of 28.36% in the Intermediate phase. The results of the chi-square test do not reveal any association between these two factors, i.e., the activity sector is independent of the progress phase of the implementation of Lean methodology.

Table 6. Lean progression stage by sector of activity.

	Initial	Intermediate	Advanced	TOTAL
Secondary	14.93%	28.36%	17.91%	61.20%
Tertiary	8.96%	14.93%	7.46%	31.34%
Mixed	1.49%	1.49%	2.99%	5.97%
TOTAL	25.38%	44.78%	28.36%	98.52% *

* Validated responses.

The Mann–Whitney test was applied to identify differences between representatives of companies linked to the tertiary and secondary sectors, since the mixed sector had a very small portion of representatives of companies that apply Lean methodology. Considering these two independent groups, "Secondary" and "Tertiary", no significant differences were observed in the items related to the results of the implementation of the Lean methodology, or in the difficulties of its implementation.

It appears that there are no significant differences between the perceptions of representatives from different sectors of activity in relation to the most significant results, difficulties, and factors considered important in the implementation of the Lean methodology.

3.4.4. Improvement of Lean Practices

Most respondents think the company needs to further enhance Lean practices (97%), which makes perfect sense for the Lean philosophy as it advocates for continuous improvement. In the study by Moreira [35], 57% of companies operating in Portugal thought they needed to intensify their current Lean practices. Bearing in mind that at that time, in most companies using Lean, the principles and tools had been introduced in the previous 3 years. Furthermore, 37% of respondents thought that there was no need to improve, and the rest replied that they did not know. Thus, we start to see some changes in the mindset of employees from then until the present study.

3.4.5. Lean and Company Competitiveness

Regarding the question "Do you believe that Lean helps to increase the company's competitiveness?" the answer was unanimous; everyone agreed that Lean helps to increase the company's competitiveness. In the above-mentioned study [35], 89% answered "yes" and the rest disagreed.

3.4.6. Models to Support the Implementation of the Lean Methodology

As for the models that help in the implementation of the Lean Methodology, it appears that 61.2% is done by the company's own staff and 32.8% subcontract specialists in the area, which does not invalidate the inclusion of company employees in the help. This was an open question; thus, respondents could describe another model, with "Both" being the answer provided by the remaining participants.

3.4.7. Factors Contributing to the Success of Lean Implementation

Respondents expressed their opinion about the important factors influencing the success of Lean implementation in the company they worked for. They rated each item on a Likert-type scale, where 1 meant a minor factor and 6 meant a high importance factor. A classification score was assigned to each item. This score represents the proportion between the number of points obtained in a given item and the maximum they could obtain (Equation (1)). Figure 3 shows the scores obtained by each item.

$$Score = \frac{total\ of\ given\ points}{max\ of\ points} \times 100\% \quad (1)$$

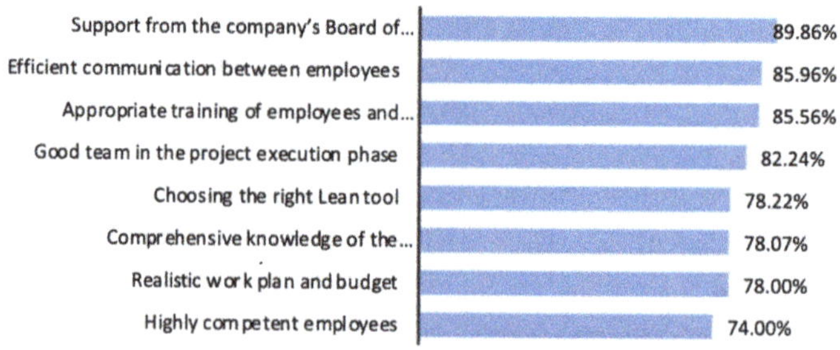

Figure 3. Factors supporting successful Lean implementation.

The item "Support from the company's Board of Directors" is considered by respondents as the most important factor for the successful implementation of the Lean methodology, obtaining a score of 89.86%. In fact, a large percentage of respondents rated this item as "Extremely important". This is followed in importance by "Efficient communication between employees" (85.96%) and "Appropriate training of employees and cultural change" (85.56%). Respondents considered the factor "Highly competent employees" with a lesser degree of importance (74%). It should be noted that all items analyzed have mode and median values between 5 and 6, which reflects their significant role for a successful Lean implementing strategy. The results obtained in this questionnaire are in line with the study by Kumar et al. [36] which refers to the added value of the involvement and commitment of the company's management for the successful implementation of the Lean methodology.

"Support from the company's Board of Directors" is the only item that presents significant differences between representatives of companies with and without Lean training, as observed from the results of the Mann–Whitney test (U = 103.000 and $p = 0.010 < 0.05$).

Employees of companies that implement Lean or Lean tools consider this to be a very important topic, with a mean grade of 5.39 and a median of 6. This is significantly higher than the values attributed by representatives of companies that do not apply Lean, who provided a mean grade of 4.43 and a median of 5.

The only factor considered important for the successful implementation of Lean that is sensitive to the fact that a company has been putting it into practice for some time is "Realistic work plan and budget". This presents significant differences between the groups with different implementation times (H = 7.063 and $p = 0.029$). Again, the group of companies stands out where the methodology has been implemented for more than 10 years, with the highest mean grade being 5.25.

3.4.8. Results of Lean Implementation

Figure 4 presents the respondents' opinions on the results obtained with the implementation of Lean in the company where they worked. These responses refer to the question "What are the results of implementing Lean in your company?". The respondents again used a Likert-type rating scale, and the classification score assigned to each item is defined by Equation (1). It is observed that "Improvement in process quality" with a score of 82%, and "Increased productivity" with a score of 81.58% are the items that present the most relevant results. This is supported by the median value of 5, mode of 6, and means of 4.92 and 4.89, respectively. The items "Efficiency in attracting new customers", "Improved knowledge of customers", and "Creation custom products or services" are those that present less significant results, with scores of 63.89%, 63.93%, and 64.29%, respectively. It should be noted that the standard deviation varies between 1.24 and 1.41, reflecting a differentiated variability of scores from item to item.

Figure 4. Results of Lean implementation.

The Kruskal–Wallis H-test has been also applied to evaluate the effect of the Lean practice time in the company as well as its results. The practice time was distributed into three categories that gave rise to three independent groups (samples): "less than 5 years ago", "between 5–10 years ago" and "more than 10 years ago". Statistically significant differences have been observed, with significances lower than 0.05 and/or 0.01 in some of the items, namely, "Efficiency in obtaining new customers " (H = 6.477 and $p = 0.039$), "Improved knowledge of customers" (H = 6.320 and $p = 0.042$), "Quality of customer service" (H = 8.169 and $p = 0.017$), "Company profit increase" (H = 6.078 and $p = 0.048$), and "Improved decision-making process" (H = 9.543 and $p = 0.008$). Effectively, the Lean practice time reveals itself as one of the fundamental points for obtaining significant results. Figure 5 presents the means of these items regarding the three mentioned categories. It is noteworthy that highest means are observed for an implementation time greater than 10 years.

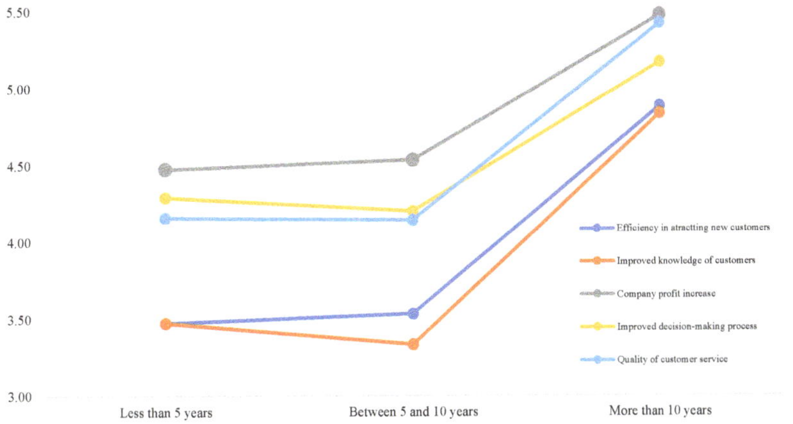

Figure 5. Mean values of "significant results" by Lean practice time (Likert scale 1–6).

Furthermore, the fact that a company in this study is national or international is not a differentiating characteristic in relation to the results obtained with the implementation of the Lean methodology.

4. SWOT Analysis of Lean Implementation in Portugal

In this section, the S(trengths), W(eaknesses), O(pportunities), and T(hreats) of the Lean philosophy implementation in companies operating in Portugal are analyzed. To the best of our knowledge, few studies have performed a SWOT evaluation of the pathway of Lean implementation in companies operating in Portugal. However, taking into account the size of the present sample, it is evident that this methodology is increasingly used. Thus, an adaptation of the SWOT language was applied to reflect strengths (advantages), weaknesses (disadvantages), opportunities, and threats (ongoing challenges), similar to [47].

4.1. Strengths (Advantages)

As can be seen in the previous section, there are important factors for the success of Lean implementation. In Table 7, the median, mode, mean, and standard deviation of the factors identified by the respondents are depicted. The factor with the highest median and mode is "Support from the company's Management". The variability in the answers is similar to all identified factors.

Table 7. Classification of important factors contributing to successful Lean implementation.

Factors	Median	Mode	Mean	Standard Deviation
Support from the company's Management	6	6	5.39	1.10
Good team in the project execution phase	5	6	5.16	1.06
Choosing the right Lean tool	5	6	5.13	1.04
Highly competent employees	5	5	4.93	1.12
Appropriate training of employees and cultural change	5	5	4.69	1.16
Efficient communication between employees	5	5	4.68	1.10
Realistic work plan and budget	5	5	4.68	1.21
Comprehensive knowledge of the company's reality	5	5	4.44	1.03

All factors expressed in Table 8 are internal; they are strengths, forces that companies have or can develop that will allow them to be successful in implementing Lean. In Table 8, the median, mode, mean, and standard deviation of the results indicated by the respondents with respect to the implementation of Lean in their company are shown. The results can impart strength or intensify company's strengths to accomplish an effective Lean implementation. Good results are an additional motivation for management and employees. Increased productivity and improvement in process quality have the best mode and mean values. This suggests that employees need to see these results to continue to be motivated by the Lean philosophy.

Table 8. Classification of the Lean implementation results.

Results	Median	Mode	Mean	Standard Deviation
Improvement in process quality	5	6	4.92	1.24
Increased productivity	5	6	4.89	1.24
Quality of customer service	5	5	4.76	1.18
Increased visibility of the company	5	5	4.63	1.24
Greater flexibility	5	5	4.53	1.30
Improved decision-making process	5	5	4.53	1.31
Company profit increase	5	5	4.50	1.25
Improved customer retention	5	5	4.26	1.32
Creating custom products or services	4	4	3.86	1.37
Improved knowledge of customers	4	4	3.84	1.37
Efficiency in attracting new customers	4	4	3.83	1.41

It is known that the main Lean strength or advantage is to reduce waste, which, consequently, helps to generate more profit and better productivity for the company. It was exactly the latter, the increase in productivity, that was one of the factors presenting the best classification in the current study, as mentioned above. Another positive factor observed is that most of participants (97%) express that they need to increase the implementation of Lean in the company. One of the main reasons for Toyota's success is never being satisfied and always wanting to evolve [48]. In terms of the implementation phase, about 45% of the respondents claim to be in an intermediate phase and 28% in an advanced phase of Lean implementation, which shows a good progression in general, when compared to the 2011 study [35]. Multinational companies are the great majority at an advanced stage of implementation. As to the sector of activity, the intermediate progress phase is predominant in companies in the secondary sector.

Another factor with a good score is the improvement in the quality of the process. Lean makes the process very flexible with better standardization, which makes it more agile, balanced (Heijunka), and has less execution time, rationalizing human effort [49,50]. In terms of quality, it makes the process more robust and with fewer errors or defects. The argument is to stop whenever there is a problem and solve it immediately (Jidoka). Visual control is very important to detect problems. With the best choice and best tool practices, the Lean Methodology helps to prevent the eight types of waste [51].

One of the great advantages of Lean is that people become versatile, with a better critical capacity and willing to solve the problems that arise in their daily lives. This characteristic ultimately increases the flexibility of companies [52], which was one of the factors with an average rating of 4.53 in the present study, with a median and mode of 5. The study found a high percentage of respondents from companies in the automobile industry. Almost all admitted having the Lean concept implemented in the company (just over 90%), which is in line with the study carried out by Wu [53]. This study states that most auto-industry companies have already adopted, to some extent, some aspects of Lean.

Among the factors that were given less importance by respondents in this study were "Efficiency in obtaining new customers" and "Improvement in customer knowledge". These are relevant items to better serve the stakeholders. As a future suggestion for improvement, companies should try to understand and implement a customer relationship framework that can best meet their needs.

Creating a culture of continuous improvement is the best way to achieve success in an organization [54]. The analysis of the responses to the questionnaire allowed us to observe a great evolution in the thinking of Lean practitioners. In particular, regarding the need to improve Lean practices in Portugal, an increased trend was perceived from Lean companies, from 57% in the study of Moreira [35] to 97% of respondents in this study.

In the performed non-parametric tests, it was found that almost all respondents from companies that have had Lean for a longer period attributed a better classification in all factors related to the results of the Lean implementation, which indicates further evolution in these companies.

4.2. Weaknesses (Disadvantages)

In this item, the study started by summarizing the main issues revealed by the responses to the question on the main difficulties experienced in Lean implementation (Figure 6 and Table 9). The presented score is calculated using Equation (1). The analysis of Table 9 shows that "Employees' resistance" is highly scored, at 71.11%. On the other hand, the "Choice of tools", with a score of 51.78%, is considered an item of less difficulty in the implementation of the Lean philosophy.

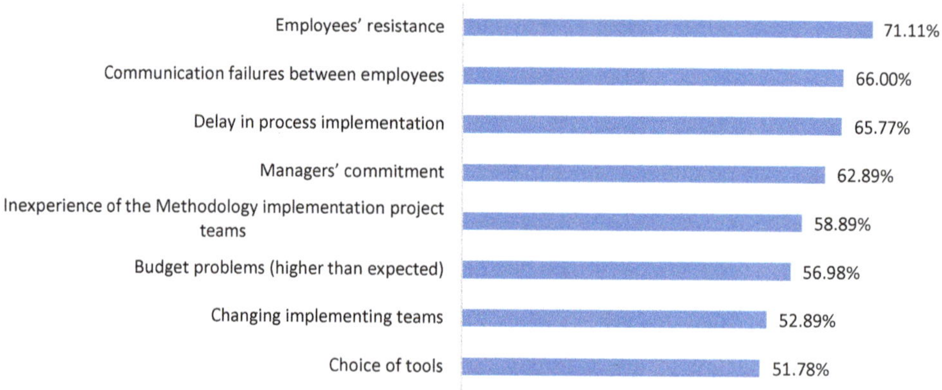

Figure 6. Difficulties experienced in Lean implementation.

Table 9. Main difficulties experienced in Lean implementation.

Difficulties	Score (%)	Median	Mode	Mean	Standard Deviation
Employees' resistance	71.11%	5	5	4.27	1.28
Communication failures between employees	66.00%	4	5	3.96	1.35
Delay in process implementation	65.77%	4	5	3.95	1.28
Managers' commitment	62.89%	4	5	3.77	1.43
Inexperience of the methodology implementation project teams	58.89%	4	4	3.53	1.42
Budget problems (higher than expected)	56.98%	3	3	3.42	1.31
Changing implementing teams	52.89%	3	3	3.17	1.46
Choice of tools	51.78%	3	4	3.11	1.26

The resistance of employees is the factor with the greatest impact on the difficulties experienced, as reported by the professionals who participated in the study. These perceptions are in agreement with most studies carried out in other countries [55–57]. This factor has a greater impact on SMEs and national companies.

Changing the mindset or culture of employees is extremely challenging [55], especially those who have been performing certain tasks for many years and/or have a long life in the company. These workers have an aversion to change, insisting that Lean will not work and no benefits will come from it, either for them or for the company. The most heard excuse is having too much work and not being able to spare time with Lean implementation [56]. It is essential to realize that the shift to a Lean culture does not happen overnight; it requires a holistic, consistent, long-term engagement and a deep organizational transformation most of the time [58]. These points require a good leadership that promotes cultural change in workers. Otherwise, it is proven that this will be one of the paths to failure [48].

In certain cultures, Lean implementation may be easier than in others [55]. It is therefore essential to constantly motivate employees and try to integrate them in the best possible way. Lean principles have been readapted, since its creation, to different requests and organizational cultures, in order to adjust to various realities, and it is its plasticity that makes Lean powerful [56].

It is also added that respondents from companies that did not have Lean implemented claimed that one of the main reasons was the lack of an adequate mindset or culture of employees towards Lean methodology. This reason was also identified in other studies from other countries [59]. According to the authors in [60], the success of Lean implementation depends strongly on employees and their mindsets. The main factor for Toyota's success with the Lean concept was its highlighted importance of the involvement of its employees and, above all, of what motivated them. Despite the negative weight inherent to this aspect, this does not mean that there cannot be behavioral changes. For a proper implementation, all employees must be integrated into the philosophy, from top management to front-line technicians [48].

Second, in terms of implementation difficulties, respondents mentioned, in similar proportion (66%), the delay in implementing the process and the lack of communication between employees. It is urgent to develop methods that promote greater interaction between employees. Teamwork can help to identify the capabilities of employees, which above all avoids the eighth waste (underutilized people). Another viable solution that has been used in several companies, to break down barriers between departments and facilitate access to information, is the application of a single data-access system for all employees, in a centralized manner. This can be implemented with software like ERP—Enterprise Resource Planning [61].

Another weakness is the commitment of managers, with a score of 62.89%. The good relationship that top management has with its employees is considered very important. Everyone must be part of the continuous improvement process to be successful, which reveals the great importance of the aforementioned Gemba Walk [62]. Only 56% of the respondents admit having practiced Lean in their workplace. This may be related, in part, with the sector of activity of the companies involved in the study—the services sector. They do not feel the need to walk on the shop floor. As the companies live in the digital age, problems tend to be solved off-site. The inclusion of top management focuses a lot on employee motivation, creating conditions for everyone to feel that they can do better every day [48]. SMEs indicate having more difficulties, highlighting the inexperience of leaders in implementing the Lean methodology, which can translate into future Lean failure. These companies must invest in more adequate training methods in order to have the aforementioned change in employees' mindsets.

Another reported weakness, with a 56.98% score, is budget issues, with SMEs being most deeply affected. This may be related to the turnover, which is usually higher in international companies, or even the lack of experience in Lean presented by SMEs.

Furthermore, results suggest that the groups "SME" and "LE" present significant differences in the items "Resistance of Employees" and "Inexperience of the project team to implement the Methodology", as shown by the results of the Mann–Whitney test (U = 347.000 e $p = 0.033 < 0.05$) and (U = 304.000 and $p = 0.007 < 0.01$), respectively. "Employee resistance" is significantly more difficult for "SME" representatives, with a mean (4.68) that is higher than the mean of "LE" representatives (4.03). In line with this item is the "Inexperience of the Methodology Implementation Project Team" as an equally significant difficulty for the

"SME" representatives; with an average score of 4.12, this is also higher than the mean of the "LE representatives", at 3.15.

Therefore, the size of the company that applies Lean is a differentiating point in the difficulties felt during the implementation, arising from the "Resistance of employees" and "Inexperience of the project team to implement the Methodology" with more significant results in "SMEs".

4.3. Opportunities

As mentioned by Taiichi Ohno, it is during economic crises that companies with better management methodologies can be distinguished [54]. The Lean Methodology has proven time and time again that it is a very efficient philosophy for improving the quality of management. The COVID-19 pandemic has once again put everyone and everything at stake. Many companies were forced to restructure, making the necessary adaptations to boost their businesses. Sales companies have become more digital to make online shopping easier. As an example, Amazon stands out, registering historical records in purchases [63]. There are companies in the production sector manufacturing products that are considered indispensable at the moment. One example is Zara, one of the well-known Lean companies, which has been making hospital gowns to help fight COVID-19 [64]. Another great example is the wine producer Adega Cooperativa de Vidigueira, Cuba and Alvito, Portugal, which is producing alcohol-based antiseptic gel from wine spirit [65]. The perfume company, Nortempresa Perfume LAB, adapted its factory in Braga, Portugal, to produce 2400 liters of disinfectant gel per day [66]. Many companies have re-adapted to the manufacture of masks, gloves, disinfectants, among other products that could, at the moment, generate added value.

Likert [48] argues that the strategy of using Lean in-product development consolidates a strong capacity for reactivity to market requests and the company's competitive positioning as a leader in its segment of operation. With the results obtained in this study, it is noted that for the area of creating personalized products or services, Lean has not been so well used, with a score of 64.29% and a median of 4 on a scale from 1 to 6; thus, this COVID-19 pandemic may be an opportunity to be explored.

The Lean application experience allows companies to turn detected weaknesses and threats into opportunities for improvement. The deeper the Lean philosophy is in an organization, the more solutions will be found for the problems that arise and in a shorter period of time [54].

4.4. Threats (Ongoing Challenges)

Threats are external factors, similar to opportunities. Companies have little control over them. In [67], the authors point to liberalization, which relaxes government restrictions, removing international investment. This might cause closure of units and retrenchment due to free flow of capital. Another threat is privatization, when a company's ownership is moved from the government to the private sphere and there is a profit orientation. The latter might cause reduction of manpower, lay-off, voluntary retirement, transfers, and other kind of measures to reduce the number of employees. Economic recession is another burden to the companies, as well as globalization, promoting an interconnected world, involving people, companies, and governments. Globalization without development induces poverty and unemployment in a country.

In 2018, Prasad et al. [68] performed a SWOT analysis with hybrid modified TOPSIS, to evaluate Lean strategy in the Indian foundry industry. The detailed threats included looming shortage of skilled workers and trained engineers in the field, R&D investment by the companies to increase productivity and reduce pollution, production of sustainable castings during technological and economic challenges, attention to the existence of competitors better equipped with technology and infrastructure to satisfy the market, and fluctuations in raw materials prices in the national and global markets.

In 2018, Jiang [69] described the threats to offsite construction in China, as incomplete policies and standards, lack of acceptance of the market, and poor development conditions for offsite construction.

More recently, in 2021, Abu and co-authors [31] performed a bibliometric and systematic review of the pathways of Lean manufacturing in wood and furniture industries. The identified threats (ongoing challenges) were characterized in different areas, namely (i) lack of competitive strength, (ii) deadlines and pressures, (iii) lack of knowledge (technical, training and/or tangible benefits) and know-how, (iv) small resources (time, capital, labor resources), environmental effects (weather, political, legislative), market demand, obstacles (sustainable financial backing, insurmountable weakness, new regulations, increased trade barriers, emergence of substitute products), and cultural and human attitudinal issues (lack of employee commitment, lack of management support and interest, difficulty in the implementation, backsliding to old ways of work).

To overcome the effect of these factors on the company, one can think of minimizing the company's weaknesses to respond better to the quest. For example, top management's short-term thinking, due to lack of knowledge of long-term gains, is an internal weakness that can be minimized. On the other hand, those employees who are more resistant to the Lean philosophy, as already mentioned, may fail due to lack of delivery and affect other employees. It is necessary to overcome entrenched and worn-out positions in processes [56]. Employees unfamiliar with the Lean methodology may think that by improving performance with fewer resources, the company can reduce the number of workers and they may feel their jobs are threatened [70]. Most studies determine that nine of the top ten barriers to change are people-related, including poor communication and employee opposition [71]. This internal weakness can and must be addressed by good communication between management and employees. Any sudden change could induce greater resistance, hence the importance of implementing an incremental process.

Another negative internal factor that could be minimized is the lack of proper training of employees. People with poor training are always a threat due to their difficulty in integrating and accepting cultural changes. Since Lean is a philosophy that defends teamwork, it is evident that this can lead to failure [57]. This internal weakness can be minimized by focusing on training workers who will thus respond more openly and effectively to external threats.

At the moment, the world is still living in a post-COVID-19 pandemic state, with most countries still struggling to vaccinate their populations. Nevertheless, the world is opening up, as is the economy, and companies are awakening from almost 2 years of low or no labor. Opportunities were created during the pandemic, which were recognized by many companies and taken advantage of. Flexibility and speed of response in an organization is crucial, in order to better help in times of need for change and in responding to threats [52].

5. Conclusions

Lean methodology is a management philosophy focused exclusively on process efficiency. The aim is to have better, cheaper, and more agile production or services systems. This methodology contains several frameworks, and each company is responsible for deciding which one is best suited to it and for making the respective adaptations according to the company's reality. Having a leadership committed to the implementation of Lean is one of the most important factors for its success.

In this work, we performed a SWOT analysis of the Lean methodology implementation in national and international companies operating in Portugal. A questionnaire was prepared and sent to several companies. The questions raised relate to concepts about the application of Lean, the sectors of the company involved in the study, the size of the company, difficulties and threats found, results obtained in terms of products and services, and the implementation phase, among others. A total of 119 individual responses were received from employees from 98 companies, and from different sectors of activity.

The SWOT analysis provided a basis to characterize the companies that apply Lean, in terms of turnover, size, implementation status, and opinion on results obtained, among others. The results are summarized below.

The strengths identified in this study are "Support from the company's management", with the highest score (89.86%) being considered the most important factor for the successful implementation of the Lean philosophy. "Efficient communication between employees" (85.96%) is the second most relevant factor, according to respondents. The item "Highly competent employees" presented a lower score of 74%. Regarding causes that intensify these internal forces, the increase in productivity and improvement in the process were the ones that stood out the most in the study. They were given the highest rating by most participants (a mode of 6). The company's evaluation, greater flexibility, increased company profit, and improvement in the decision-making process are other factors that also scored well, with a median and mode of 5.

The main weakness indicated by the respondents in the questionnaire was the employees' resistance to the change process, with a score of 71.11%. Workers resist changing their behavioral patterns, which can be explained by the threat of unemployment. The lack of communication between employees and the delay in implementing the process were the second and third factors with a higher score. Another weakness is the commitment of managers, with a score of 62.89%. The good relationship between top management and its employees is relevant. "Choice of Tools" is considered an item of less difficulty in the implementation of the Lean philosophy (51.78%), preceded by the "Changing implementing teams", with a score of 52.89%. The results of the non-parametric tests highlighted differences between LE and SMEs in terms of "Employee Resistance" and "Inexperience of the methodology implementation project team", with SMEs being more negatively affected. Another factor in which differences were observed is management support. This is more relevant in companies that already have Lean implemented. The years of Lean practice are also revealed as one of the fundamental points for obtaining significant results. Finally, a realistic work plan and budget are essential factors for success, namely in the group of companies in which Lean has been practiced for the greatest number of years.

Opportunities—with the COVID-19 pandemic, companies were forced to restructure, making the necessary adaptations to leverage their core businesses. Sales companies became more digital to facilitate online shopping, with Amazon being the most distinguished. Other companies in the production sector, namely clothing, perfume, and wine manufacturing/processing companies, among others, changed their products in order to meet market needs (gloves, masks, etc.). There was an opportunity that was quickly recognized, and companies turned it into assets. The Lean application experience allows companies to turn detected weaknesses and threats into opportunities for improvement. The deeper the Lean philosophy is in an organization, the faster solutions to the problems are found [54].

As threats are external factors, the company has little control over them. The accumulated experience resulting from the application of Lean makes it possible to transform the detected weaknesses and threats into opportunities for improvement. The effects of threats, such as the emergence of competition, changes in norms, or other unforeseen events can be diminished, increasing strengths (internal) and decreasing weaknesses (internal). It should be noted that flexibility and speed of response in an organization is essential to better help in times of need for change and in responding to threats [52].

With the recent economic crisis already disclosed by the European Commission, due to the COVID-19 pandemic, with a forecast of a recession of the Portuguese GDP of 9.8% in 2020, with the eurozone as a whole to fall 8.7% [72], it would make sense to study the effects of this same crisis on the implementation of Lean. Will companies strive to enter a new era of product development with Lean? Or will this opportunity be missed? Living in the Industry 4.0 era, the companies must be focused on the total digitization of physical assets and their integration into digital ecosystems with the various suppliers in the value chain. In [73], a survey is conducted on the challenges of Industry 4.0 in the automotive sector. The authors identify 36 challenges, grouped into four dimensions, i.e., Manufacturing Management (MM), Manufacturing Technology (MT), Manufacturing Strategy (MS), and Workforce & Organizational Management (WOM). The authors identify two pivotal dimensions that need to be considered in Industry 4.0 implementation, namely MT and WOM. Thus, the perfect moment to start applying Lean and move to Industry 4.0 is now!

According to a survey promoted by CIP (Confederation of the Portuguese Industry) and ISCTE's (High School for Enterprises Sciences and Technologies) Marketing Future Cast Lab, about two-thirds of companies that diversified products and services in response to the pandemic will maintain the changes made in the future. The survey indicates that 19% of companies diversified their offer, especially in the industrial sector, and in 87% of these cases, they did so without recourse to public funding [69]. This flexibility that many companies are adopting may also be an indicator of the implementation of Lean Management.

The approach followed in the study may encourage Portuguese Industry practitioners who still have issues regarding the implementation of Lean or any other continuous improvement tool into their company, to be willing to take a step forward. This study also reviews the importance of the implementation of Industry 4.0 and how the continuous improvement methodologies, IoT, and CPS are crucial for a successful execution. The practitioners must to perform a SWOT analysis for their company and plan their steps towards continuous improvement tools and Industry 4.0 integrations according to their companies' prioritized weaknesses and threats. Thus, the current study has practical relevance.

From an academic point of view, the methodology used in the current study may provide new insights to future researchers on how to design studies in applications of continuous improvements and Industry 4.0 adoption from other perspectives.

The main limitation is to be able to have a significant sample to study. The companies are somewhat reluctant to reply to questionnaires from academia. This was frustrating and proved fruitless. The next step was to try to select the target audience using the social network of 224 professionals and companies, LinkedIn. Here, we received some replies, but in a total of 1045 messages and emails sent, we only received 119 responses (11.4%). This was a big limitation and was time consuming. In our opinion, though there have been substantial improvements, the need for a tighter relationship between academia and companies is imperative.

Future research will consider assessing the reasons associated with the implementation of Lean in certain departments of the companies, thereby belittling others, shown empirically to be adequate for it. The performed analysis reveals that Lean is applied only in certain sectors, which irretrievably entails the loss of its potential as a whole. Other future research questions may consider the characterization of the implementation of Industry 4.0 in Portugal.

Author Contributions: All authors contributed equally for the paper. All authors have read and agreed to the published version of the manuscript.

Funding: The author CP was partially supported by CMUP, which is financed by national funds through FCT—Fundação para a Ciência e a Tecnologia, I.P., under the project with reference UIDB/00144/2020.

Institutional Review Board Statement: Not applicable.

Informed Consent Statement: Not applicable.

Data Availability Statement: The data presented in this study are available on request from the corresponding author.

Acknowledgments: F.J.G.S. would like to thank INEGI Research Center due to its continuous support, namely to Jorge Seabra, as well as all remaining CETRIB/INEGI/LAETA teal.

Conflicts of Interest: The authors declare no conflict of interest.

References

1. Hardcopf, R.; Liu, G.J.; Shah, R. Lean Production and Operational Performance: The Influence of Organizational Culture. *Int. J. Prod. Econ.* **2021**, *235*, 108060. [CrossRef]
2. Blume, S.; Herrmann, C.; Thiede, S. Increasing Resource Efficiency of Manufacturing Systems Using a Knowledge-Based System. *Procedia CIRP* **2018**, *69*, 236–241. [CrossRef]
3. Womack, J.P.; Jones, D.T.; Roos, D. *The Machine That Changed the World*; Simon & Schuster: New York, NY, USA, 1990.
4. Rodrigues, J.; Sá, J.C.; Silva, F.J.G.; Ferreira, L.P.; Jimenez, G.; Santos, G. A Rapid Improvement Process through "Quick-Win" Lean Tools: A Case Study. *Systems* **2020**, *8*, 55. [CrossRef]
5. Elkhairi, A.; Fedouaki, F.; El Alami, S. Barriers and Critical Success Factors for Implementing Lean Manufacturing in SMEs. *IFAC-PapersOnLine* **2019**, *52*, 565–570. [CrossRef]
6. Alhuraish, I.; Robledo, C.; Kobi, A. A Comparative Exploration of Lean Manufacturing and Six Sigma in Terms of Their Critical Success Factors. *J. Clean. Prod.* **2017**, *164*, 325–337. [CrossRef]
7. Cherrafi, A.; Garza-Reyes, J.A.; Belhadi, A.; Kamble, S.S.; Elbaz, J. A Readiness Self-Assessment Model for Implementing Green Lean Initiatives. *J. Clean. Prod.* **2021**, *309*, 127401. [CrossRef]
8. Vasanthakumar, C.; Vinodh, S.; Vishal, A.W. Application of Analytical Network Process for Analysis of Product Design Characteristics of Lean Remanufacturing System: A Case Study. *Clean Technol. Environ. Policy* **2017**, *19*, 971–990. [CrossRef]
9. Costa, T.; Silva, F.J.G.; Pinto Ferreira, L. Improve the Extrusion Process in Tire Production Using Six Sigma Methodology. *Procedia Manuf.* **2017**, *13*, 1104–1111. [CrossRef]
10. Cai, W.; Lai, K.-h.; Liu, C.; Wei, F.; Ma, M.; Jia, S.; Jiang, Z.; Lv, L. Promoting Sustainability of Manufacturing Industry through the Lean Energy-Saving and Emission-Reduction Strategy. *Sci. Total Environ.* **2019**, *665*, 23–32. [CrossRef] [PubMed]
11. Alves, A.C.; Ferreira, A.C.; Maia, L.C.; Leão, C.P.; Carneiro, P. A Symbiotic Relationship between Lean Production and Ergonomics: Insights from Industrial Engineering Final Year Projects. *Int. J. Ind. Eng. Manag.* **2019**, *10*, 243–256. [CrossRef]
12. Brito, M.; Vale, M.; Leão, J.; Ferreira, L.P.; Silva, F.J.G.; Gonçalves, M.A. Lean and Ergonomics Decision Support Tool Assessment in a Plastic Packaging Company. *Procedia Manuf.* **2020**, *51*, 613–619. [CrossRef]
13. Pinto, B.; Silva, F.J.G.; Costa, T.; Campilho, R.D.S.G.; Pereira, M.T. A Strategic Model to Take the First Step towards Industry 4.0 in SMEs. *Procedia Manuf.* **2019**, *38*, 637–645. [CrossRef]
14. Vinodh, S.; Antony, J.; Agrawal, R.; Douglas, J.A. Integration of Continuous Improvement Strategies with Industry 4.0: A Systematic Review and Agenda for Further Research. *TQM J.* **2021**, *33*, 441–472. [CrossRef]
15. Rosa, C.; Silva, F.J.G.; Ferreira, L.P. Improving the Quality and Productivity of Steel Wire-Rope Assembly Lines for the Automotive Industry. *Procedia Manuf.* **2017**, *11*, 1035–1042. [CrossRef]
16. Rosa, C.; Silva, F.J.G.; Ferreira, L.P.; Campilho, R. SMED Methodology: The Reduction of Setup Times for Steel Wire-Rope Assembly Lines in the Automotive Industry. *Procedia Manuf.* **2017**, *13*, 1034–1042. [CrossRef]
17. Antoniolli, I.; Guariente, P.; Pereira, T.; Ferreira, L.P.; Silva, F.J.G. Standardization and Optimization of an Automotive Components Production Line. *Procedia Manuf.* **2017**, *13*, 1120–1127. [CrossRef]
18. Nakajima, S. *Introduction to TPM: Total Productive Maintenance*; Productivity Press: Boca Raton, FL, USA, 1988.
19. Sivaraman, P.; Nithyanandhan, T.; Lakshminarasimhan, S.; Manikandan, S.; Saifudheen, M. Productivity Enhancement in Engine Assembly Using Lean Tools and Techniques. *Mater. Today Proc.* **2020**, *33*, 201–207. [CrossRef]
20. Jia Yuik, C.; Puvanasvaran, P. Development of Lean Manufacturing Implementation Framework in Machinery and Equipment SMEs. *Int. J. Ind. Eng. Manag.* **2020**, *11*, 157–169. [CrossRef]
21. Pinto, G.; Silva, F.J.G.; Fernandes, N.O.; Casais, R.; Baptista, A.; Carvalho, C. Implementing a Maintenance Strategic Plan Using TPM Methodology. *Int. J. Ind. Eng. Manag.* **2020**, *11*, 192–204. [CrossRef]
22. Klašnja, N.; Sremčev, N.; Vukelić, D.; Simeunović, N.; Lazarević, M. Optimization of Cable Harness Assembly Systems Based on Lean Concept Application. *Int. J. Ind. Eng. Manag.* **2019**, *10*, 115–123. [CrossRef]
23. Correia, D.; Silva, F.J.G.; Gouveia, R.M.; Pereira, T.; Ferreira, L.P. Improving Manual Assembly Lines Devoted to Complex Electronic Devices by Applying Lean Tools. *Procedia Manuf.* **2018**, *17*, 663–671. [CrossRef]
24. Neves, P.; Silva, F.J.G.; Ferreira, L.P.; Pereira, T.; Gouveia, A.; Pimentel, C. Implementing Lean Tools in the Manufacturing Process of Trimmings Products. *Procedia Manuf.* **2018**, *17*, 696–704. [CrossRef]
25. Martins, M.; Godina, R.; Pimentel, C.; Silva, F.J.G.; Matias, J.C.O. A Practical Study of the Application of SMED to Electron-Beam Machining in Automotive Industry. *Procedia Manuf.* **2018**, *17*, 647–654. [CrossRef]
26. Zhou, J.; He, P.; Qin, Y.; Ren, D. A Selection Model Based on SWOT Analysis for Determining a Suitable Strategy of Prefabrication Implementation in Rural Areas. *Sustain. Cities Soc.* **2019**, *50*, 101715. [CrossRef]
27. Rocha, M.S.R.; Caldeira-Pires, A. Environmental Product Declaration Promotion in Brazil: SWOT Analysis and Strategies. *J. Clean. Prod.* **2019**, *235*, 1061–1072. [CrossRef]
28. Cui, J.; Allan, A.; Lin, D. SWOT Analysis and Development Strategies for Underground Pedestrian Systems. *Tunn. Undergr. Sp. Technol.* **2019**, *87*, 127–133. [CrossRef]
29. Phadermrod, B.; Crowder, R.M.; Wills, G.B. Importance-Performance Analysis Based SWOT Analysis. *Int. J. Inf. Manage.* **2019**, *44*, 194–203. [CrossRef]
30. Sodhi, H.S.; Singh, D.; Singh, B.J. SWOT Analysis of Waste Management Techniques Quantitatively. *Int. J. Adv. Oper. Manag.* **2020**, *12*, 103–121. [CrossRef]

31. Abu, F.; Gholami, H.; Saman, M.Z.M.; Zakuan, N.; Sharif, S.; Streimikiene, D. Pathways of Lean Manufacturing in Wood and Furniture Industries: A Bibliometric and Systematic Review. *Eur. J. Wood Wood Prod.* **2021**, *79*, 753–772. [CrossRef]
32. Tomioka, A.M.; de Souza Leite, J.; Neves, J.M.S.; Silva, M.L.P. A Filosofia Lean Na Indústria Brasileira: Revisão Da Literatura. *Braz. J. Dev.* **2020**, *6*, 11823–11843. [CrossRef]
33. De, D.; Chowdhury, S.; Dey, P.K.; Ghosh, S.K. Impact of Lean and Sustainability Oriented Innovation on Sustainability Performance of Small and Medium Sized Enterprises: A Data Envelopment Analysis-Based Framework. *Int. J. Prod. Econ.* **2020**, *219*, 416–430. [CrossRef]
34. Mishra, R.P.; Chakraborty, A. Strengths, Weaknesses, Opportunities and Threats Analysis of Lean Implementation Frameworks. *Int. J. Lean Enterp. Res.* **2014**, *1*, 162. [CrossRef]
35. Moreira, F.J.T. Estudo Da Implementação Da Filosofia Lean Na Indústria Portuguesa (In Portuguese). Master's Thesis, School of Engineering, Polytechnic of Porto, Porto, Portugal, 2011.
36. Kumar, M.; Antony, J.; Douglas, A. Does Size Matter for Six Sigma Implementation? Findings from the Survey in UK SMEs. *Total Qual. Manag.* **2009**, *21*, 623–635. [CrossRef]
37. Pinheiro, L.M.P.; De Toledo, J.C. Aplicação Da Abordagem Lean No Processo de Desenvolvimento de Produto: Um Survey Em Empresas Industriais Brasileiras. *Gest. Prod.* **2016**, *23*, 320–332. [CrossRef]
38. Livermore, J.A. Factors That Significantly Impact the Implementation of an Agile Software Development Methodology. *J. Softw.* **2008**, *3*, 31–36. [CrossRef]
39. Zahraee, S.M. A Survey on Lean Manufacturing Implementation in a Selected Manufacturing Industry in Iran. *Int. J. Lean Six Sigma* **2016**, *7*, 136–148. [CrossRef]
40. Madariaga, L.; Nussbaum, M.; Burq, I.; Marañón, F.; Salazar, D.; Maldonado, L.; Alarcón, C.; Naranjo, M.A. Online Survey: A National Study with School Principals. *Comput. Human Behav.* **2017**, *74*, 35–44. [CrossRef]
41. Gill, F.J.; Leslie, G.D.; Grech, C.; Latour, J.M. Using a Web-Based Survey Tool to Undertake a Delphi Study: Application for Nurse Education Research. *Nurse Educ. Today* **2013**, *33*, 1322–1328. [CrossRef]
42. Mircioiu, C.; Atkinson, J. A Comparison of Parametric and Non-Parametric Methods Applied to a Likert Scale. *Pharmacy* **2017**, *5*, 26. [CrossRef]
43. Field, A. *Discovering Statistics Using IBM SPSS Statistics*, 4th ed.; Carmichael, M., Ed.; Sage Publications Ltd.: Thousand Oaks, CA, USA, 2013. [CrossRef]
44. Jasiulewicz-Kaczmarek, M. SWOT Analysis for Planned Maintenance Strategy-a Case Study. *IFAC-PapersOnLine* **2016**, *49*, 674–679. [CrossRef]
45. Economia Finanças. Definição de Grande Empresa, Média Empresa, Pequena Empresa e Microempresa (In Portuguese). Available online: https://economiafinancas.com/2016/definicao-grande-media-pequena-microempresa (accessed on 13 June 2020).
46. Alves, A.C.; Dinis-Carvalho, J.; Sousa, R.M.; Moreira, F.; Lima, R.M. Benefits of Lean Management: Results Form Some Industrial Cases in Portugal. In *6 Congresso Luso-Moçambicano de Engenharia (CLME'2011)*; Edições INEGI: Porto, Portugal, 2011; p. 9.
47. Stoller, J.K. A Perspective on the Educational "SWOT" of the Coronavirus Pandemic. *Chest* **2021**, *159*, 743–748. [CrossRef]
48. Liker, J. *The Toyota Way: Fourteen Management Principles from the World's Greatest Manufacturer*; McGraw-Hill: New York, NY, USA, 2004.
49. Teixeira, P.; Sá, J.C.; Silva, F.J.G.; Ferreira, L.P.; Santos, G.; Fontoura, P. Connecting Lean and Green with Sustainability towards a Conceptual Model. *J. Clean. Prod.* **2021**, *322*, 129047. [CrossRef]
50. Ribeiro, P.; Sá, J.C.; Ferreira, L.P.; Silva, F.J.G.; Pereira, M.T.; Santos, G. The Impact of the Application of Lean Tools for Improvement of Process in a Plastic Company: A Case Study. *Procedia Manuf.* **2019**, *38*, 765–775. [CrossRef]
51. Gay, C. 8 Wastes of Lean Manufacturing. Available online: https://www.machinemetrics.com/blog/8-wastes-of-lean-manufacturing (accessed on 27 June 2021).
52. Schonberger, R.J. Japanese Production Management: An Evolution—With Mixed Success. *Pap. Knowl. Towar. Media Hist. Doc.* **2007**, *25*, 403–419. [CrossRef]
53. Wu, Y.C. Lean Manufacturing: A Perspective of Lean Suppliers. *Int. J. Oper. Prod. Manag.* **2003**, *23*, 1349–1376. [CrossRef]
54. Liker, J.; Hoseus, M. *Toyota Culture: The Heart and Soul of the Toyota Way*; McGraw-Hill: New York, NY, USA, 2008.
55. Parkes, A. Cultural Conditioning of Lean Management in Great Britain. *J. Contemp. Manag.* **2016**, *15*, 49–65.
56. Bhasin, S. An Appropriate Change Strategy for Lean Sucess. *Manag. Decis.* **2012**, *50*, 439–458. [CrossRef]
57. Gupta, G.; Mishra, R.P. A SWOT Analysis of Reliability Centered Maintenance Framework. *J. Qual. Maint. Eng.* **2016**, *22*, 130–145. [CrossRef]
58. Mutingi, M.; Monageng, R.; Mbohwa, C. Lean Healthcare Implementation in Southern Africa: A SWOT Analysis. *Lect. Notes Eng. Comput. Sci.* **2015**, *2218*, 866–869.
59. Eswaramoorthi, M.; Kathiresan, G.R.; Prasad, P.S.S.; Mohanram, P.V. A Survey on Lean Practices in Indian Machine Tool Industries. *Int. J. Adv. Manuf. Technol.* **2011**, *52*, 1091–1101. [CrossRef]
60. Drew, J.; McCallum, B.; Roggenhofer, S. *Journey to Lean: Making Operational Change Stick*; Springer: Berlin/Heidelberg, Germany, 2016.
61. Jacobs, F.R.; Weston, F.C.T. Enterprise Resource Planning (ERP)-A Brief History. *J. Oper. Manag.* **2007**, *25*, 357–363. [CrossRef]
62. Dalton, J. Gemba Walks. In *Great Big Agile*; Apress: Berkeley, CA, USA, 2019; pp. 173–174. [CrossRef]

63. Faria, L.M. COVID-19. Fortuna de Jeff Bezos Aumentou 23.6 mil milhões com a Pandemia (In Portuguese). Available online: https://expresso.pt/coronavirus/2020-04-16-COVID-19.-Fortuna-de-Jeff-Bezos-aumentou-236-mil-milhoes-com-a-pandemia (accessed on 14 June 2020).
64. Diário de Notícias. Batas protectoras da Zara já chegaram a hospitais (In Portuguese). Available online: https://www.dn.pt/dinheiro/-batas-protectoras-da-zara-ja-chegaram-a-hospitais-12120973.html (accessed on 27 June 2021).
65. Catarino, A. Produtores de vinho fazem gel desinfetante para hospital de Beja (In Portuguese). Available online: https://www.tsf.pt/portugal/sociedade/produtores-de-vinho-fazem-gel-desinfetante-para-hospital-de-beja-12028081.html (accessed on 22 July 2020).
66. Pinheiro, S. COVID 19: Antes a Nortempresa fazia perfume, agora produz gel desinfectante (In Portuguese). Available online: https://www.publico.pt/2020/04/08/impar/noticia/COVID-19-nortempresa-fazia-perfume-produz-gel-desinfectante-1911541 (accessed on 26 June 2021).
67. Rane, A.B.; Sunnapwar, V.K.; Rane, S. Strategies to Overcome the HR Barriers in Successful Lean Implementation. *Int. J. Procure. Manag.* **2016**, *9*, 223–247. [CrossRef]
68. Prasad, S.; Khanduja, D.; Sharma, S.K. Integration of SWOT Analysis with Hybrid Modified TOPSIS for the Lean Strategy Evaluation. *Proc. Inst. Mech. Eng. Part B J. Eng. Manuf.* **2018**, *232*, 1295–1309. [CrossRef]
69. Jiang, R.; Mao, C.; Hou, L.; Wu, C.; Tan, J. A SWOT Analysis for Promoting Off-Site Construction under the Backdrop of China's New Urbanisation. *J. Clean. Prod.* **2018**, *173*, 225–234. [CrossRef]
70. Achanga, P.; Shehab, E.; Roy, R.; Nelder, G. Critical Success Factors for Lean Implementation within SMEs. *J. Manuf. Technol. Manag.* **2006**, *17*, 460–471. [CrossRef]
71. Shook, J. How to Change a Culture: Lessons From NUMMI. *MIT Sloan Manag. Rev.* **2010**, *51*, 63–68.
72. CIP. Comunicados De Imprensa CIP (In Portuguese). Available online: http://cip.org.pt/:http://cip.org.pt/COVID-19-informacoes-as-empresas (accessed on 14 July 2020).
73. Wankhede, V.A.; Vinodh, S. Analysis of Industry 4.0 Challenges Using Best Worst Method: A Case Study. *Comput. Ind. Eng.* **2021**, *159*, 107487. [CrossRef]

sustainability

Article

Improving Operational and Sustainability Performance in a Retail Fresh Food Market Using Lean: A Portuguese Case Study

Pedro Alexandre Marques [1,*], André M. Carvalho [2] and José Oliveira Santos [1]

[1] EIGeS—Research Centre in Industrial Engineering, Management and Sustainability, Lusófona University, Campo Grande, 376, 1749-024 Lisboa, Portugal; jose.oliveira.santos@ulusofona.pt
[2] Polytechnic Institute of Cávado and Ave, 4750-810 Funchal, Portugal; andrecarvalho@ipca.pt
* Correspondence: pedro.alexandre.marques@ulusofona.pt; Tel.: +351-21-751-5500

Abstract: This paper presents a real application of a lean–green improvement initiative conducted at a large Portuguese hypermarket store. It explores how lean tools and techniques may be used to not only improve the operational performance, but also sustainability. A case study was carried out in one store of a multinational retail enterprise, with the aim of enhancing both the operational and sustainability performance in the cold meat section, one of the most relevant areas of the fresh food markets. The Gemba Kaizen event approach, which comprises three main stages, was adopted. During the workshop stage, the structured problem-solving methodology was followed, and was recorded in an A3 format. As a consequence of this project, food waste in the cold meat market was reduced by half, whereas the out-of-stock index decreased by a third. In addition, the pilot store hit top performance within all stores of the company in Portugal, ranking first in all key indicators for the cold meat market. The lean–green scope and performance improvement procedures developed and implemented in the pilot store were later deployed to other stores of the company. This is one of the first publications regarding the application of lean management in the food retail sector for improving both the operational and sustainability performance.

Keywords: food waste; fresh products markets; lean; out-of-stocks; shrinkage; sustainability

Citation: Marques, P.A.; Carvalho, A.M.; Santos, J.O. Improving Operational and Sustainability Performance in a Retail Fresh Food Market Using Lean: A Portuguese Case Study. *Sustainability* **2022**, *14*, 403. https://doi.org/10.3390/su14010403

Academic Editors: José Carlos Sá, Francisco J. G. Silva, Gilberto Santos, Luís Pinto Ferreira and Manuel Pereira Lopes

Received: 20 November 2021
Accepted: 28 December 2021
Published: 31 December 2021

1. Introduction

Food retailers offer a wide range of fresh products, including fruits and vegetables, meat, fish, baked goods, or cold meats and cheese, among others. Fresh categories typically account for up to 40 percent of grocery chains' revenues and, in addition to this, they are also strong drivers of store traffic and customer loyalty [1]. However, they also represent a major risk, derived from lost sales, food waste, and the resulting perception of unsustainable operations.

Retail companies continuously face stiff competition, a reduction in sales margins, and increasingly demanding customers [2]. At the same time, regarding sustainability, the impact of fresh food waste is immense. Retailers, governments, and associations worldwide have taken different approaches to curb the problem. Price reductions or offerings of fresh produce, legislation preventing the dumping of food perishables, and the development of new business models are some the solutions that have risen in the face of this problem [3–5]. Several examples can be provided. Food producers and retailers have redesigned labels and packaging in an effort to prevent loss and to avoid food waste [6,7]. Countries forbade food retailers from dumping fresh food products [8]. Associations, communities, and start-ups around the world are coming up with new solutions to reduce food waste [3,9].

Accordingly, the performance of fresh food departments and the reduction in waste are two major objectives of food retailers and markets in the present day [10]. It has been argued that the performance of these markets can be improved by improving freshness, increasing sales, or reducing "shrinkage" [11]. In a fresh food market, the shrinkage rate is able to measures the proportion of food wasted due to spoilage, mishandling, a loss of quality, and the expiration date, thus also leading to the loss of potential sales of a store. From a sustainability point of view, the term "shrinkage" is key, as it does not demand further incentives to consumption or require faster transportation and/or further energy usage. Furthermore, from an operational point of view, the focus on shrinkage has a benefit. Being seen as "waste", its reduction is easily framed within traditional operational improvement initiatives already used in the industry, such as lean operations. In fact, shrinkage has been loosely applied to encapsulate some of the areas that generate loss [12].

The benefits of matching lean and sustainability—or "green"—improvement initiatives may be explored from multiple perspectives. There have been some approaches joining the operational and sustainability perspectives in the retail sector, but the research trend is still emerging [13]. Examples include the improved use of technology [14], the inclusion of shrinkage as a variable for optimization [15], or the promotion of sustainability-oriented management [16,17]. However, clear descriptions of the methods used to join the sustainability and operational performance improvement in the retail sector—especially those based on real cases—are still scarce.

In this work, we focus on "shrinkage" as an indicator of sustainability, once it is applied in a fresh food market with the intention to measure and reduce food waste. By being traditionally recognized within the lean practices of the industry (usually to gauge lost sales), we expect that it will more easily be accepted by both leaders and the workforce. By using a familiar lean indicator to tackle sustainability issues, we hope to align the pursuit of the "green" perspective with that of the operational performance, reducing misconceptions and resistance to sustainability initiatives.

In light of this reality, this paper aims to understand if and how lean tools and methods may be used to efficiently tackle sustainability issues related to food waste, in the same way that they provided an effective approach to tackle operational and process waste in the retail sector. It presents a real lean–green project conducted at a Portuguese food retail store that aimed to reduce food shrinkage and the resulting waste in one of the most important fresh products markets: the cold meat section. In this paper, we focus on shrinkage events that can lead to food waste, thus not including events of theft, nor accounting errors. As a result, it was observed that these methods were tackling shrinkage in order to improve the sustainability and operational performance indicators in the fresh food market. While it aimed primarily to reduce food waste, it also intended to decrease the number of out-of-stocks (OOS) in the same market—an objective that was initially perceived as a possible trade-off, but was later seamlessly integrated within the first one.

The methodology used in this work was based on the use of traditional lean tools and methods. Three stages comprising the Gemba Kaizen event approach were adopted by the team to perform the improvement initiative. An A3 problem-solving report provided the structured roadmap to record and display all steps of the improvement cycle. As a result, shrinkage was reduced, and although some variability has been observed, the weekly average has been cut to almost half (from 4.4% to 2.4%). At the same time, OOS events were also consistently reduced.

This work provides important insights for retail managers in pursuit of a balance between sustainability and operational strategies. It shows how the use of process and operations improvement methods, especially in the scope of a lean management program, can deliver effective solutions to reduce food waste while still providing operational improvements. The article is organized around five sections. After a careful and extensive literature review about the retail industry, related sustainability issues, and the adoption of lean practices in the sector, the case is presented and described, and is followed by the discussion of the results and a summary of the conclusions. Finally, the limitations of the research are identified and suggestions for future research are proposed.

2. Literature Review

2.1. Retail Store Operations: Research Summary and Opportunities

The retail sector comprises an important portion of the economy [18]. A report issued by the European Physical Society [19] concludes that retail made a net contribution of around 4.5% to the European Union GDP, while the Price Waterhouse Coopers [20] reports that, in the United States, the direct impact of retail in the national total GDP is 7.7%. It is thus natural that the interest on the part of the academic community in this sector has increased over the years [21], as evidenced by the substantial increment of the number of papers published [22], especially from 2015 [23].

Research on retail management encompasses many topics [24], including, among others, shelve replenishment, inventory management, food waste management, product promotions, product positioning, pricing, or online commerce. Given the wide variety of themes, Caro et al. [25] propose a classification of the topics representing the central challenges in retail management, considering eight categories: (1) Inventory, (2) Pricing, (3) Assortment, (4) Incentives, (5) Online retail, (6) Industry Studies, (7) Returns, and (8) Other topics.

Similarly, Mou et al. [18] provide a complete review of the current state of research on retail store operations using a classification of six research items/themes and seven operational decisions. In their literature review, the authors classified a total of 255 journal articles, published from 2008 to 2016. The authors concluded that more than two thirds of all papers fall in the "Inventory Management Decision" category, and, within this category, that more than 95% focus on the themes of "Uncertainty", "Perishability", and "Availability".

A review of these themes quickly and clearly establishes their importance for retail operations management. The theme of "Uncertainty" comprises three perspectives: purchase quantity, purchase timing, and purchase preferences of customers. According to Bouzaabia et al. [26], the in-store operations performance significantly contributes to reducing the impact of uncertainty in customer-perceived indicators. Reducing uncertainty thus positively impacts both perishability and availability. Several studies have been made on how to manage uncertainty. Among other relevant findings, it is curious to note that high inventory levels make in-store logistics more prone to execution errors, causing a higher risk for product shrinkage [27] and shelf stockouts [28].

In fresh products categories, "Perishability" is the biggest concern of food retailers, not only because it negatively affects the customer perception of quality, but also because of the monetary loss it creates [15]. Perishable products account for over 40% of sales in grocery chains [1], and, with factors such as a limited lifetime, high safety and quality requirements, and short lead time, they are requirements that are highly complex to manage [29].

Finally, "Availability" is key among all customer-perceived indicators of customer service in retailing, including in the online channels [30]. Product availability, or on-shelf availability (OSA), is defined as the probability of having a product in stock when a customer order arrives [31]. The higher the stock level on the shelves, the lower the likelihood of an out-of-stock (OOS) situation; nevertheless, this would lead to an increase in the capital holding costs, as well as the risk for product shrinkage [27]. According to Aastrup and Kotzab [32], both retailers and producers suffer significant losses due to poor OSA. They identify two main research streams in the area of OOS: demands issues and supply issues. Both are described below:

- Demand side issues, including the study of consumer responses to stockouts [33–38]. These issues have been proven to affect store the image and brand loyalty [34]; and it has been concluded that more than 15% of customers usually decide to quit their purchase and go elsewhere to purchase a stockout product [39];
- Supply side issues, including the analyses of root-causes that explain the occurrence of OOS situations, as well as of countermeasures to improve the performance [35,40–45]. Reasons that contribute to OOS events include situations of a poor in-store operations performance [35], poor backroom inventory handling procedures [42], misplaced SKUs [43], late delivery by a supplier [44], large product variety [28], and promotional events

that lead to more uncertainty on demand [45]; some authors also claim that higher levels of the inventory make in-store logistics more prone to execution errors [28,39].

Given the current business environment, it is no surprise that "Uncertainty", "Perishability", and "Availability" rank as the top themes in the literature on retail operations management. However, and looking at the research published until 2016, the low frequency of the topic of "Sustainability" stands out. It is true that the topic has grown considerably in the past few years, with themes as wide as in-store operations and food waste [24,46], reverse logistics [47], or marketing and communication [48,49]. Recently, de Moraes et al. [46] conducted a systematic review to map the causes of food waste to practices adopted by retailers to reduce the magnitude of that phenomenon, whereas Huang et al. [50] developed a systematic literature review to understand how food retailers from 27 different countries deal with food waste both internally and externally. However, the topic still offers large room for exploration.

While there are some older works on the strategies and effect of the move towards green retailing [51,52], the recent nature of the interest in this research topic is patent in the numerous works focused on understanding and reporting on the causes and reduction practices of sustainability problems [16,46,53,54].

Some of the greatest operational challenges faced by retailers are related to the need to ensure adequate on-shelf availability, suitable inventory levels, and acceptable amounts of product shrinkage. These goals are actually interdependent among each other [27] and strongly depend on how effective the processes of a store are designed and executed. For these reasons, many retailers have been attempting to adopt operational excellence programs based on lean principles, methods, and tools to improve performance, productivity, and customer satisfaction [55–57]. While the origins of lean management are traced to the Toyota production system (TPS) [58], over time, it has seen several transformations [59] and applications [60], establishing it as a wide-ranging management system [61] that can be effectively adopted to eliminate wastes in any organization or industry [62–65]. Examples in the retail industry include how lean principles and concepts are successfully applied in 7-Elevens in Japan [66]. The promotion of a mindset at Tesco's supply chain and store operational flows [67,68], and the building of a lean continuous improvement (Kaizen) culture, was built at Amazon [69]. Regarding this topic, it is to highlight the research conducted by Domingo [70], who classifies OOS on shelves, in the context of lean, as waste in retailing.

In recent years, the concept of lean retail has been popularized—gaining naming variations that include "lean logistics", "lean distribution", and "lean consumption" [55,70]. Sowards [71] describe how lean can be applied to improve the productivity of a shop, attending to the continuous identification and elimination of the seven types of waste. Evans and Lindsay [72] report a Kaizen event conducted at the retail services of Magnivision to investigate problems that continuously plagued employees. Jaca et al. [73] described how a distribution center enhanced its productivity by removing inefficiencies at the warehouse. Noda [56] explained the sustained adoption of standardized work and process improvement practices based on lean principles by a mid-size Japanese retailer that sold foods, consumables, apparel, and general merchandise goods.

2.2. *Retail Food Waste, Lean and Sustainability*

Within the fresh products markets, perishability is a critical issue. As reported by Kor et al. [74], approximately 45% of all fruits and vegetables, 35% of fish and seafood, 30% of cereals, and 20% of meat and dairy products are wasted by suppliers, retailers, and consumers every year. In the retail management disciplines, the issue of retail food waste has been related to the "rate of shrinkage", a performance indicator that represents the gap between inventories and sales and is commonly used as an indicator of the performance of retail stores [1,75].

The management of perishable products is not only exposed to the stigma of food waste, but demands a high cost in its prevention [11]. Given their limited life span, the freshness of perishable items is quickly lost with time. Every perishable item has a certain date before which it needs to be sold to the consumer. As a result, the handling and transportation of these products is subject to several requirements, strict norms, and sanitary regulations. On the sustainability side, their fast perishability—which affects storage conditions, transportation, order frequency and times, temperature control, packaging, and origin and traceability [76]—originates additional constraints and requirements. Grocery retailers are thus facing increasing pressure to tackle sustainability issues, promoting the transparency of their food supply chains and reducing food waste across them [77]. To Srivastava et al. [76], the fresh food retail industry is particularly vulnerable to supply chain risks.

In direct pursuit of a sustainability performance, or due to compliance, reputational gains, or as an opportunity to enhance business efficiency, retail companies are increasingly aware of the need to take actions to reduce product loss [11], including situations of food waste. Research in this scope has been conducted in various countries—where movements against food waste are emerging, asking the food chain actors, including retailers, for specific interventions to face this issue [78–81]. However, research on the effects of lean programs on sustainability metrics has been limited [82,83].

The prevailing literature describes lean as an approach that has a main objective of systematically identifying and eliminating waste in the organizational processes [84]. In fact, the entire lean toolbox is focused on the elimination of waste, purging all actions where there is no addition of value [85–87]. Food production takes a considerable part of the research on the relationship [88–90]. Another significant part of the studies on the balance between lean and green is focused on the supply chain, often influenced by approaches and perspectives previously used in the manufacturing sector [88,91,92]. Nevertheless, there is limited research on how to balance lean and green in in-store operations [93,94], and thus a gap is especially evident in the collection of empirical evidence. In this scope, we advance a case study showing the results that may be achieved by using lean methods and tools in pursuit of both the operational and sustainability performance.

Given that operational improvement efforts in the retail sector have frequently been addressed by the use of lean management, an opportunity presents itself to understand if and how lean may be used to tackle issues that pertain to both the operational and sustainability performance—and, as desired by this article, to explore the results that may be achieved in both dimensions when lean methods are applied to fresh products' markets.

Food waste is firstly and foremost a sustainability issue, since it generates significant economic, social, and environmental impacts [78], as is pointed out by the 2030 United Nations Agenda [46]. Nevertheless, the research conducted by Teller et al. [24] shows that the root-causes for the occurrence of food waste in the retail sector depend deeply on operational issues, such as the store format and product category, but also customer behaviors, demand variability, a poor efficiency within in-store operations, replenishment procedures, and high-demanding requirements for the quality of the products from both retail organizations and customers.

In the specific problem of food waste, the very basic principles of lean management provide a clear alignment with the challenges faced by fresh products' retailers. Retailers need to strike a balance between maximizing product availability on the shelves while minimizing the wastage of perishable products [35]. Factors such as a limited lifetime, high safety, and quality requirements, together with short lead time requirements, make them highly complex to manage [26]. In the fresh food categories, product deterioration is the biggest concern of food retailers, not only because it negatively affects the customer perception of quality, but also because of the monetary loss it creates [16].

The relationship between lean and green is thus balanced between benefits and misalignments. Sanchez-Rodrigues and Kumar [82] show, for example, that the implementation of a lean program in a particular company has been especially beneficial in food supply chains in terms of achieving a significant food waste reduction. However, the authors also highlight some misalignments, including superior vehicle usage, increased emissions, and swollen costs due to fleet renewal towards electric vehicles, caused by JIT deliveries and tensions between time and stock and multiple deliveries within the same time window. Looking at the efforts used to reduce food waste, we found that metrics such as OOS are often used and may provide both operational and suitability gains. However, they may also present trade-offs—and isolating one single metric will offer just one piece in a broader picture [27]. In this sense, the adaptation of the "shrinkage" metric to include only situations of food waste due to poor handling, storage and display, exceeded expiration date, or similar factors offers a better opportunity to jointly tackle operational and sustainability issues.

3. Case Study

The case study focuses on one store of a multinational corporation in the retail sector located nearby Lisbon, in Portugal. With more than 70 stores in the country, the corporation employs around 9000 people. The hypermarket store is a medium-large format type with an area of around 7000 square meters. In 2019, the company was developing a strategic pilot initiative with the aim of implementing a lean green daily management program. This program aimed to improve operational key performance indicators, and, in parallel, to increase sustainability metrics. Due to their importance for the overall sustainability results, the improvement efforts targeted the markets under the fresh products department. The metrics studied in this particular study were the "shrinkage rate"—directly responsible for both food and operational waste—and, as a complementary metric, due to its strong correlation, the "out-of-stocks index".

3.1. Problem Statement

At the start of this study, the company faced poor performances in the shrinkage metrics, as well as in the product availability indicators. The term "shrinkage", in this context, is applied with a different perspective, going beyond its traditional scope of a measure of lost sales. In this work, it includes situations of food waste due to poor handling, storage and display, exceeded expiration date, and similar factors. Given our focus on sustainability, it does not include events of theft, nor accounting errors, which are otherwise normally considered in this metric.

The problems addressed by this study were not exclusively felt in the pilot store; they were also transversal to the majority of the corporation's stores. The shrinkage problem is particularly severe in most of the fresh product's markets.

3.2. Methodology

Given our interest in exploring the ability of lean tools and methods to improve both operational and sustainability metrics, a typically lean approach was followed. The methodology followed is outlined in Figure 1. It is coherent with the Gemba Kaizen event multi-stage approach described by Martin and Osterling [95] and Hamel [96]. The activities that took place during the Kaizen event execution followed the six-step structured-problem solving method provided by the A3 framework.

The event herein described was conducted in the cold meat market, which includes both the counter and as a self-service area. The whole project took approximately four and a half months to be completed. The specific steps involved in each of the three phases, depicted in Figure 1, as well as their duration, are detailed below:

1. Kaizen Event Planning (duration: 2 weeks):
    a. Definition of the event scope;
    b. Selection of team members;
    c. Gathering of relevant data;
    d. Planning of the workshop;
2. Kaizen Event Execution (duration: two-day workshop):
    a. Ground rules, agenda and methodology;
    b. Definition of the problem (step 1 of A3 Problem Solving);

c. Characterization of the problem and determination of the baseline performance, or current condition (step 2 of A3 Problem Solving);
d. Goal statement, or desired future state (step 3 of A3 Problem Solving);
e. Root-causes analysis (step 4 of A3 Problem Solving);
f. Definition and implementation of an action plan (step 5 of A3 Problem Solving);
g. Definition of a control plan (step 6 of A3 Problem Solving);

3. Kaizen Event Follow-up—stabilization (duration: 4 months):
   a. Monitoring of the actions' effectiveness;
   b. Standardization, training and process stability.

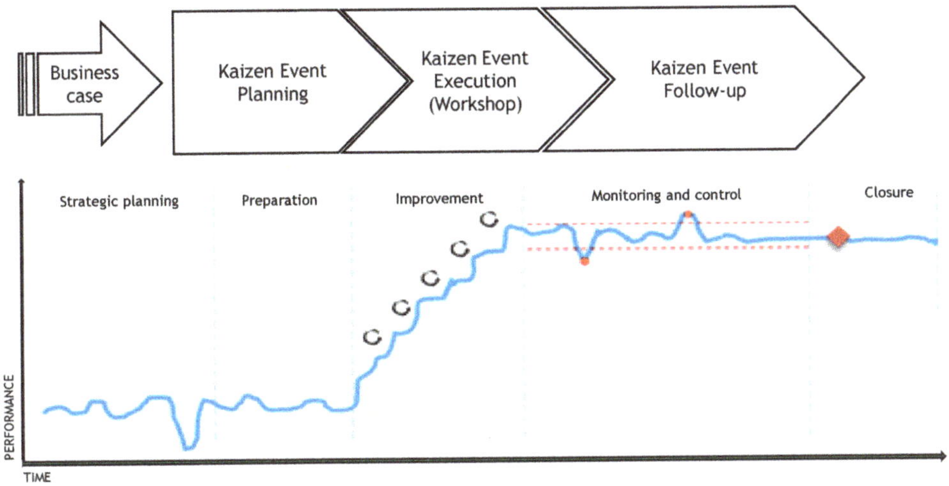

Figure 1. The stages of the Gemba Kaizen event and their relationship with the phases of process management and its performance.

3.3. *Kaizen Event Planning*

Data from the first quarter of 2019 revealed that the overall product shrinkage in the store that served as a pilot was higher compared with the same period of the previous year. In addition to this, a negative trend asserted itself over several months in the food waste indicators. It was thus considered important to conduct a set of Kaizen events in order to decrease the store shrinkage rate, a metric that relates the total monetary value lost due to food waste factors with the overall store sales.

The shrinkage rate usually correlates with the out-of-stocks (OOS) index. In facts, stockouts tend to increase when products that would be available for sale have to be removed from the shelves due to inadequate conditions. For this reason, the project was also a good opportunity to increase the product availability, so the team decided that it was important to consider the OOS index too.

As part of the preparation for the workshop stage, the facilitator of the event, together with staff from the store, gathered quantitative and qualitative (through direct observations) data, not only to be able to estimate baseline performances, but also to support a careful diagnosis of the current situation of the processes involved in the shrinkage creation, as well as to help the team with factual evidences during the root-cause analysis.

The project directly involved the director of the store, who acted as a sponsor; a member of the lean management team of the company, who provided the expertise in the principles, methods, and tools; the manager of the cold meat market, who participated as process owner in the project; and all of the four employees from the permanent operational staff belonging to that market. The store operators of the market and their manager were invited to participate in the workshop, whose duration was scheduled to last two days. An experienced member from the internal lean team was selected to facilitate the workshop. The agenda, objectives, and outcomes for the event—both in terms of the sustainability and operational performance—were defined and communicated to participants and interested parties along with other relevant resources; in particular, the material needed to conduct the workshop and other logistics matters.

3.4. Kaizen Event Execution

The methodology followed in the workshop was completely aligned with the structured problem-solving mindset. The A3 report, exhibited in Figures 2 and 3, summarizes the steps followed by the team during the two-day event. The detail of each sequence of steps are presented in the following subsections.

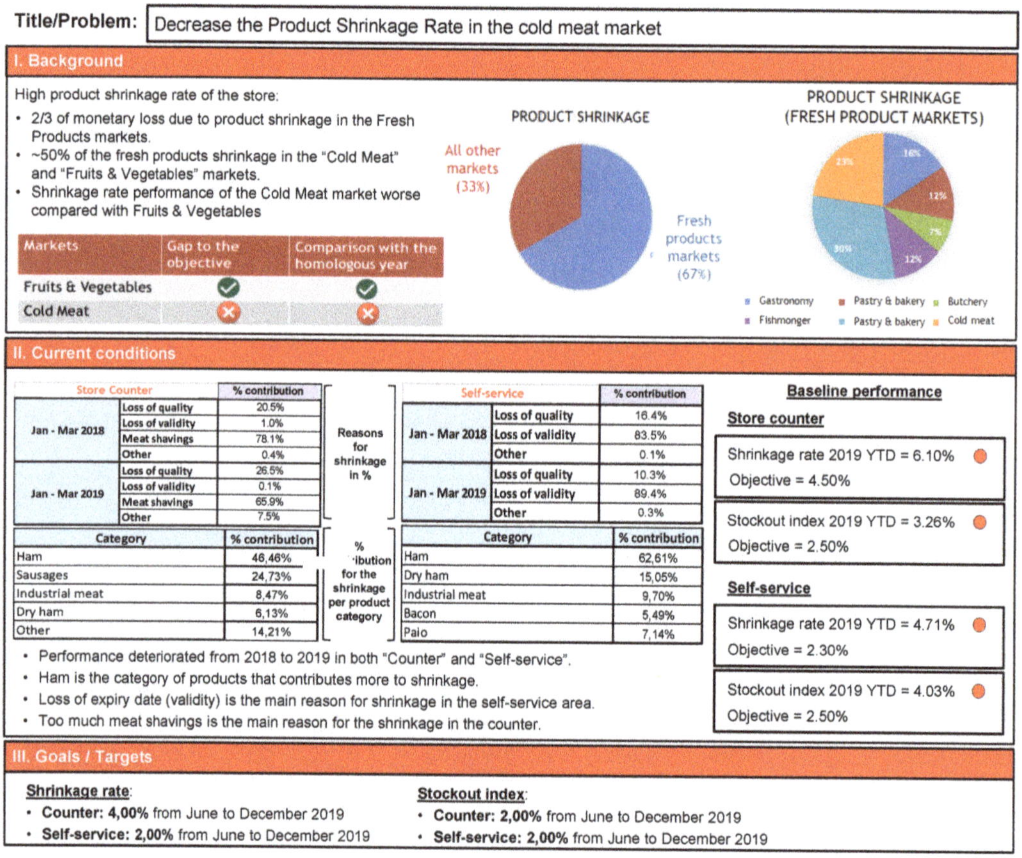

Figure 2. A3 problem-solving for day 1 of the Kaizen workshop.

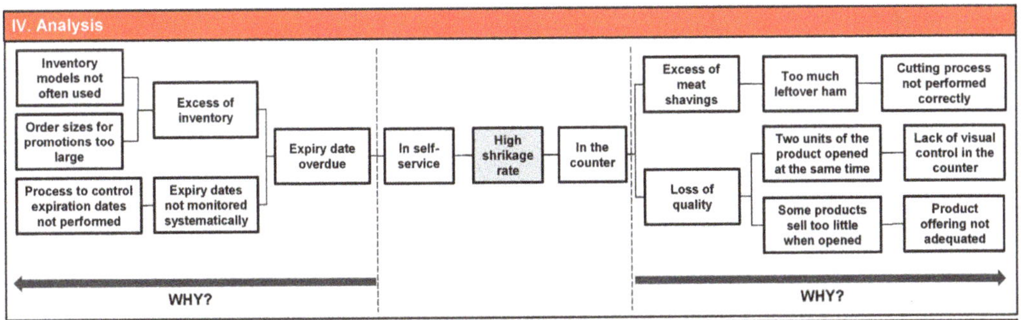

V. Countermeasures

	ROOT CAUSE	COUNTERMEASURE DESCRIPTION	DUE DATE	ACCOUNTABILITY
COUNTER	Cutting process not performed correctly	• Describe the correct cutting procedure in a visual standard • Train operators of the counter in the cutting procedure • Specify 100g as the minimum weight for the leftover ham	End of May 2019	Market Manager
COUNTER	Lack of visual control in the counter	• Standardization of the cold meat case and implementation of visual control	End of May 2019	Market Manager
COUNTER	Product offering not adequated	• Do not offer the least sold items in the counter, offer them in self-service • Adjust the order size in the self-service to consider the expected demand from customers	End of May 2019	Market Manager
SELF-SERVICE	Inventory models not often used	• Train the the Market Manager in the usage of the inventory models	7th June 2019	Operational Effectiveness Manager
SELF-SERVICE	Order sizes for promotions too large	• Forecast demand before ordering a certain quantity of a product that will be in promotion	End of May 2019	Market Manager
SELF-SERVICE	Process to control expiration dates not performed	• Perform daily work planning and allocate one person to perform this process	End of May 2019	Market Manager

VI. Follow-up actions

TASK	FREQUENCY	ACCOUNTABILITY	KPI	GOAL
Monitor the laftover ham weight	Daily	Supervisor	Leftover ham weight	≤100 g/unit
Monitor the shrinkage rate in the counter	Daily	Market manager	Shrinkage rate	≤4%
Monitor the shrinkage rate in self-service	Daily	Market manager	Shrinkage rate	≤2%
Monitor the stockout index in the counter	Daily	Market manager	Stockout index	≤2%
Monitor the stockout index in self-service	Daily	Market manager	Stockout index	≤2%
Confirm implementation of countermeasures	Weekly	Director of Store	% implementation	100%

Figure 3. A3 problem-solving for day 2 of the Kaizen workshop.

3.4.1. Definition of the Problem

Collected data allowed for the store to understand that the highest levels of product shrinkage were in its fresh products markets, particularly in two areas: fruits and vegetables (F&V) and cold meat. The F&V market performed satisfactorily, since the shrinkage results were better than both the objective and in comparison with the previous year. On the contrary, the store performed poorly regarding this food waste indicator in the cold meat market, with a shrinkage rate far from the stated objective, as well as from the previous year's performance. These facts were recorded in advance in the "Background" field of the A3 report (Figure 2). The scope was therefore defined: decrease the shrinkage rate in the cold meat market.

3.4.2. Characterization of the Problem and Estimation of the Baseline Performance

Despite the shrinkage rate being the main indicator to be impacted by this event, it was also considered important by the team to follow the evolution of the stockouts index, which measures the percentage of non-available products on the shelves, because these indicators are usually correlated with each other. The baseline for these KPIs was determined, and, for both metrics, the team realized a large gap between the results and objectives.

As one can observe in Figure 2, the cold meat market comprises two areas:

1. Counter;
2. Self-service.

Data from 2019 led to the conclusion that the end of the expiration date was identified as having by far the largest impact on the shrinkage value reported in the self-service area. The same conclusion could be drawn based on the data values of the same period of the previous year. Furthermore, "ham" was the product category that contributed the most for the overall shrinkage in both areas: self-service and counter.

In the counter, the excess of meat shavings was identified as having the most impact on both the food waste and monetary loss. Meat shavings correspond to resold trimmings that result from slicing the ham, including the remaining piece after cutting. They contribute to the reduction in the overall food waste; however, the fact that the product is sold as shavings and not as ham implies an economic loss for the store. A loss of quality is also a relevant reason for waste in the counter: because some products do not have sufficient demand, they are likely to overpass the required due time once opened.

3.4.3. Setting of Improvement Goals

The third step was the setting of SMART (specific, measurable, achievable, relevant to business, time-bounded) goals for both KPIs. The goals, included in the third room of the A3 exhibited in Figure 2, were aligned with the annual objectives established and communicated by the enterprise to employees.

3.4.4. Root-Causes Analysis

A root-cause analysis exercise was performed during the workshop to determine the underlying reasons for the shrinkage occurrence in both the counter and the self-service area. By consecutively asking "why", it was possible to construct a branched tree diagram of causes (Figure 3). Each possible deployed sub-cause was only validated and then only recorded if evidence (e.g., through "go and observe" walks) and factual data proved their relevance. The conjoint exercise in the workshop regarding the determination of potential root causes and their prioritization benefited from the work carried out during the Kaizen event planning stage, where quantitative and qualitative data were gathered in advance and a preliminary diagnosis could be performed.

In the counter, it was possible to gather evidence that could confirm the following root causes:

- The cutting process is not performed correctly, which tends to generate an excess of cold meat shavings. It was possible to verify that the way the ham cutting process was usually performed did not follow the best practices, thus producing an excess of trimmings;
- A lack of existing visual controls in the counter do not prevent the two pieces of the same product reference from being inadvertently opened;
- The range of products offered in the counter is not adequate. It was found that there is around a dozen products whose level of demand is quite low. The risk that these products, once opened, may have their quality degraded is significant. This means that such products should not have been available at the counter.

The root-causes determined in the self-service area were the following:

- Poor inventory management practices; in particular, two situations:
    - Available inventory models are not used or not correctly utilized by the market manager;
    - Very large order sizes when product promotions occur. It was found that, very often, the quantities of products ordered for promotion events were too high, resulting in too high inventory levels;

- The envisaged process to control the products' expiry dates were not consistently performed, mainly due to a lack of an effective daily planning and management.

3.4.5. Definition of Planned Countermeasures (Action Plan)

All of the root-causes determined in the previous step were validated by the team members participating in the workshop. It allowed for people to appropriate the improvement actions or countermeasures that were defined to respond to each root-cause. They are described in the fifth section of the A3 report. For example, a visual standard (Figure 4) was developed to assist the personnel at the counter in better performing the ham-cutting process. In addition to this, the standard also specified the acceptable weight of ham that should be left.

Figure 4. Visual standard for the "ham-cutting" operation created during the Kaizen event.

3.4.6. Definition of a Control Plan and Planning for Follow-Up Actions

Finally, a set of follow-up actions was defined, where key metrics would be measured, monitored, and evaluated on a daily or weekly basis. Due to the fact that the shrinkage rate is a lagging indicator, and is only available on the following day, the team tried to define a leading indicator to track the performance. For example, in the counter, the team monitored during the first weeks of the follow-up the percentage of cut ham items whose weight was above the acceptable weight indicated in the standard illustrated in Figure 4. This also allowed the team to confirm that the leftover ham was a relevant cause for the shrinkage rate, since the higher the percentage of items above the upper limit, the higher the shrinkage.

3.5. Kaizen Event Follow-Up

The follow-up actions were determined and put in place to verify if the countermeasures produced the expected result. This phase also aimed to maintain the new procedures and routines in order to stabilize processes. Run charts were used to visually monitor the evolution of KPIs on a daily basis. Figure 5 illustrates the chart used to monitor the accumulated overall shrinkage rate generated in the counter, where it was possible to visualize a positive trend towards the objective. Similar charts were created and used to control the evolution of the shrinkage rate regarding the main product categories: ham and sausages.

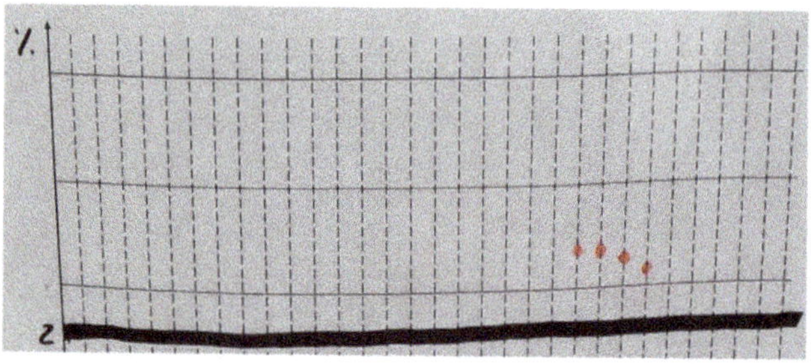

Figure 5. Run chart to monitor the overall shrinkage rate in the counter.

The already-in-place shrinkage control procedures contemplates the measurement and recording of the amount/weight of wasted product, stratifying by product category and by the type or cause of the shrinkage.

The general trend of the indicators was positive. It was interesting to notice that stockouts also decreased, thus meaning that shrinkage reduction was not achieved at the expense of a lack of product; in fact, quite the contrary. Figure 6 shows the evolution of the shrinkage rate and of the stockouts index in the counter. On average, the percentage values for the shrinkage rate were reduced to around a half, thus much more positively impacting the reduction in the food waste generated by this market. Moreover, the countermeasures were also revealed to be effective in reducing the out-of-stock percentage, which was decreased by one third. The results in the self-service area were similar.

Objective= 2,00%

	Period	Shrinkage rate [%]	Shrinkage rate in 2018 [%]	Stockouts index [%]
2019	Until May 2019	4.71%	3.82%	4.03%
	1-14 June 2019	4.18%	5.24%	3.55%
	15-30 June 2019	2.01%	5.49%	3.50%
	1-14 July 2019	1.27%	4.07%	3.06%
	15-31 July 2019	2.33%	3.63%	2.96%
	1-14 August 2019	1.35%	4.80%	3.14%
	15-31 August 2019	4.43%	3.00%	2.41%
	1-14 September 2019	1.74%	4.19%	2.64%
	15-30 September 2019	2.75%	4.97%	2.82%
	1-14 October 2019	1.54%	4.19%	2.60%

Comply with the objective and it is better when compared with the previous year

Do not comply with the objective but it is better when compared with the previous year

Do not comply with the objective and it is worse when compared with the previous year

Figure 6. Evolution of the two main KPIs in the counter.

From Figure 6, one can also notice a deterioration of the shrinkage rate performance during the summer period in August. It reveals a higher variability during this period, mainly due to the vacation period of some of the permanent staff and their replacement by temporary and untrained employees who are hired during this period. In order to sustain the achieved performance results, a set of procedures were standardized and the people from the cold meat market were trained in these standards, including the temporary personnel.

The results demonstrated the overall effectiveness of the countermeasures in reducing the food waste measured by the shrinkage rate and, at the same time, decreasing the proportion of stockouts, hence contributing to increased sales. This integrated perspective provides an interesting notion of the strong relationship existing in food retail between social and environmental sustainability and economic and business sustainability.

4. Discussion and Conclusions

This paper described the case study of a Gemba Kaizen event conducted at a food retail store in Portugal with the main purpose of reducing the "shrinkage rate" in the cold meat market, a sustainability indicator related to food waste. As was explained in the Kaizen event planning stage, a secondary aim concerned the diminishing of OOS measured by an operational indicator called the "stockouts index". The event comprised three stages—preparation, improvement workshop, and follow-up—and it was conducted as part of a daily Kaizen program that was being implemented in the store, which served as a pilot for a broader and strategic project of the company called a "lean store".

Improvement actions were defined and implemented in the two areas that comprise the cold meat market (self-service and counter) after determining the root-causes for the shrinkage problem. During the weeks that followed the Kaizen workshop, the pilot store was able to reach the top-ranking position regarding the shrinkage indicator in the cold meat market, among all of the stores that the retail company own in Portugal. The national coordinator for the cold meat market has estimated overall savings of more than EUR 100,000 a year if the measures introduced by this project are adopted country-wide.

The main performance indicator classified in the sustainability category monitored by the company is the percentage of cardboard and plastic waste resulting from the replenishment operations that are recycled. Typically, the shrinkage rate was not regarded as an indicator of sustainability, but mainly as a metric that is indicative of the amount of unsold product. This project introduced the theme of lean green within the company. One of the things that attracted attention from this initiative was that the reduction in the shrinkage rate not only contributed to decreasing the amount of wasted food, but also contributed to diminishing the OOS occurrences, thus contributing to increase the sales of the store. For these reasons, and from a business point of view, it was possible for managers to establish a linkage between sustainability and the financial results of the company.

Two foundational success factors were revealed to be critical for the outcome of this project: firstly, the visible commitment from the leadership of the store, and, secondly, the involvement of the operational staff from the very beginning during the pre-diagnosis activities, which allowed them to understand and be aware of the importance of reducing the amount of wasted food, while contributing to increase sales.

5. Limitations and Future Research

The case study described in this paper was conducted at a single store with a specific format dimension. The operations and other considerations made regarding the cold meat market described for the pilot store are very applicable to other hypermarket formats too; however, we do recognize the situation as a limitation of the study. In addition, the researchers were not able to conduct a study on the reduction in the shrinkage rate in other relevant markets from the fresh product's department. To benchmark the learning derived from the lean case herein presented, there is a need to extend this research to another store format in order to validate the applicability of the lean and green methodologies, not only in the cold meat areas, but also in other fresh products' markets.

Author Contributions: Conceptualization, P.A.M. and A.M.C.; methodology, P.A.M. and A.M.C.; formal analysis, P.A.M. and A.M.C.; investigation, P.A.M.; resources, P.A.M. and A.M.C.; writing—original draft preparation, P.A.M. and A.M.C.; writing and editing, P.A.M. and A.M.C.; project administration, P.A.M. and A.M.C.; funding acquisition, J.O.S. All authors have read and agreed to the published version of the manuscript.

Funding: This research was funded by the HORIZON 2020 EU Programme PhytoAPP project, grant number 101007642 H2020-MSCA-RISE-2020.

Institutional Review Board Statement: Not Applicable.

Informed Consent Statement: Not Applicable.

Data Availability Statement: Restrictions apply to the availability of the data.

Conflicts of Interest: The authors declare no conflict of interest.

References

1. Buck, R.; Minvielle, A. A fresh take on food retailing. *Perspect. Retail Consum. Goods* **2014**, 71–84. Available online: https://www.mckinsey.com/~/media/mckinsey/dotcom/client_service/retail/articles/perspectives%20-%20winter%202013/3_fresh_take_on_food_retailing_vf.ashx (accessed on 19 November 2021).
2. Hirsch, S.; Lanter, D.; Finger, R. Profitability and Competition in EU Food Retailing. In Proceedings of the 2018 Agricultural & Applied Economics Association Annual Meeting, Washington, DC, USA, 5–7 August 2018.
3. Ribeiro, R.; Sobral, P.; Peças, P.; Henriques, E. A sustainable business model to fight food waste. *J. Clean. Prod.* **2018**, *177*, 262–275. [CrossRef]
4. Buisman, M.E.; Haijema, R.; Bloemhof-Ruwaard, J.M. Discounting and dynamic shelf life to reduce fresh food waste at retailers. *Int. J. Prod. Econ.* **2019**, *209*, 174–184. [CrossRef]
5. Lohnes, J.D. Regulating surplus: Charity and the legal geographies of food waste enclosure. *Agric. Hum. Values* **2021**, *38*, 351–363. [CrossRef]
6. Wilson, N.L.W.; Bradley, J.R.; Saputo, R.; Ho, S. Food waste: The role of date labels, package size, and product category. *Food Qual. Prefer.* **2017**, *55*, 35–44. [CrossRef]
7. Kavanaugh, M.; Quinlan, J.J. Consumer knowledge and behaviors regarding food date labels and food waste. *Food Control* **2020**, *115*, 107285. [CrossRef]
8. Arcuri, S. Food poverty, food waste and the consensus frame on charitable food redistribution in Italy. *Agric. Hum. Values* **2019**, *36*, 263–275. [CrossRef]
9. Massari, S.; Antonelli, M.; Principato, L.; Pratesi, C.A. Design Thinking to engage consumers in achieving zero waste food experiences: The CEASE framework. In *Design Thinking for Food Well-Being*; Batat, W., Ed.; Springer: Cham, Switzerland, 2021; pp. 247–260.
10. Álvarez-Rodríguez, C.; Martín-Gamboa, M.; Iribarren, D. Sustainability-oriented efficiency of retail supply chains: A combination of life cycle assessment and dynamic network data envelopment analysis. *Sci. Total Environ.* **2020**, *795*, 135977. [CrossRef]
11. Broekmeulen, R.A.C.M.; Van Donselaar, K.H. Quantifying the potential to improve on food waste, freshness and sales for perishables in supermarkets. *Int. J. Prod. Econ.* **2019**, *209*, 265–273. [CrossRef]
12. Beck, A. Reconceptualising loss in retailing: Calling time on shrinkage. *Int. Rev. Retail Distrib. Consum. Res.* **2017**, *27*, 407–423. [CrossRef]
13. Gomez-Conde, J.; Lunkes, R.J.; Rosa, F.S. Environmental innovation practices and operational performance: The joint effects of management accounting and control systems and environmental training. *Account. Audit. Account. J.* **2019**, *32*, 1325–1357. [CrossRef]
14. Bertolini, M.; Romagnoli, G.; Weinhard, A. Proposing a value-added indicators framework for the apparel and fashion sector: Design and empirical evaluation. *Int. J. RF Technol.* **2017**, *3*, 143–164. [CrossRef]
15. Saucède, F.; Fenneteau, H.; Codron, J.M. Department upkeep and shrinkage control: Two key variables in optimizing the performance of fruit and vegetables departments. *Int. J. Retail Distrib. Manag.* **2014**, *42*, 733–758. [CrossRef]
16. Naidoo, M.; Gasparatos, A. Corporate environmental sustainability in the retail sector: Drivers, strategies and performance measurement. *J. Clean. Prod.* **2018**, *203*, 125–142. [CrossRef]
17. Álvarez-Rodríguez, C.; Martín-Gamboa, M.; Iribarren, D. Sustainability-oriented management of retail stores through the combination of life cycle assessment and dynamic data envelopment analysis. *Sci. Total Environ.* **2019**, *683*, 49–60. [CrossRef]
18. Mou, S.; Robb, D.J.; DeHoratius, N. Retail store operations: Literature review and research directions. *Eur. J. Oper. Res.* **2018**, *265*, 399–422. [CrossRef]
19. EPS—European Physical Society. *The Importance of Physics to the Economies of Europe—A Study by Cebr for the Period 2011–2016*; European Physical Society: Mulhouse, France, 2019; pp. 5–13.
20. PwC—Price Waterhouse Coopers. *The Economic Impact of the US Retail Industry*; National Retail Federation: Washington, DC, USA, 2020; pp. 1–19.
21. Fredriksson, A.; Liljestrand, K. Capturing food logistics: A literature review and research agenda. *Int. J. Logist. Res. Appl.* **2015**, *18*, 16–34. [CrossRef]
22. Fisher, M. Foreword: Special issue on retail operations. *Prod. Oper. Manag.* **2013**, *22*, 755–757. [CrossRef]
23. Lagorio, A.; Pinto, R. Food and grocery retail logistics issues: A systematic literature review. *Res. Transp. Econ.* **2020**, *87*, 100841. [CrossRef]
24. Teller, C.; Holweg, C.; Reiner, G.; Kotzab, H. Retail store operations and food waste. *J. Clean. Prod.* **2018**, *185*, 981–997. [CrossRef]

25. Caro, F.; Kök, A.G.; Martínez-de-Albéniz, V. The future of retail operations. *Manuf. Serv. Oper. Manag.* **2020**, *22*, 47–58. [CrossRef]
26. Bouzaabia, O.; Van Riel, A.C.R.; Semeijn, J. Managing in-store logistics: A fresh perspective on retail service. *J. Serv. Manag.* **2013**, *24*, 112–129. [CrossRef]
27. Bertolini, M.; Ferreti, G.; Vignali, G.; Volpi, A. Reducing out of stock, shrinkage and overstock through RFID in the fresh food supply chain: Evidence from an Italian retail pilot. *Int. J. RF Technol.* **2013**, *4*, 107–125. [CrossRef]
28. Chuang, H.; Oliva, R.; Liu, S. Examining the link between retailer inventory leanness and operational efficiency: Moderating roles of firm size and demand uncertainty. *Prod. Oper. Manag.* **2019**, *28*, 2338–2364. [CrossRef]
29. Kirci, M.; Biçer, I.; Seifert, R.W. Optimal replenishment cycle for perishable items facing demand uncertainty in a two-echelon inventory system. *Int. J. Prod. Res.* **2019**, *57*, 1250–1264. [CrossRef]
30. Calvo, E.; Cui, R.; Wagner, L. Disclosing product availability in online retail. *Manuf. Serv. Oper. Manag.* **2020**. [CrossRef]
31. Chopra, S.; Meindl, P. *Supply Chain Management: Strategy, Planning and Operations*, 3rd ed.; McGraw-Hill: New York, NY, USA, 2008.
32. Aastrup, J.; Kotzab, H. Forty years of Out-of-Stock research—And shelves are still empty. *Int. Rev. Retail Distrib. Consum. Res.* **2010**, *20*, 147–164. [CrossRef]
33. Sloot, L.M.; Verhoef, P.C.; Franses, P.H. The impact of brand equity and the hedonic level of products on consumer stock-out reactions. *J. Retail.* **2005**, *81*, 15–34. [CrossRef]
34. Kucuk, S.U. Reducing out-of-stock costs in a developing retailer sector. *J. Int. Consum. Mark.* **2004**, *16*, 75–104. [CrossRef]
35. Mckinnon, A.C.; Mendes, D.; Nababteh, M. In-store logistics: An analysis of on-shelf availability and stockout responses for three product groups. *Int. J. Logist. Res. Appl.* **2007**, *10*, 251–268. [CrossRef]
36. Ku, H.; Kuo, C.; Fang, W.; Yu, Y. The impact of retail out-of-stock options on preferences: The role of consumers' desire for assimilation versus differentiation. *Mark. Lett.* **2014**, *25*, 53–66. [CrossRef]
37. Nguyen, D.H.; De Leeuw, S.; Dullaert, W.E.H. Consumer behaviour and order fulfilment in online retailing: A systematic review. *Int. J. Manag. Rev.* **2018**, *20*, 255–276. [CrossRef]
38. Ali, D.H. Assessment of the effect of out-of-stock on customers purchasing behavior. *Int. J. Transp. Eng. Technol.* **2019**, *5*, 25–29.
39. Benrqya, Y. An examination of the effects of cross-docking on retail out of stock. *Int. J. Retail Distrib. Manag.* **2021**, *49*, 636–651. [CrossRef]
40. Fisher, M.; Raman, A. *The New Science of Retailing: How Analytics are Transforming the Supply Chain and Improving Performance*; Harvard Business Press: Boston, MA, USA, 2010; pp. 9–28.
41. Ta, H.; Esper, T.L.; Tokar, T. Appealing to the crowd: Motivation message framing and crowdsourcing performance in retail operations. *Prod. Oper. Manag.* **2021**, *30*, 3192–3212. [CrossRef]
42. Pires, M.; Pratas, J.; Liz, J.; Amorim, P. A framework for designing backroom areas in grocery stores. *Int. J. Retail Distrib. Manag.* **2017**, *45*, 230–252. [CrossRef]
43. Rizwan, M.; Hensen, E.B. On the Shelf Availability in Food Based Discount Stores Molde Grocery Environment. Master's Thesis, Molde University College, Molde, Sweden, 2009.
44. Moorthy, R.; Behera, S.; Verma, S. On-shelf availability in retailing. *Int. J. Comput. Appl.* **2015**, *116*, 47–51. [CrossRef]
45. Van Donselaar, K.H.; Peters, J.; De Jong, A.; Broekmeulen, R.A.C.M. Analysis and forecasting of demand during promotions for perishable items. *Int. J. Prod. Econ.* **2016**, *172*, 65–75. [CrossRef]
46. De Moraes, C.C.; Costa, F.H.O.; Pereira, C.R.; Silva, A.L.; Delai, I. Retail food waste: Mapping causes and reduction practices. *J. Clean. Prod.* **2020**, *256*, 120124. [CrossRef]
47. Mangla, S.K.; Govindan, K.; Luthra, S. Critical success factors for reverse logistics in Indian industries: A structural model. *J. Clean. Prod.* **2016**, *129*, 608–621. [CrossRef]
48. Cho, Y.; Soster, R.L.; Burton, S. Enhancing environmentally conscious consumption through standardized sustainability information. *J. Consum. Aff.* **2018**, *52*, 393–414. [CrossRef]
49. Upadhyaya, S.; Hughes, M.; Houston, H.R. Using sustainability as a framework for marketing curricula and pedagogy. *J. Sustain. Educ.* **2019**, *20*, 19.
50. Huang, I.Y.; Manning, L.; James, K.L.; Grigoriadis, V.; Millington, A.; Wood, V.; Ward, S. Food waste management: A review of retailers' business practices and their implications for sustainable value. *J. Clean. Prod.* **2021**, *285*, 125484. [CrossRef]
51. Thompson, B. Green retail: Retailer strategies for surviving the sustainability storm. *J. Retail Leis. Prop.* **2007**, *6*, 281–286. [CrossRef]
52. Sinha, R. Green retailing: An exploratory study examining the effects of sustainability on global retail landscape. In Proceedings of the Conference on Inclusive and Sustainable Growth Role of Industry, Government and Society, Nagpur, India, 15–16 July 2011.
53. ElAlfy, A.; Palaschuk, N.; El-Bassiouny, D.; Wilson, J.; Weber, O. Scoping the evolution of corporate social responsibility (CSR) research in the sustainable development goals (SDGs) era. *Sustainability* **2020**, *12*, 5544. [CrossRef]
54. Van Woensel, T.; Van Donselaar, K.; Broekmeulen, R.; Fransoo, J. Consumer responses to shelf out-of-stocks of perishable products. *Int. J. Phys. Distrib. Logist. Manag.* **2007**, *37*, 704–718. [CrossRef]
55. Lukic, R. The effects of application of Lean concept in retail. *Econ. Ser. Manag.* **2012**, *15*, 88–98.
56. Noda, T. Integration of Lean operation and pricing strategy in retail. *J. Mark. Dev. Compet.* **2015**, *9*, 50–60.
57. Madhani, P.M. Performance optimisation of retail industry: Lean Six Sigma approach. *ASBM J. Manag.* **2020**, *13*, 74–91.
58. Arnheiter, E.D.; Maleyeff, J. The integration of Lean management and Six Sigma. *TQM Mag.* **2005**, *17*, 5–18. [CrossRef]

59. Bhamu, J.; Sangwan, K.S. Lean manufacturing: Literature review and research issues. *Int. J. Oper. Prod. Manag.* **2012**, *34*, 876–940. [CrossRef]
60. Stone, K.B. Four decades of Lean: A systematic literature review. *Int. J. Lean Six Sigma* **2012**, *3*, 112–132. [CrossRef]
61. Ohno, T. *Toyota Production System: Beyond Large Scale Production*; Productivity Press: New York, NY, USA, 1988; pp. 75–92.
62. Gutpa, S.; Jain, S.K. A literature review on Lean manufacturing. *Int. J. Manag. Sci. Eng. Manag.* **2013**, *8*, 241–249.
63. Alaskari, O.; Ahmad, M.M.; Pinedo-Cuenca, R. Development of a methodology to assist manufacturing SMEs in the selection of appropriate Lean tools. *Int. J. Lean Six Sigma* **2016**, *7*, 62–84. [CrossRef]
64. Myerson, P. Lean retail. In *The Routledge Companion to Lean Management*; Netland, T.H., Powell, D.J., Eds.; Routledge: New York, NY, USA, 2017; pp. 413–421.
65. Morcillo-Bellido, J.; Duran, A. Supply chain sustainability in Spanish major retailer through strategic alliances and Lean practices. In *Closing the Gap between Practice and Research in Industrial Engineering*; Viles, E., Ormazábal, M., Lleo, A., Eds.; Springer: Cham, Switzerland, 2018; pp. 11–18.
66. Naruo, S.; Toma, S.G. From Toyota production system to Lean retailing: Lessons from Seven-Eleven Japan. In *Advances in Production Management System*; Olhager, J., Perrson, F., Eds.; Springer: New York, NY, USA, 2017; pp. 387–395.
67. Womack, J.P.; Jones, D.T. *Lean Solutions: How Companies and Customers can Create Value and Wealth Together*; Free Press: New York, NY, USA, 2005.
68. Smith, L.; Sparks, L. Tesco's supply chain management. In *Logistics & Retail Management: Emerging Issues and New Challenges in the Retail Supply Chain*, 3rd ed.; Fernie, J., Sparks, L., Eds.; Kogan Page: London, UK, 2009; pp. 143–171.
69. Onetto, M. When Toyota met e-commerce. *McKinsey Q.* **2014**, *2*, 1–7.
70. Domingo, T.M. The Adoption of Lean Techniques to Optimize the On-Shelf Availability of Products and Drive Business Performance in the Food Industry: A South African Manufacturing and Retail Case Study. Master's Thesis, Cape Town University, Cape Town, South Africa, 2013.
71. Sowards, D. How Lean improves shop productivity. *Contractor Magazine*, 15 August 2017; p. 36.
72. Evans, J.R.; Lindsay, W.M. *An Introduction to Six Sigma & Process Improvement*, 2nd ed.; CENGAGE Learning: Stamford, CT, USA, 2015; pp. 199–221.
73. Jaca, C.; Santos, J.; Errasti, A.; Viles, E. Lean thinking with improvement teams in retail distribution: A case study. *Total Qual. Manag.* **2012**, *23*, 449–465. [CrossRef]
74. Kor, Y.Y.; Prabhu, J.; Esposito, M. How large food retailers can help solve the food waste crisis. *Harv. Bus. Rev.* **2017**. Available online: https://hbr.org/2017/12/how-large-food-retailers-can-help-solve-the-food-waste-crisis (accessed on 19 November 2021).
75. Buzby, J.C.; Bentley, J.T.; Padera, B.; Ammon, C.; Campuzano, J. Estimated fresh produce shrink and food loss in U.S. supermarkets. *Agriculture* **2015**, *5*, 626–648. [CrossRef]
76. Srivastava, S.K.; Chaudhuri, A.; Srivastava, R.K. Propagation of risks and their impact on performance in fresh food retail. *Int. J. Logist. Manag.* **2015**, *26*, 568–602. [CrossRef]
77. Vadakkepatt, G.G.; Winterich, K.P.; Mittal, V.; Zinn, W.; Beitelspacher, L.; Aloysius, J.; Ginger, J.; Reilman, J. Sustainable retailing. *J. Retail.* **2021**, *97*, 62–80. [CrossRef]
78. Cicatiello, C.; Franco, S.; Pancino, B.; Blasi, E. The value of food waste: An exploratory study on retailing. *J. Retail. Consum. Serv.* **2016**, *30*, 96–104. [CrossRef]
79. Filimonau, V.; Gherbin, A. An exploratory study of food waste management practices in the UK grocery retail sector. *J. Clean. Prod.* **2017**, *167*, 1184–1194. [CrossRef]
80. Kulikovskaja, V.; Aschemann-Witzel, J. Food waste avoidance actions in food retailing: The case of Denmark. *J. Int. Food Agribus. Mark.* **2017**, *29*, 328–345. [CrossRef]
81. Hermsdorf, D.; Rombach, M.; Bitsch, V. Food waste reduction practices in German food retail. *Br. Food J.* **2017**, *119*, 2532–2546. [CrossRef] [PubMed]
82. Sanchez-Rodrigues, V.; Kumar, M. Synergies and misalignments in Lean and Green practices: A logistics industry perspective. *Prod. Plan. Control* **2019**, *30*, 369–384. [CrossRef]
83. Kouhizadeh, M.; Zhu, Q.; Sarkis, J. Blockchain and the circular economy: Potential tensions and critical reflections from practice. *Prod. Plan. Control* **2020**, *31*, 950–966. [CrossRef]
84. Bhasin, S. *Lean Management beyond Manufacturing: A Holistic Approach*; Springer: Cham, Coventry, UK, 2015; pp. 11–26.
85. Bhuiyan, N.; Baghel, A. An overview of continuous improvement: From the past to the present. *Manag. Decis.* **2005**, *43*, 761–771. [CrossRef]
86. Womack, J.P.; Jones, D.T. *Lean Thinking: Banish Waste and Create Wealth in Your Corporation—Expanded and Revised*; Free Press: New York, NY, USA, 2003; pp. 29–36.
87. Bicheno, J.; Holweg, M. *The Lean Toolbox: A Handbook for Lean Transformation*, 5th ed.; PICSIE Books: Buckingham, UK, 2016; pp. 18–31.
88. Amani, P.; Lindbom, I.; Sundström, B.; Östergren, K. Green-Lean synergy-root-cause analysis in food waste prevention. *Int. J. Food Syst. Dyn.* **2015**, *6*, 99–109.
89. Van der Goot, A.J.; Pelgrom, P.J.M.; Berghout, J.A.M.; Geerts, M.E.J.; Jankowiak, L.; Hardt, N.A.; Keijer, J.; Schutyser, M.A.I.; Nikiforidis, C.V.; Boom, R.M. Concepts for further sustainable production of foods. *J. Food Eng.* **2016**, *168*, 42–51. [CrossRef]

90. Jagtap, S.; Bhatt, C.; Thik, J.; Rahimifard, S. Monitoring potato waste in food manufacturing using image processing and internet of things approach. *Sustainability* **2019**, *11*, 3173. [CrossRef]
91. Lindberg, E.; Sohlin, T. *Food Recalls in the Food Supply Chain: A Qualitative Study of Different Product Flows in a Retail Context, Degree Project*; Umeå University: Umeå, Sweden, 2021.
92. Pejić, V.; Lerher, T.; Jereb, B.; Lisec, A. Lean and Green paradigms in logistics: Review of published research. *Promet Traffic Transp.* **2016**, *26*, 593–603. [CrossRef]
93. Jónsdóttir, S. Innovation Processes in Retailing: A Way to Green Food Retail? Master's Thesis, Lund University, Lund, Sweden, 2012.
94. Petljak, K.; Zulauf, K.; Štulec, I.; Seuring, S.; Wagner, R. Green supply chain management in food retailing: Survey-based evidence in Croatia. *Supply Chain Manag.* **2018**, *23*, 1–15. [CrossRef]
95. Martin, K.; Osterling, M. *The Kaizen Event Planner: Achieving Rapid Improvement in Office, Service, and Technical Environments*; Productivity Press: New York, NY, USA, 2007; pp. 3–32.
96. Hamel, M.R. *Kaizen Event Fieldbook: Foundation, Framework, and Standard Work for Effective Events*; Society of Manufacturing Engineers: Dearborn, MI, USA, 2010; pp. 85–196.

MDPI
St. Alban-Anlage 66
4052 Basel
Switzerland
www.mdpi.com

Sustainability Editorial Office
E-mail: sustainability@mdpi.com
www.mdpi.com/journal/sustainability

www.ingramcontent.com/pod-product-compliance
Lightning Source LLC
Chambersburg PA
CBHW040752250926
51850CB00007BB/1481